T0366356

Non-linear Modeling and Analysis of Solids and Structures

Steen Krenk

CAMBRIDGE
UNIVERSITY PRESS

University Printing House, Cambridge CB2 8BS, United Kingdom

Cambridge University Press is part of the University of Cambridge.

It furthers the University's mission by disseminating knowledge in the pursuit of education, learning and research at the highest international levels of excellence.

www.cambridge.org
Information on this title: www.cambridge.org/9780521830546

First published 2009

A catalogue record for this publication is available from the British Library

ISBN 978-0-521-83054-6 Hardback

To Jette

Contents

Preface *page* ix

1 Introduction 1

1.1 A simple non-linear problem 2

1.1.1 Equilibrium 3

1.1.2 Virtual work and potential energy 6

1.2 Simple non-linear solution methods 7

1.2.1 Explicit incremental method 8

1.2.2 Newton–Raphson method 9

1.2.3 Modified Newton–Raphson method 13

1.3 Summary and outlook 14

1.4 Exercises 15

2 Non-linear bar elements 17

2.1 Deformation and strain 18

2.2 Equilibrium and virtual work 20

2.3 Tangent stiffness matrix 24

2.4 Use of shape functions 26

2.5 Assembly of global stiffness and forces 31

2.6 Total or updated Lagrangian formulation 36

2.7 Summing up the principles 39

2.8 Exercises 43

3 Finite rotations 47

3.1 The rotation tensor 49

3.2 Rotation of a vector into a specified direction 53

3.3 The increment of the rotation variation 55

3.4 Parameter representation of an incremental rotation 60

3.5 Quaternion parameter representation 63

3.5.1 Representation of the rotation tensor 64

3.5.2 Addition of two rotations 65
3.5.3 Incremental rotation from quaternion parameters 67
3.5.4 Mean and difference of two rotations 68
3.6 Alternative representation of the rotation tensor 69
3.7 Summary of rotations and their virtual work 72
3.8 Exercises 73

4 Finite rotation beam theory 76
4.1 Equilibrium equations 77
4.2 Virtual work, strain and curvature 78
4.3 Increment of the virtual work equation 81
4.3.1 Constitutive stiffness 82
4.3.2 Geometric stiffness 83
4.3.3 The load increments 85
4.4 Finite element implementation 86
4.4.1 Element stiffness matrix 87
4.4.2 Loads and internal forces 89
4.4.3 Shear locking 91
4.5 Summary of 'elastica' beam theory 98
4.6 Exercises 99

5 Co-rotating beam elements 100
5.1 Co-rotating beams in two dimensions 101
5.1.1 Co-rotation form of the tangent stiffness 104
5.1.2 Element deformation stiffness 107
5.1.3 Total tangent stiffness 110
5.1.4 Finite element implementation 112
5.2 Co-rotating beams in three dimensions 117
5.2.1 Co-rotation form of the tangent stiffness 120
5.2.2 Element deformation stiffness 127
5.2.3 Total tangent stiffness 130
5.2.4 Finite element implementation 133
5.3 Summary and extensions 139
5.4 Exercises 141

6 Deformation and equilibrium of solids 145
6.1 Deformation and strain 146
6.1.1 Non-linear strain 148
6.1.2 Decomposition into deformation and rigid body motion 151
6.2 Virtual work and stresses 154
6.2.1 Piola–Kirchhoff stress 155
6.2.2 Cauchy and Kirchhoff stresses 158

6.2.3 Stress rates 160
6.3 Total Lagrangian formulation 165
6.3.1 Equilibrium and residual forces 166
6.3.2 Tangent stiffness 167
6.3.3 Finite element implementation 170
6.4 Updated Lagrangian formulation 174
6.4.1 Transformation from total to updated format 174
6.4.2 Virtual work in the current configuration 176
6.4.3 Finite element implementation 180
6.5 Summary of non-linear motion of solids 185
6.6 Exercises 186

7 Elasto-plastic solids 189
7.1 Elastic solids 190
7.1.1 Stress invariants 192
7.1.2 Strain invariants and small strain elasticity 198
7.1.3 Isotropic elasticity at finite strain 200
7.2 General plasticity theory 203
7.2.1 Reversible deformation 204
7.2.2 Maximum plastic dissipation rate 207
7.2.3 Evolution equations 212
7.2.4 Isotropic and kinematic hardening 216
7.3 Von Mises plasticity models 218
7.3.1 Yield surface and flow potential 219
7.3.2 Explicit integration 222
7.3.3 Radial return algorithm 225
7.4 General aspects of plasticity models 229
7.4.1 Combined isotropic and kinematic hardening 230
7.4.2 Internal variables and non-associated flow 234
7.4.3 General computational procedure 237
7.5 Models for granular materials 241
7.5.1 Flow potential and yield surface 242
7.5.2 Elasticity and hardening 247
7.6 Finite strain plasticity 249
7.7 Summary 252
7.8 Exercises 253

8 Numerical solution techniques 256
8.1 Iterative solution of equilibrium equations 257
8.1.1 Non-linear iteration strategies 259

8.1.2 Direction and step-size control 260
8.2 Orthogonal residual method 263
8.3 Arc-length methods 270
8.3.1 General constraint formulation 272
8.3.2 Hyperplane constraints 274
8.3.3 Hypersphere constraint 278
8.4 Quasi-Newton methods 283
8.5 Summary 287
8.6 Exercises 288

9 Dynamic effects and time integration 290
9.1 Newmark algorithm for linear systems 292
9.1.1 Energy balance and stability 295
9.1.2 Numerical accuracy and damping 300
9.2 Non-linear Newmark algorithm 304
9.3 Energy-conserving integration 309
9.3.1 State-space formulation 310
9.3.2 Non-linear kinematics for Green strain 311
9.3.3 Energy-conserving algorithm 315
9.4 Algorithmic energy dissipation 323
9.4.1 Spectral analysis of linear systems 323
9.4.2 Linear algorithm with energy dissipation 325
9.4.3 Non-linear algorithm with energy dissipation 327
9.5 Summary and outlook 331
9.6 Exercises 333

References 336
Index 345

Preface

The aim of this book is to take the reader on a concentrated tour of some of the central issues of non-linear modeling and analysis of structures and solids. Traditionally, the non-linear theories of solids have been treated in books on continuum mechanics, while the questions of analysis have formed the focus of books on finite element techniques. The idea of the present book is to place the emphasis on modeling with a view to its numerical implementation right from the outset. Two guiding principles have determined the main style of the book: the story should be told in the form of concentrated chapters, each giving the central ideas of a specific aspect such as 'finite rotations' or 'elasto-plastic solids', and the reader should have the possibility of getting a feel for the numerical implementation by access and use of simple high-level implementations of the basic algorithms. A text based on these principles cannot provide exhaustive coverage, but aims at giving an interesting introduction to the basic ideas, which can then be studied elsewhere in greater detail as needed. It is hoped that the combination of a concise theoretical presentation in plain language supported by specific algorithms will make the text of interest to graduate students as well as professionals.

The book contains nine chapters: a brief introductory chapter setting the scene by use of elementary arguments, four chapters on structures, two chapters on non-linear deformation and material behavior of solids, and finally two chapters on numerical techniques for non-linear quasi-static and dynamic analysis. The theory is combined with demonstrations and exercises using a small MATLAB toolbox FEMFILES providing routines for creation and assembly of element matrices and permitting the solution of non-linear finite element problems in a fairly simple script file format. The toolbox FEMFILES is available from the author via the internet. Exercises that require the use of a high-level program like FEMFILES are marked $*$.

The text started as a draft manuscript prepared for a short introductory course on non-linear aspects of the finite element method at Aalborg University in the fall of 1992. A visit to Lund Institute of Technology sponsored by NorFA provided an opportunity to include additional material on the numerical aspects. The text was later extended with material on finite rotations, co-rotating formulation of elements, potential theory of plasticity theory and plasticity models for geotechnical materials, and conservation algorithms for numerical integration of dynamic problems. Several parts of this work have been sponsored by the Danish Technical Research Council. The work on bringing it all together was initiated during a visiting appointment as Melchor Professor at the University of Notre Dame, Indiana, in the fall of 2001. The final stage has been combined with courses at Helsinki University of Technology 2004, and at Aalborg University and Lund Institute of Technology 2005.

1

Introduction

Many problems of practical interest involve non-linear behavior of solids and structures. In the present context a solid means a body with a firm shape, as opposed to a fluid, while a structure refers to a solid composed of slender elements such as beams, plates and shells. Typical problems are the motion of robots, collapse scenarios of structures, metal forming processes in industrial production, and material deformation and failure in geotechnical engineering. These problems typically involve a considerable change of shape, often accompanied by non-linear material behavior.

The finite element method is an important tool for the analysis of non-linear problems, such as geometrical and material non-linear behavior of solids and structures. The solution of non-linear problems by the finite element method involves modeling, leading to the formulation of an appropriate set of non-linear equations describing the problem, followed by an appropriate strategy for the numerical solution of these equations. In contrast to linear problems, where the solution strategy reduces to the solution of a system of linear equations, the solution phase in a non-linear problem typically involves an iterative procedure.

Non-linear modeling and analysis is a very active research area with many engineering applications. The many different aspects involved are not covered in any single text. However, some central references to general texts should be given here. A brief introduction to some of the basic problems of non-linear finite element analysis of solids and structures is included in the book by Cook *et al.* (1989). A general state-of-the-art presentation of the finite element method, including the non-linear aspects of solids, structures and fluids, has been given in Zienkiewicz and Taylor (2000). A presentation with main emphasis on incremental formulation of geometrically non-linear problems, including details of implementation, has been given by Bathe (1996). The books by Crisfield (1991, 1997) and Belytschko *et al.*

(2000) are entirely devoted to non-linear analysis of solids and structures, combining illustrative examples with specific finite element procedures.

The present text is an introduction to some of the central ideas of non-linear modeling and finite element analysis. It covers theoretical aspects of geometric and material non-linearity and associated numerical techniques. The text proceeds from the elementary level to a fairly rigorous presentation of ideas used in current research. Only the main ideas can be covered, and the references should be consulted according to need. This first chapter gives an illustration of geometric non-linear behavior with reference to a simple two-element truss model. The example serves as a vehicle for an informal introduction to a non-linear load–displacement relation, the tangent stiffness, and the relation between the equilibrium and the virtual work approach to the problem. The example also provides a simple realistic non-linear equation on which to try different variants of the Newton–Raphson solution technique.

1.1 A simple non-linear problem

The simple two-element truss model shown in Fig. 1.1 has often been used to illustrate some basic features of geometric non-linear behavior, see e.g. Bathe (1996, p. 494) and Crisfield (1991, pp. 2–13). The structure consists of two identical truss elements, loaded with a vertical force f at the center and simply supported at the other ends. The vertical displacement at the center is called u. In the initial configuration the length of the bars is l_0.

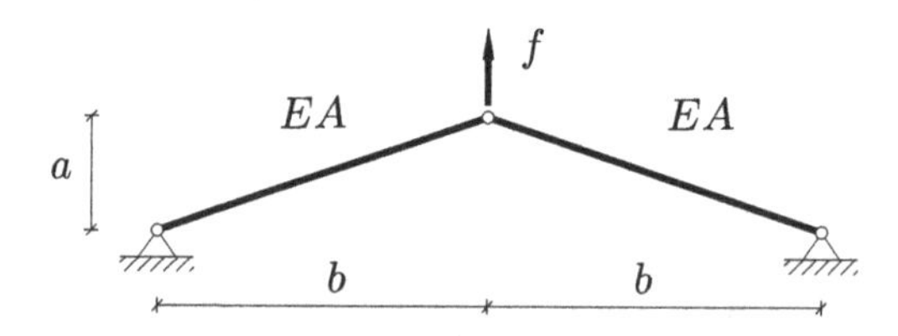

Fig. 1.1. Two-element truss model.

Application of the load leads to a deformed state with vertical displacement u of the central node, Fig. 1.2. The structure is assumed to be shallow, i.e. $a \ll b$. This permits series expansion of the square roots defining the original bar length l_0 and the bar length l corresponding to the current deformed state:

$$l_0 = \sqrt{b^2 + a^2} \simeq b\Big(1 + \frac{1}{2}\frac{a^2}{b^2}\Big), \tag{1.1}$$

$$l = \sqrt{b^2 + (a+u)^2} \simeq b\Big[1 + \frac{1}{2}\Big(\frac{a+u}{b}\Big)^2\Big]. \tag{1.2}$$

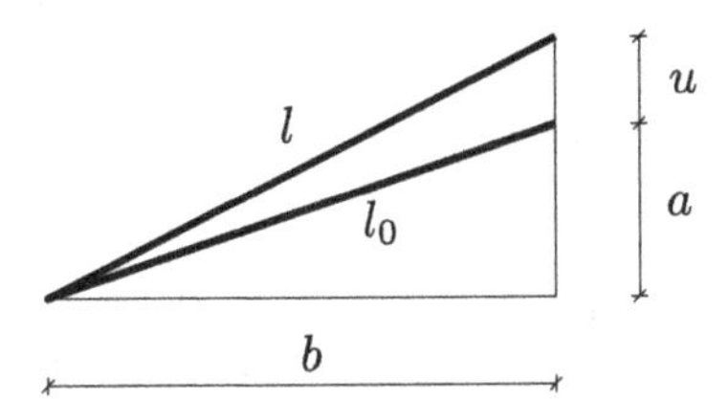

Fig. 1.2. Initial length l_0 and current length l.

The deformation of the bars is described by their elongation. A non-dimensional measure of deformation is the engineering strain, defined as the elongation relative to the original length,

$$\varepsilon = \frac{l - l_0}{l_0} \simeq \frac{a}{l_0}\frac{u}{l_0} + \frac{1}{2}\Big(\frac{u}{l_0}\Big)^2. \tag{1.3}$$

The first term is the linear part of the strain, while the second term is non-linear. A true measure of deformation must not be influenced by any rigid body motion of the bar, and thus a true deformation measure must be a non-linear function of the displacement component(s). If the displacement u is small relative to all characteristic lengths of the geometry – l_0 and a – the linear term will constitute a fair approximation, but if this approximation is used, some of the characteristic non-linear features of the problem are lost.

1.1.1 Equilibrium

The two bars are assumed to be linear elastic with axial stiffness EA, where E is the elastic modulus and A is the cross-section area. Thus, the axial force in each bar is expressed in terms of the strain as

$$N = EA\,\varepsilon \simeq EA\Big[\frac{a}{l_0}\frac{u}{l_0} + \frac{1}{2}\Big(\frac{u}{l_0}\Big)^2\Big]. \tag{1.4}$$

Equilibrium of the central node in the deformed state requires that the external force f is equal to the internal force $g(u)$ generated by deformation of the structure. Projection of the normal force gives

$$g(u) = 2\,N\,\frac{a+u}{l} \simeq \frac{2EA}{l_0^3}\big(au + \tfrac{1}{2}u^2\big)(a+u). \tag{1.5}$$

In non-dimensional form this is

$$g(u) = 2\,EA\left(\frac{a}{l_0}\right)^3\left[\frac{u}{a} + \frac{3}{2}\left(\frac{u}{a}\right)^2 + \frac{1}{2}\left(\frac{u}{a}\right)^3\right], \tag{1.6}$$

where the normalized displacement is u/a. The load–displacement relation (1.6) is shown in Fig. 1.3 corresponding to a downward load.

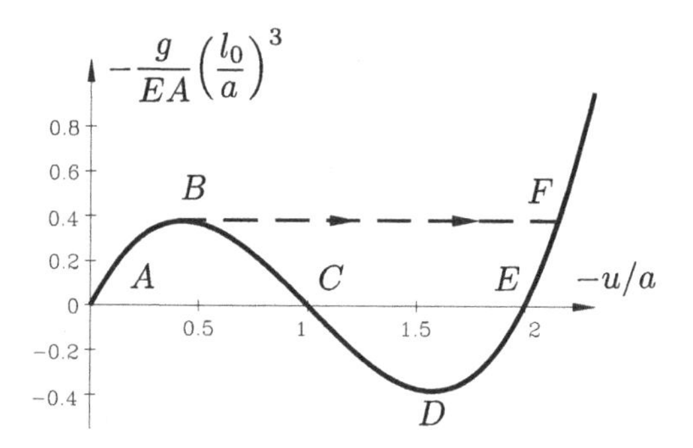

Fig. 1.3. Load–displacement curve for two-element truss.

From the unloaded state A an increasing downward load leads to a local maximum B. In this state the structure cannot support a further increase of the load. Thus, further increase of the load from B would lead to snap-through to F. The snap-through is an unstable dynamic process, and thus the load–displacement curve in Fig. 1.3 is not fully representative. Alternatively the structure may be loaded in displacement control, in which the central node is given a controlled downward displacement $-u$. This would require an increasing load from A to B, and then a decreasing load from B to C, where $u = -a$ and the two bars form a straight line. An upward force is now required to proceed to D and E, where the structure is stress-free, forming an angle symmetric to the original configuration with respect to the base line. Further downward load leads through F with increasing stiffness of the structure.

For a structure with one degree of freedom, the stiffness is a measure of the change in force for a given change in displacement. Thus, the tangent stiffness K is defined as the stiffness corresponding to infinitesimal changes in u and g:

$$K = \frac{dg}{du}. \tag{1.7}$$

In the present case the tangent stiffness K follows from straightforward differentiation of (1.6):

$$K = \frac{2EA}{l_0}\Big(\frac{a}{l_0}\Big)^2\Big[1+3\Big(\frac{u}{a}\Big)+\frac{3}{2}\Big(\frac{u}{a}\Big)^2\Big]. \tag{1.8}$$

Although this expression defines the tangent stiffness K, it does not convey the physics of the problem very clearly. This is better accomplished by differentiation of the equilibrium equation (1.5):

$$K = \frac{d}{du}\Big(2N\frac{a+u}{l_0}\Big) = 2\frac{EA}{l_0}\Big(\frac{a+u}{l_0}\Big)^2 + 2\frac{N}{l_0}. \tag{1.9}$$

Here $a+u$ is the height of the structure in the current state, while N is the current value of the axial force. The first term is due to changes in the normal force N, while the second term is due to changes in the geometric configuration with constant normal force N. Sometimes the first term is separated into a constant corresponding to $u = 0$ and the rest, whereby (1.9) takes the form

$$\begin{aligned} K &= 2\frac{EA}{l_0}\Big(\frac{a}{l_0}\Big)^2 + 2\frac{EA}{l_0}\frac{2au+u^2}{l_0^2} + 2\frac{N}{l_0} \\ &= \quad K_0 \quad + \quad K_u \quad + K_\sigma, \end{aligned} \tag{1.10}$$

where K_0 is the linear stiffness, K_u is the initial displacement stiffness, and K_σ is the initial stress stiffness. In an incremental procedure, where the geometry is updated, the current value of u is absorbed in the updated value of a, and in that case the initial displacement stiffness K_u vanishes.

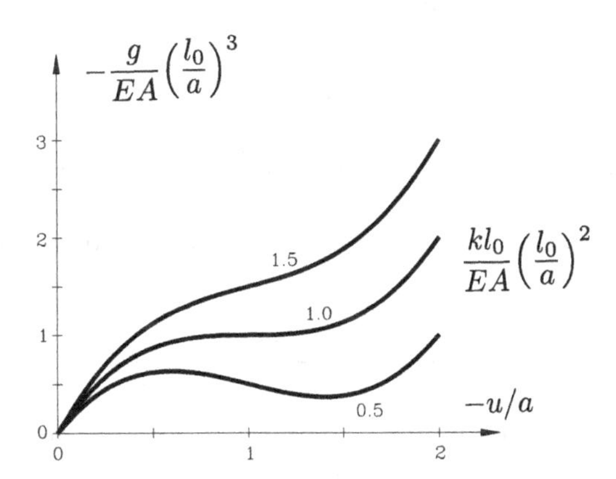

Fig. 1.4. Load–displacement curve for two-element truss with spring.

A family of load–displacement curves with different degrees of non-linearity can be obtained by introducing a vertical linear elastic spring with stiffness

k at the central node of the structure. The load–displacement relation (1.6) is changed to

$$g(u) = 2EA\left(\frac{a}{l_0}\right)^3\left[\frac{u}{a} + \frac{3}{2}\left(\frac{u}{a}\right)^2 + \frac{1}{2}\left(\frac{u}{a}\right)^3\right] + ku \tag{1.11}$$

and the tangent stiffness (1.8) to

$$K = \frac{2EA}{l_0}\left(\frac{a}{l_0}\right)^2\left[1 + 3\left(\frac{u}{a}\right) + \frac{3}{2}\left(\frac{u}{a}\right)^2\right] + k. \tag{1.12}$$

Figure 1.4 shows the load–displacement curve for different values of the spring stiffness k. For $k \geq EAa^2/l_0^3$ the variation of load with displacement is monotonic, corresponding to $K \geq 0$.

1.1.2 Virtual work and potential energy

The load–displacement relations (1.6) and (1.11) were obtained from equilibrium of the center node. For structures with more degrees of freedom or more complicated elements it is often convenient to make use of the principle of virtual work. Essentially, the principle of virtual work is a restatement of a set of equilibrium equations, where each equation is multiplied by a corresponding infinitesimal virtual displacement component. With an appropriate definition of the force and displacement components summation of their products forms a scalar invariant, known as the virtual work.

In the particular example of the two-element truss with an elastic spring the equilibrium equation can be written as

$$2\,N\,\frac{a+u}{l} + ku - f = 0. \tag{1.13}$$

Multiplication by a virtual displacement δu gives the virtual work equation

$$\delta V = 2\,N\,\frac{a+u}{l}\,\delta u + (ku)\delta u - f\delta u = 0. \tag{1.14}$$

The displacement factor in the first term is similar to the first variation of the strain (1.3):

$$\delta\varepsilon = \frac{\partial}{\partial u}\left[\frac{a}{l_0}\frac{u}{l_0} + \frac{1}{2}\left(\frac{u}{l_0}\right)^2\right]\delta u = \frac{a+u}{l_0}\frac{\delta u}{l_0}. \tag{1.15}$$

If, for the time being, the difference between l_0 and l is neglected, the virtual work equation (1.14) can now be written as

$$\delta V \simeq 2\int_0^{l_0} N\,\delta\varepsilon\,ds + (ku)\,\delta u - f\,\delta u = 0. \tag{1.16}$$

The integral is the internal virtual work of the bar elements, the second term is the virtual work of the elastic spring, while the last term is the external virtual work.

Apart from the factor l_0/l that is somehow missing, the use of virtual work in the present case where $\delta\varepsilon$ is constant within the elements is almost trivial. However, for more general problems with more degrees of freedom and non-trivial displacement fields within the elements, the principle of virtual work is an important tool for establishing the balance equations of the discretized model. The question of the factor l_0/l is discussed in Chapter 2, where the theory of non-linear bar elements is discussed more rigorously. Here, the relation between virtual work and potential energy is discussed briefly before turning to elementary numerical solution methods for non-linear equilibrium equations.

When the internal forces such as the axial force N and the spring force ku are functions of the state of displacement given by u, and the external load is also a function of u, the virtual work δV can be considered as the differential of an energy function $\Phi(u)$ – the potential energy. In the present case (1.16) is written as

$$\delta\Phi(u) = 2\int_0^{l_0} EA\,\varepsilon\delta\varepsilon\, ds + ku\,\delta u - f\,\delta u. \tag{1.17}$$

This relation can be integrated with respect to the displacement u, giving the following expression for the potential energy:

$$\Phi(u) = 2\int_0^{l_0} \tfrac{1}{2}EA\varepsilon^2\, ds + \tfrac{1}{2}k\,u^2 - fu. \tag{1.18}$$

The potential energy is the internal strain energy of the structure, including the spring, minus the external work represented by fu. For linear elastic structures it may be simpler to derive the equilibrium equations from the potential energy by considering an incremental change δu of the displacements. However, the principle of virtual work is valid irrespective of the specific material behavior, and thus the principle of virtual work has become the method of choice for setting up equilibrium equations.

1.2 Simple non-linear solution methods

For a system with only one degree of freedom non-linear behavior can often be described explicitly as a function of the displacement u, and the problem may then be considered as one of displacement control. However, in the case of several degrees of freedom the use of displacement control is non-trivial, and most problems are formulated in terms of a load history, for which the corresponding displacement history is to be calculated. This requires the solution of a system of non-linear equations. Here some of the simpler methods for solving non-linear equations are briefly introduced, leaving more specialized techniques to Chapter 8. The methods are illustrated for a single degree of freedom and then generalized to matrix form.

1.2.1 Explicit incremental method

An explicit incremental method, often called the Euler explicit method, is obtained by replacing the differentials in the definition (1.7) of the tangent stiffness with finite increments Δf and Δu:

$$\Delta u = K^{-1} \Delta f. \tag{1.19}$$

The load–displacement history is described by a number of increments Δf_n, Δu_n, $n = 1, 2, \ldots$ defining the states

$$f_n = f_{n-1} + \Delta f_n, \quad u_n = u_{n-1} + \Delta u_n, \quad n = 1, 2, \ldots \tag{1.20}$$

In the explicit incremental method the tangent stiffness K corresponds to the state at the beginning of the increment. Thus, the precise form of (1.19) is

$$\Delta u_n = K^{-1}(u_{n-1}) \Delta f_n, \quad n = 1, 2, \ldots \tag{1.21}$$

This procedure is illustrated in Fig. 1.5.

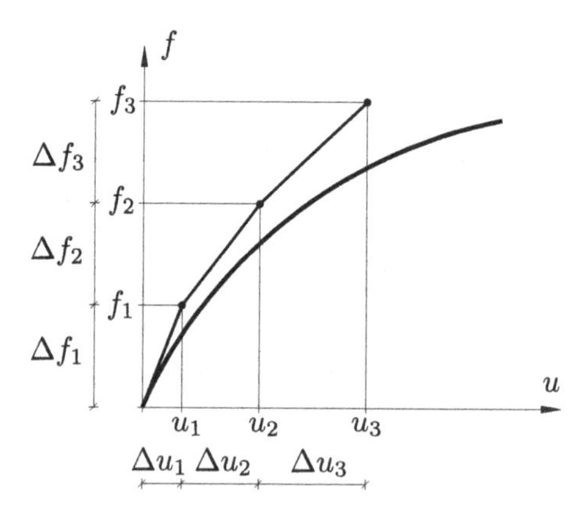

Fig. 1.5. Explicit incremental method.

It is seen that the computed states deviate more and more from the exact load–displacement curve. There are two reasons for this: the tangent stiffness of each increment is taken at the left end-point and in this particular case overestimates the stiffness, and deviations from the exact curve are added to a cumulative error. While it is difficult to use an exact representation for the stiffness corresponding to the full increment, the problem of increasing deviations can be countered by introducing equilibrium iterations as discussed in the following.

The explicit incremental method is easily generalized to multi-degree of freedom systems. Let the displacement vector be $\mathbf{u}$ and the corresponding load vector $\mathbf{f}$. The tangent stiffness matrix $\mathbf{K}$ is then defined by

$$d\mathbf{f} = \mathbf{K}(\mathbf{u})\, d\mathbf{u}. \tag{1.22}$$

The corresponding explicit incremental method is

$$\Delta\mathbf{u}_n = \mathbf{K}^{-1}(\mathbf{u}_{n-1})\, \Delta\mathbf{f}_n, \quad n = 1, 2, \ldots \tag{1.23}$$

The use of the inverse matrix $\mathbf{K}^{-1}$ in (1.23) should not be taken literally. In practice the matrix $\mathbf{K}$ is factored and the product $\mathbf{K}^{-1}\Delta\mathbf{f}$ found by back substitution.

1.2.2 Newton–Raphson method

In order to avoid accumulating errors in each additional load step, equilibrium iterations may be used to establish equilibrium to a desired degree of accuracy at each load step. This procedure is a special instance of the Newton–Raphson method, well known from numerical analysis. In principle, the method works by applying two steps intermittently: (i) check if equilibrium is satisfied to within the desired accuracy; (ii) if not, make a suitable adjustment of the state of deformation.

The first step consists in checking the equilibrium equation. This is done by forming the difference between the external load f and internal force $g(u)$,

$$r(u, f) = f - g(u) = 0, \tag{1.24}$$

where $r(u, f)$ is called the residual force. In a state of equilibrium the internal force $g(u)$ is equal to the external load f, and thus the residual vanishes. In practice, lack of equilibrium will be produced at the beginning of each load increment, where the load f is increased, while no new displacement estimate u is yet available. Thus, the need arises for obtaining an improved estimate of the state of displacement u.

In the absence of equilibrium an improved estimate of the displacement u is obtained from a linearized form of the residual $r(u + \delta u, f)$ around the known residual $r(u, f)$,

$$r(u + \delta u, f) = r(u, f) + \delta r(u, f) + \cdots = 0. \tag{1.25}$$

The dots indicate higher-order terms, because δr is only a linearized form of the increment of the residual. In the classic form of equilibrium iterations the load f is assumed fixed within the given load step, and thus the increment of the residual only depends on the internal force $g(u)$. The linearized increment is then given by the first derivative of the internal force as

$$\delta r = -\frac{dg(u)}{du}\,\delta u = -K(u)\,\delta u. \tag{1.26}$$

Here the tangent stiffness K, introduced in (1.7), has been introduced. The displacement increment is now obtained from the linearized form of (1.25) by substitution of the tangent stiffness relation (1.26). When rearranging the terms in (1.25), the linearized equation becomes

$$K(u)\,\delta u = r. \tag{1.27}$$

In this equation the residual $r(u, f)$ is known, as it relates to the current state of load f and displacement u. The tangent stiffness $K(u)$ at the current displacement state u can also be calculated. Thus, this equation permits determination of the displacement increment δu,

$$\delta u = K^{-1}(u)\,r. \tag{1.28}$$

Once the displacement increment δu is determined, the current displacement state is updated to

$$u^i = u^{i-1} + \delta u^i. \tag{1.29}$$

In this equation the superscript is used to indicate that the iteration i changes the estimated displacement from u^{i-1} to u^i. In a computer program the iteration superscript i is not needed, as the register containing u^{i-1} is simply overwritten by the new value u^i according to the assignment statement

$$u := u + \delta u. \tag{1.30}$$

Here, $:=$ is the assignment operator, implying that the variable u is assigned a new value. In this book many of the algorithms are presented in the form of pseudocode – i.e. a code format that appears like high-level programs such as MATLAB. In the pseudocode presented here assignments are indicated by the normal equality sign, as all equalities are assignment statements.

The Newton–Raphson equilibrium iteration procedure is illustrated in Fig. 1.6. The figure shows load step n. This load step starts from a state of equilibrium already established at the previous load f_{n-1} with displacement u_{n-1}. The load step is initiated by increasing the load by Δf_n to f_n. This generates the first residual $r_n^1 = \Delta f_n$. This residual and the tangent stiffness

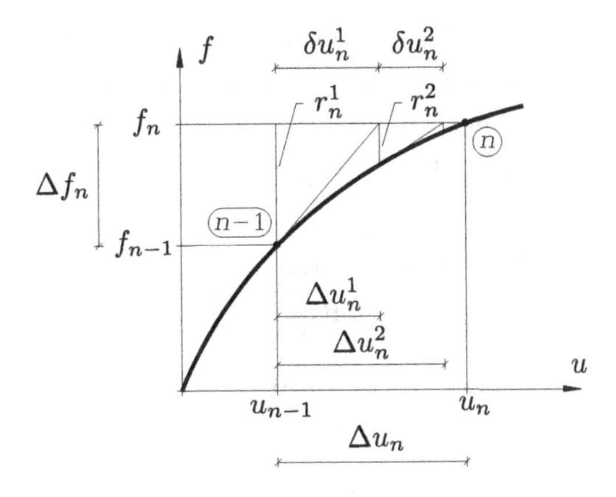

Fig. 1.6. Newton–Raphson equilibrium iterations.

$K(u_{n-1})$ lead to the displacement increment δu_n^1, shown in the figure. At the new displacement $u_{n-1}+\delta u_n^1$, the internal force g – represented by the curve – is still smaller than the imposed load. The difference forms the residual r_n^2, and the procedure is continued. It should be noted that the use of sub- and superscripts to indicate load step and iteration number, respectively, is merely for illustration in relation to the figure. These indices are not needed when programming the algorithm.

The iteration process needs a termination criterion. This may be taken as the requirement that the current residual force r_n should be less than a prescribed fraction ϵ of the load increment Δf of the present load step,

$$|r| < \epsilon\,|\Delta f|. \tag{1.31}$$

The value of ϵ could be on the order of say 10^{-4}–10^{-6}. For structures developing very small stiffness, the criterion (1.31) may be supplemented by the displacement criterion

$$|\delta u| < \epsilon\,|\Delta u|, \tag{1.32}$$

where Δu is the total displacement increment accumulated in the present load step.

In the corresponding multi-component problem with displacement vector $\mathbf{u}$ and load vector $\mathbf{f}$, the residual force vector is

$$\mathbf{r}(\mathbf{u},\mathbf{f}) = \mathbf{f} - \mathbf{g}(\mathbf{u}). \tag{1.33}$$

The tangent stiffness matrix is defined by the incremental change of the

ALGORITHM 1.1. Newton–Raphson method.

Load steps $n = 1, 2, \ldots, n_{\max}$

$$\mathbf{f}_n = \mathbf{f}_{n-1} + \Delta\mathbf{f}_n$$

$$\mathbf{u}_n = \mathbf{u}_{n-1}$$

Iterations $i = 1, 2, \ldots, i_{\max}$

$$\mathbf{K}_n = \frac{d\mathbf{g}(\mathbf{u}_n)}{d\mathbf{u}}$$

$$\mathbf{r}_n = \mathbf{f}_n - \mathbf{g}(\mathbf{u}_n)$$

$$\delta\mathbf{u}_n = \mathbf{K}_n^{-1}\mathbf{r}_n$$

$$\mathbf{u}_n = \mathbf{u}_n + \delta\mathbf{u}_n$$

Stop iteration when $\|\mathbf{r}_n\| < \epsilon\,\|\Delta\mathbf{f}_n\|$

End of load step

---

internal forces,

$$\mathbf{K}(\mathbf{u}) = \frac{d\mathbf{g}(\mathbf{u})}{d\mathbf{u}}. \tag{1.34}$$

The tangent stiffness is now introduced into a linearized form of the equilibrium condition, whereby the following vector equation is obtained for the displacement sub-increment $\delta\mathbf{u}$:

$$\mathbf{K}(\mathbf{u})\,\delta\mathbf{u} = \mathbf{r}. \tag{1.35}$$

In contrast to the one-dimensional case, the solution of these equations may be a non-trivial part of the procedure. The termination criteria will typically make use of the 'length' of the corresponding vectors, whereby

$$(\mathbf{r}_n^T\mathbf{r}_n)^{1/2} < \epsilon\,(\Delta\mathbf{f}^T\Delta\mathbf{f})^{1/2}, \tag{1.36}$$

$$(\delta\mathbf{u}^T\delta\mathbf{u})^{1/2} < \epsilon\,(\Delta\mathbf{u}^T\Delta\mathbf{u})^{1/2}. \tag{1.37}$$

The Newton–Raphson procedure is summarized as Algorithm 1.1. Note that in the iteration loop the computer overwrites quantities like $\mathbf{r}_n$ by their new value in the same register. Therefore, the superscript i does not appear explicitly in the algorithm. Similarly, the load step subscript n is only used to avoid the explicit indication of storing the result of each completed load step. The actual algorithm is conveniently programmed without the use of indexed variables in the iteration loops.

1.2.3 Modified Newton–Raphson method

In the original Newton–Raphson method the current tangent stiffness matrix $\mathbf{K}(\mathbf{u})$ is computed and factored in each iteration. For non-linear problems with a single or a few degrees of freedom this is usually not a problem, but for problems with many degrees of freedom the computational cost involved in forming the stiffness matrix $\mathbf{K}(\mathbf{u})$ and solving the corresponding equations for $\delta\mathbf{u}$ in each iteration may be considerable. It is seen from Fig. 1.6 and Algorithm 1.1 that $\mathbf{K}_n$ appears within the inner loop. A simple modification of the Newton–Raphson method consists in moving the stiffness matrix $\mathbf{K}$ outside the iteration loop. Then $\mathbf{K} = \mathbf{K}_{n-1}$ is only computed and factored once for each load step on the basis of the previous state of displacement $\mathbf{u}_{n-1}$,

$$\mathbf{K}_{n-1} = \frac{d\mathbf{g}(\mathbf{u}_{n-1})}{d\mathbf{u}}. \tag{1.38}$$

This simplifies the iteration loop as shown in Fig. 1.7 and Algorithm 1.2.

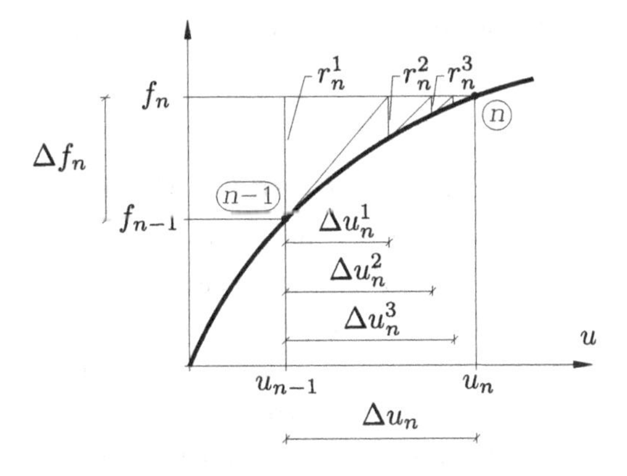

Fig. 1.7. The modified Newton–Raphson method.

The asymptotic convergence of the modified Newton–Raphson method is slower than that of the Newton–Raphson method, and this may offset some of its computational efficiency. A different, more refined type of modification makes use of a secant approximation of $\mathbf{K}$. This requires a non-trivial generalization of the secant concept to multi-degree of freedom systems. The corresponding methods, called quasi-Newton methods, are described in Chapter 8.

The classic Newton methods encounter problems at a load maximum. Several methods have been developed to deal with this problem. A common feature of these methods is that the load increment is also subject to changes during the iterations, e.g. by linking load and displacement increments. This

ALGORITHM 1.2. Modified Newton–Raphson method.

Load steps $n = 1, 2, \ldots, n_{\max}$

$$\mathbf{f}_n = \mathbf{f}_{n-1} + \Delta \mathbf{f}_n$$

$$\mathbf{u}_n = \mathbf{u}_{n-1}$$

$$\mathbf{K}_{n-1} = \frac{d\mathbf{g}(\mathbf{u}_{n-1})}{d\mathbf{u}}$$

Iterations $i = 1, 2, \ldots, i_{\max}$

$$\mathbf{r}_n = \mathbf{f}_n - \mathbf{g}(\mathbf{u}_n)$$

$$\delta\mathbf{u}_n = \mathbf{K}_{n-1}^{-1}\mathbf{r}_n$$

$$\mathbf{u}_n = \mathbf{u}_n + \delta\mathbf{u}_n$$

Stop iteration when $\|\mathbf{r}_n\| < \epsilon \, \|\Delta\mathbf{f}_n\|$

End of load step

---

introduces a kind of displacement control near the maximum, as discussed in Chapter 8.

1.3 Summary and outlook

Non-linear problems of structures and solids involve processes in which neighboring states are connected by non-linear relations. This is illustrated by the simple example of a two-bar truss loaded by a quasi-static force. A model of the problem requires representation of the material behavior and the formulation of suitable equilibrium equations. In this introductory chapter these issues were dealt with in an ad hoc fashion, also introducing approximations to simplify the presentation. In the following chapters these issues are dealt with in a rigorous way within the framework of finite element analysis.

While the equilibrium equations of simple systems often can be formulated directly, it is generally advantageous to use the principle of virtual work, which retains its simplicity for larger and more complicated systems. Several aspects of this will appear in later chapters. The idea of the principle of virtual work is to consider the work done by the actual forces through an imagined – or virtual – displacement field. This changes a multi-component problem into a similar number of scalar problems. A necessary requirement is that the product of the internal forces and the virtual measures of deformation constitute work. This property of the virtual work serves to identify suitable internal force and stress measures, when a displacement and strain representation has been selected. This is illustrated in the next chapter in connection with a general discussion of bar elements, and is used to define several stress measures for solids in Chapter 6. The virtual work also plays a key role in defining the properties associated with rotations and moments as discussed in Chapter 4.

In order to obtain the solution to specific problems, the material behavior must be represented in the form of a relation between the internal forces and the corresponding measures of deformation. In the problem of the two-bar truss the bars were assumed to be linear elastic, thus simplifying the presentation. In many situations non-linear material behavior is an important part of the problem. The basic form of the equations of elasto-plastic material behavior is described in Chapter 7.

Even the simple two-bar truss exhibits a non-monotonic relation between load and displacement. The simple Newton-type solution methods briefly outlined in this chapter need to be modified in order to enable computation of non-monotonic force–displacement relations. This problem is dealt with in Chapter 8 on numerical methods. The chapter on numerical methods can be read without reference to the chapters before, and indeed reading this chapter first will enable the reader to supplement the theory of the intermediate chapters with non-trivial numerical examples. In the final chapter the numerical methods are extended to dynamic problems with inertial effects.

1.4 Exercises

Exercise 1.1 Consider the load–displacement curve of the two-element truss shown in Fig. 1.3.

(a) Determine the non-dimensional coordinates to the points B and D.

(b) Select a suitable load step magnitude and sketch the states produced by the explicit incremental method, the Newton–Raphson method, and the modified Newton–Raphson method. Note in particular the passage of the maximum at B.

Exercise 1.2 Consider the two-element truss shown in Fig. 1.1 and add a vertical spring at the central node with spring stiffness $k = 1.2EAa^2/l_0^3$.

(a) Introduce the non-dimensional displacement $v = -u/a$ and the non-dimensional load $p = -fl_0^3/(EAa^3)$. Give the relation between p and v and the corresponding tangent stiffness.

(b) Organize the explicit incremental method, the Newton–Raphson method, and the modified Newton–Raphson method in tabular form and compute the p–v relation with a suitable load step, e.g. 0.2–0.4. Sketch the result for each of the three methods.

Exercise 1.3* Implement the Newton–Raphson algorithm shown as Algorithm 1.1 in MATLAB for analysis of the two-bar truss treated in Section 1.1. Organize the implementation in four m-files:

(a) `data.m`: script file containing the model parameters and the load increment, e.g. $b = 1.0$, $a = 0.1$, $EA = 1.0$, $k = 0$, and $\Delta f = -0.0001$.

(b) `g_bar.m`: the internal force in (1.24), obtained as a function of the displacement u from the expression (1.11).

(c) `kt_bar.m`: the function $K(u)$ in (1.12), giving the tangent stiffness as a function of the displacement u.

(d) `nr_bar`: script file containing the Newton–Raphson algorithm from Algorithm 1.1. (Use limited loops, e.g. $n = 1 : n_{\max}$ and $i = 1 : i_{\max}$, with $n_{\max} = 20$ and $i_{\max} = 8$.)

Use the program to study the behavior of the Newton–Raphson algorithm via plots of f as a function of displacement u, particularly near the turning point of the load–displacement curve. Use e.g. $\Delta f = -1.0 \times 10^{-4}$ and $\Delta f = -0.5 \times 10^{-4}$.

Exercise 1.4* Make a modified version `mnr_bar.m` of the driver routine `nr-bar.m` from Exercise 1.3 using the modified Newton–Raphson algorithm. Use the two routines to study the behavior of the algorithms for different values of the spring constant k. Note in particular that the modified Newton–Raphson algorithm has convergence problems for *increasing* stiffness. Sketch this problem.

Exercise 1.5* Both the full and the modified form of the Newton–Raphson algorithm are based on prescribed load increments, and therefore require monotonically changing load. In the two-bar truss problem the load will increase monotonically if a vertical spring with stiffness $k > (EA/l_0)(a/l_0)^2$ is introduced. The actual force transferred to the truss is $f_{\text{eff}} = f - ku$.

Introduce a spring of sufficient stiffness and solve the problem with the program developed in Exercise 1.3. Plot the effective load f_{eff} against the displacement u to demonstrate recovery of the full curve from Fig. 1.3. For a very stiff spring equal load increments correspond to equal displacements, and the method is equivalent to the use of displacement control. This method is only directly applicable in the case of a single load component. General solution methods are discussed in Chapter 8.

2

Non-linear bar elements

The finite element method has been the method of choice for modeling and analysis of structures and solids for several decades. The basic idea is that the structure (or solid) is considered as an assembly of elements, and that each element is modeled in a standard format that permits repetitive use of the individual element formats. Bar elements only contain a single internal degree of freedom – the elongation – and they are therefore a convenient means for introducing and illustrating the basic features of geometrically non-linear finite element analysis.

In a geometrically non-linear problem the first question to arise is the definition of a non-linear measure of deformation, the strain. This is addressed in Section 2.1. When a structure is assembled from the individual elements, use is generally made of the principle of virtual work. The principle of virtual work is a restatement of the equilibrium equations in scalar form. It turns out that once a non-linear strain definition has been adopted, the corresponding definition of stress follows from the formulation of the principle of virtual work. This is the subject of Section 2.2. The tangent stiffness matrix of a geometrically non-linear bar element is derived in Section 2.3 in global coordinates.

The derivation of the equilibrium and stiffness relations of the bar element is quite simple because the strain is constant within the element. In order to illustrate the relation to more complex problems involving other types of elements, the tangent stiffness matrix is re-derived by use of shape functions in global coordinates. This indicates the procedure followed in isoparametric solid elements. Another alternative makes use of a local coordinate system rotating together with the element during displacement. This so-called co-rotational approach is typically used for generalized continuum elements such as beams and shells, where the displacement representation depends on the local orientation of the beam axis, middle plane, etc. Co-rotating beam elements are treated in Chapter 5. Bar elements have been given an extensive treatment by Crisfield (1991), and Mattiasson (1983) has described the co-rotational formulation of bar elements in detail.

The implementation of the finite element method involves the assembly of elements into the global structure. A brief sketch of the assembly procedure is given in Section 2.5. The actual solution of the finite element relations can follow one of two formulations. In the first the initial configuration is used as reference through the full analysis, and total displacements from this configuration are used. This is the total Lagrangian formulation. In the other formulation the reference state is updated each time an equilibrium state has been established. Thus, the displacements in this formulation refer to the last equilibrium state. This is the updated Lagrangian formulation. The simple bar element is used to illustrate these two formulations in Section 2.6.

In spite of its simplicity the bar element contains the main features of geometrically non-linear structural and solid elements, and Section 2.7 sums up these main features in a general form. These general features are encountered repeatedly in various settings in the following chapters.

2.1 Deformation and strain

In most finite element formulations for structures and solids the displacements are the primary variables of the problem. The displacements may lead to deformation of the elements, and this in turn to internal forces. It is important to use deformation measures that vanish identically for rigid body displacements. In a theory with finite displacements this requirement can be satisfied by several different strain definitions. The most common of these are briefly discussed here for a simple bar element to indicate their use and limitations.

In a bar element the deformation is characterized by the elongation. Figure 2.1 shows a bar element with initial length l_0. Deformation introduces the elongation u, whereby the new length is

$$l = l_0 + u. \tag{2.1}$$

A suitable measure of strain must give the relative deformation. The simplest definition is the engineering strain,

$$\varepsilon_E = \frac{u}{l_0} = \frac{l - l_0}{l_0}. \tag{2.2}$$

This is the traditional definition used in the theory of infinitesimal strain.

For finite deformation it is somewhat arbitrary to refer to the initial length

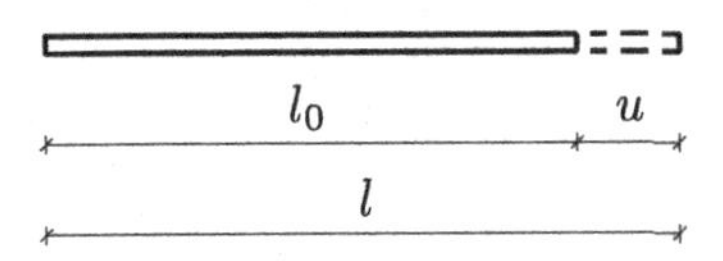

Fig. 2.1. Bar element.

l_0. Alternatively, the strain increment $\delta\varepsilon$ can be defined with reference to the current length l by the relation

$$\delta\varepsilon = \frac{\delta l}{l}. \tag{2.3}$$

Integration of this relation gives the logarithmic strain

$$\varepsilon_L = \int_{l_0}^{l} \delta\varepsilon = \ln\left(\frac{l}{l_0}\right). \tag{2.4}$$

This strain makes no distinction between initial and final length, and interchange of l_0 and l merely changes the sign of ε_L. For small strains ε_E and ε_L are nearly equal.

Many problems involve large displacements but only small to moderate strains. In those problems the important point is to use a strain definition without 'self straining', i.e. a strain definition that does not produce straining for arbitrary rigid body motion. It is then convenient to use l^2 instead of l as a basis for the strain definition. One reason for the use of l^2 is that the square of the length a of a vector $\mathbf{a} = [a_1, a_2, a_3]^T$ is the sum of the squared coordinates, $a^2 = a_1^2 + a_2^2 + a_3^2$. Another is that a definition based on l^2 is simpler to generalize to two- and three-dimensional continua. A strain definition based on l^2 can be obtained by rewriting (2.2) in the form

$$\varepsilon_E = \frac{(l - l_0)(l + l_0)}{l_0\,(l + l_0)} = \frac{l^2 - l_0^2}{l_0^2\,(2 + \varepsilon_E)}. \tag{2.5}$$

If ε_E is omitted from the denominator, the definition of the Green strain ε_G is obtained,

$$\varepsilon_G = \frac{l^2 - l_0^2}{2\,l_0^2}. \tag{2.6}$$

This strain definition, and its generalization to two and three dimensions, are often used in solid mechanics.

The different strains introduced here are related by

$$\varepsilon_L = \ln(1 + \varepsilon_E) = \tfrac{1}{2}\ln(1 + 2\varepsilon_G). \tag{2.7}$$

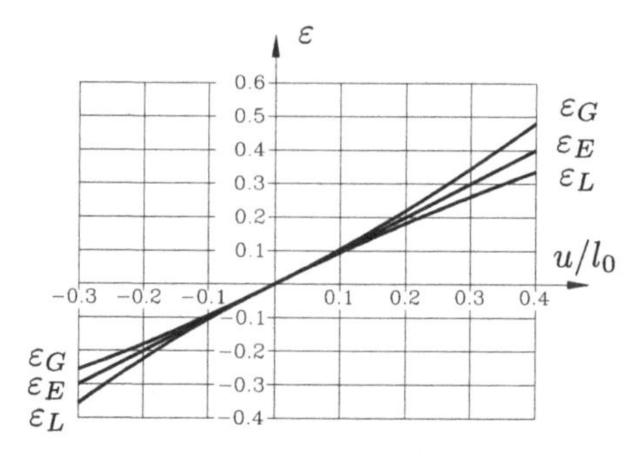

Fig. 2.2. The strain measures ε_E, ε_G and ε_L.

The strains ε_E, ε_L and ε_G are shown in Fig. 2.2 as a function of u/l_0. It is seen that $\varepsilon_G > \varepsilon_E > \varepsilon_L$. For $|u/l_0| < 0.05$ the deviation from ε_E is of the order 2–3%. For larger strains it may be necessary to distinguish between the strain to be used or account for the difference through the stress–strain relation.

2.2 Equilibrium and virtual work

Figure 2.3 shows a bar element with end-points A and B. The coordinates of A and B are referred to a global Cartesian coordinate system. In the initial configuration the coordinates are $\mathbf{x}_{A_0}$ and $\mathbf{x}_{B_0}$, respectively. Boldface letters are used to denote vectors and matrices. The dimension corresponds to the dimension of the space, i.e. 2 or 3. It turns out that this dimension is not important in the formulation of the equilibrium and stiffness relations except at the very end, when individual components are written down.

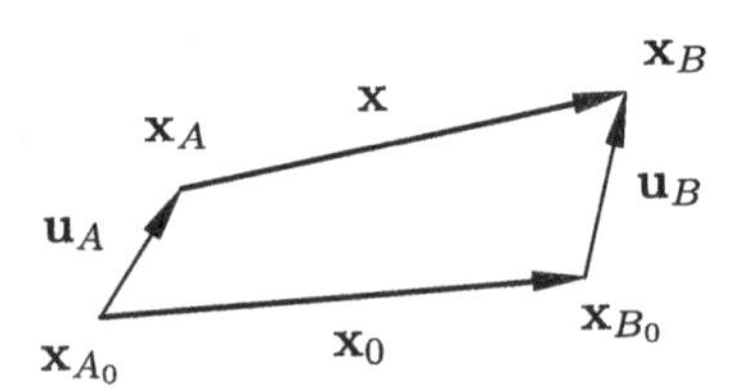

Fig. 2.3. Bar element AB in initial and displaced configurations.

The points A_0 and B_0 are now given the displacements $\mathbf{u}_A$ and $\mathbf{u}_B$, respectively. Only the direction and length of the vector $\overrightarrow{AB}$ influence the equilibrium and stiffness. The initial position is

$$\mathbf{x}_0 = \mathbf{x}_{B_0} - \mathbf{x}_{A_0}. \tag{2.8}$$

The difference between the displacement at A and B is denoted

$$\mathbf{u} = \mathbf{u}_B - \mathbf{u}_A, \tag{2.9}$$

and thus $\overrightarrow{AB}$ after displacement is

$$\mathbf{x} = \mathbf{x}_0 + \mathbf{u}. \tag{2.10}$$

It is now a simple matter to calculate the initial length l_0 and the length l after deformation:

$$l_0^2 = \mathbf{x}_0^T \mathbf{x}_0, \tag{2.11}$$

$$l^2 = \mathbf{x}^T \mathbf{x} = (\mathbf{x}_0 + \mathbf{u})^T (\mathbf{x}_0 + \mathbf{u}). \tag{2.12}$$

The strain of the bar can now be expressed in terms of l_0 and l by any of the strain measures defined in Section 2.1.

The fact that the square of the length is given in simple form suggests the use of the Green strain ε_G. From the definition (2.6) it follows that

$$\varepsilon_G = \frac{l^2 - l_0^2}{2\, l_0^2} = \frac{1}{l_0^2}(\mathbf{x}_0^T \mathbf{u} + \tfrac{1}{2}\mathbf{u}^T \mathbf{u}). \tag{2.13}$$

This formula can be interpreted as a projection of the displacement $\mathbf{u}$ on the mean vector $\mathbf{x}_{1/2} = \frac{1}{2}(\mathbf{x}_0 + \mathbf{x})$, scaled by l_0^2:

$$\varepsilon_G = \frac{1}{l_0^2}\,(\mathbf{x}_0 + \tfrac{1}{2}\mathbf{u})^T \mathbf{u} = \frac{1}{l_0^2}\,\tfrac{1}{2}(\mathbf{x}_0 + \mathbf{x})^T \mathbf{u} = \frac{1}{l_0^2}\,\mathbf{x}_{1/2}^T\, \mathbf{u}. \tag{2.14}$$

This interpretation is illustrated in Fig. 2.4. In the general formulation of the Green strain the interpretation in terms of a mean position vector may be helpful in the evaluation of ε_G. A slightly more general formulation is discussed in Exercise 2.1.

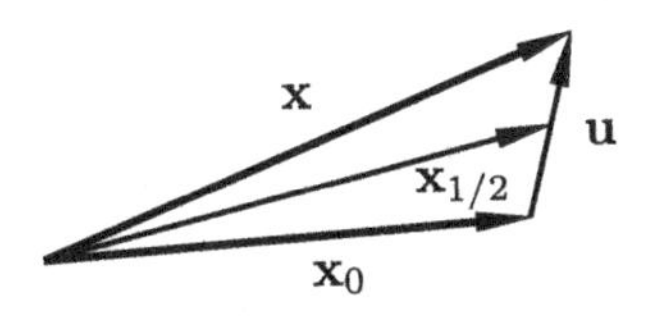

Fig. 2.4. Projection of displacement $\mathbf{u}$ on mean vector $\mathbf{x}_{1/2} = \frac{1}{2}(\mathbf{x}_0 + \mathbf{x})$.

The variation of the strain (2.13) is needed for the formulation of the principle of virtual work:

$$\delta\varepsilon_G = \frac{1}{l_0^2}(\mathbf{x}_0 + \mathbf{u})^T \delta\mathbf{u} = \frac{1}{l_0^2}\mathbf{x}^T \delta\mathbf{u}. \tag{2.15}$$

Note that the variation of the strain consists of the projection of the displacement variation $\delta\mathbf{u}$ on the current vector $\mathbf{x}$, scaled by l_0^2.

The universal method of establishing finite element relations consists in the use of the principle of virtual work. In the case of bar elements the state of deformation within the element is homogeneous, and the relations could therefore in principle be obtained directly, but in order to illustrate the general procedure and to identify precisely the internal force that is conjugate to the Green strain, the principle of virtual work is also used here. According to the principle of virtual work an arbitrary displacement variation must lead to identical internal and external work. The internal virtual work is expressed in terms of the internal force(s), while the external virtual work is expressed in terms of the external loads. Thus, the principle of virtual work establishes a relation between the internal and the external forces.

For a bar element with the external forces $\mathbf{f}_A$ and $\mathbf{f}_B$ acting at A and B, respectively, the principle of virtual work is

$$\delta V = \int N\,\delta\varepsilon\,ds - \mathbf{f}_A^T\,\delta\mathbf{u}_A - \mathbf{f}_B^T\,\delta\mathbf{u}_B = 0. \tag{2.16}$$

In order to be specific, a strain measure must be selected and an appropriate length measure must be used in the integral. Here the Green strain is used together with the initial length l_0 of the element. With this choice and $\delta\varepsilon_G$ from (2.15) the principle of virtual work takes the form

$$\delta V = \int_0^{l_0} \frac{1}{l_0^2}(N\,\mathbf{x}^T)(\delta\mathbf{u}_B - \delta\mathbf{u}_A)\,ds - \mathbf{f}_A^T\delta\mathbf{u}_A - \mathbf{f}_B^T\delta\mathbf{u}_B = 0. \tag{2.17}$$

Taking the transpose of (2.17) leads to changing the order of the factors, whereby

$$\begin{aligned}\delta V &= \delta\mathbf{u}_A^T\Big(-\int_0^{l_0}\frac{1}{l_0^2}(N\,\mathbf{x})\,ds - \mathbf{f}_A\Big)\\ &\quad + \delta\mathbf{u}_B^T\Big(\int_0^{l_0}\frac{1}{l_0^2}(N\,\mathbf{x})\,ds - \mathbf{f}_B\Big) = 0.\end{aligned} \tag{2.18}$$

This equation must be valid for any choice of the virtual displacements

$\delta\mathbf{u}_A$ and $\delta\mathbf{u}_B$, and therefore the following two equations are obtained after evaluating the integrals:

$$\mathbf{f}_A = -N\frac{1}{l_0}\mathbf{x}, \qquad \mathbf{f}_B = N\frac{1}{l_0}\mathbf{x}. \tag{2.19}$$

These equations give a precise definition of the normal force N appearing in the principle of virtual work as conjugate to the Green strain. The vector $\mathbf{x}/l_0$ is in the direction of the bar after displacement with length l/l_0. Thus Nl/l_0 is the actual force in the bar element. For moderate strain $l/l_0 \simeq 1$, and N gives the actual force directly. For large strains the difference between N and the actual force can be absorbed in the constitutive equation for N. The main point is that with this definition of N the use of the Green strain and initial length in the principle of virtual work (2.16) is an *exact* representation of the equilibrium equations. The use of engineering strain is discussed in Exercise 2.2.

For a linear elastic bar, in which N is proportional to ε_G, the constitutive relation is

$$N = EA\,\varepsilon_G, \tag{2.20}$$

where E is the modulus of elasticity and A is the cross-section area. The internal forces $\mathbf{q}_A, \mathbf{q}_B$ generated by the deformation of the element can then be expressed in terms of the displacements $\mathbf{u}_A$ and $\mathbf{u}_B$ by substitution of (2.20) into (2.19):

$$\mathbf{q}_A = -EA\,\varepsilon_G\frac{1}{l_0}\mathbf{x}, \qquad \mathbf{q}_B = EA\,\varepsilon_G\frac{1}{l_0}\mathbf{x}. \tag{2.21}$$

With the displacements $\mathbf{u}_A$ and $\mathbf{u}_B$ available the displacement vector $\mathbf{u}$ is evaluated from (2.9), the current vector $\mathbf{x}$ from (2.10), and ε_G from (2.14). This enables the evaluation of the internal forces $\mathbf{q}_A, \mathbf{q}_B$ in an iterative solution procedure. In order to find the corresponding displacement increment the current stiffness is needed.

The formulae for strain and equilibrium derived so far have been expressed in a compact form making use of the differences in position and displacement between the two ends of the bar. This form is convenient for derivation and discussion of the physical meaning of the formulae. However, in a finite element formulation it is important to have a convenient matrix format, in which *all* coordinates, displacements and forces appear in vector format. A tilde is introduced to denote an extended vector, containing the vectors from all element nodes. In this notation the extended position, displacement and internal force vectors of the bar element are

$$\tilde{\mathbf{x}}^T = [\mathbf{x}_A^T, \mathbf{x}_B^T], \quad \tilde{\mathbf{u}}^T = [\mathbf{u}_A^T, \mathbf{u}_B^T], \quad \tilde{\mathbf{q}}^T = [\mathbf{q}_A^T, \mathbf{q}_B^T] \tag{2.22}$$

for coordinates, displacements and forces, respectively. In terms of this notation the formula (2.11) for the initial length of the element is

$$l_0^2 = \mathbf{x}_0^T \mathbf{x}_0 = \tilde{\mathbf{x}}_0^T \begin{bmatrix} \mathbf{I} & -\mathbf{I} \\ -\mathbf{I} & \mathbf{I} \end{bmatrix} \tilde{\mathbf{x}}_0, \tag{2.23}$$

where $\mathbf{I}$ is the unit matrix corresponding to the dimension of the space, typically 2 or 3.

By use of (2.8) and (2.9) the formula (2.14) for the Green strain is then rewritten as

$$\varepsilon_G = \frac{1}{l_0^2} \tfrac{1}{2}(\tilde{\mathbf{x}}_0 + \tilde{\mathbf{x}})^T \begin{bmatrix} \mathbf{I} & -\mathbf{I} \\ -\mathbf{I} & \mathbf{I} \end{bmatrix} \tilde{\mathbf{u}} = \frac{1}{l_0^2} \tilde{\mathbf{x}}_{1/2}^T \begin{bmatrix} \mathbf{I} & -\mathbf{I} \\ -\mathbf{I} & \mathbf{I} \end{bmatrix} \tilde{\mathbf{u}}. \tag{2.24}$$

The strain increment (2.15) takes the form

$$\delta\varepsilon_G = \frac{1}{l_0^2} \tilde{\mathbf{x}}^T \begin{bmatrix} \mathbf{I} & -\mathbf{I} \\ -\mathbf{I} & \mathbf{I} \end{bmatrix} \delta\tilde{\mathbf{u}}. \tag{2.25}$$

The difference between the mean position $\tilde{\mathbf{x}}_{1/2}$ in ε_G and the current position $\tilde{\mathbf{x}}$ in $\delta\varepsilon_G$ is clear.

The internal element forces (2.21) have the following simple matrix form:

$$\tilde{\mathbf{q}} = \frac{N}{l_0} \begin{bmatrix} \mathbf{I} & -\mathbf{I} \\ -\mathbf{I} & \mathbf{I} \end{bmatrix} \tilde{\mathbf{x}} = \frac{EA\varepsilon_G}{l_0} \begin{bmatrix} \mathbf{I} & -\mathbf{I} \\ -\mathbf{I} & \mathbf{I} \end{bmatrix} \tilde{\mathbf{x}}, \tag{2.26}$$

where ε_G is evaluated from (2.24). Note that the strain and the internal forces $\tilde{\mathbf{q}}$ are non-linear functions of the displacements $\tilde{\mathbf{u}}$.

2.3 Tangent stiffness matrix

The tangent stiffness matrix gives the changes in the internal forces $\mathbf{q}_A$ and $\mathbf{q}_B$ corresponding to infinitesimal changes in the displacements $\mathbf{u}_A$ and $\mathbf{u}_B$. It is conveniently found from differentiation of (2.21a):

$$\begin{aligned} d\mathbf{q}_A &= -\,\mathbf{x}\,\frac{dN}{l_0} - \frac{N}{l_0}\,d\mathbf{x} \\ &= -\Big(\frac{\mathbf{x}}{l_0}\,\frac{dN}{d\mathbf{u}} + \frac{N}{l_0}\mathbf{I}\Big)\,d(\mathbf{u}_B - \mathbf{u}_A), \end{aligned} \tag{2.27}$$

where $\mathbf{I}$ is the unit matrix. It follows from (2.21) that

$$d\mathbf{q}_B = -\,d\mathbf{q}_A. \tag{2.28}$$

For a linear elastic bar with the constitutive relation (2.20), differentiation gives

$$\frac{dN}{d\mathbf{u}} = EA\,\frac{d\varepsilon_G}{d\mathbf{u}} = \frac{EA}{l_0^2}\,(\mathbf{x}_0^T + \mathbf{u}^T) = \frac{EA}{l_0^2}\,\mathbf{x}^T. \tag{2.29}$$

When this expression is inserted into (2.27) and (2.28), the result can be written in block matrix format as

$$\begin{bmatrix} d\mathbf{q}_A \\ d\mathbf{q}_B \end{bmatrix} = \left(\frac{EA}{l_0^3} \begin{bmatrix} \mathbf{x}\,\mathbf{x}^T & -\mathbf{x}\,\mathbf{x}^T \\ -\mathbf{x}\,\mathbf{x}^T & \mathbf{x}\,\mathbf{x}^T \end{bmatrix} + \frac{N}{l_0} \begin{bmatrix} \mathbf{I} & -\mathbf{I} \\ -\mathbf{I} & \mathbf{I} \end{bmatrix} \right) \begin{bmatrix} d\mathbf{u}_A \\ d\mathbf{u}_B \end{bmatrix}. \tag{2.30}$$

Note that the format of the equations (2.27) implies the identity $d\mathbf{q}_B = -d\mathbf{q}_A$ and $d\mathbf{q}_B = d\mathbf{q}_A = \mathbf{0}$ for a rigid body translation $d\mathbf{u}_B = d\mathbf{u}_A$.

The first matrix in (2.30) represents the constitutive stiffness due to material deformation, while the second matrix is the initial stress stiffness, often called geometric stiffness. The separation into these two parts can be illustrated by considering the representation of a vector $\mathbf{N}$ by its length N and a unit vector $\mathbf{e} = \mathbf{N}/N$ giving the direction, i.e.

$$\mathbf{N} = N\,\mathbf{e}. \tag{2.31}$$

The corresponding incremental form is

$$d\mathbf{N} = dN\,\mathbf{e} + N\,d\mathbf{e}. \tag{2.32}$$

The increments are illustrated in Fig. 2.5 for the case where the length of the vector $\mathbf{e}$ remains constant, i.e. $d\mathbf{e}$ corresponds to a rotation of $\mathbf{e}$. The rotation term $Nd\mathbf{e}$ corresponds to the initial stress term, while the change of length $dN\mathbf{e}$ corresponds to the constitutive stiffness term.

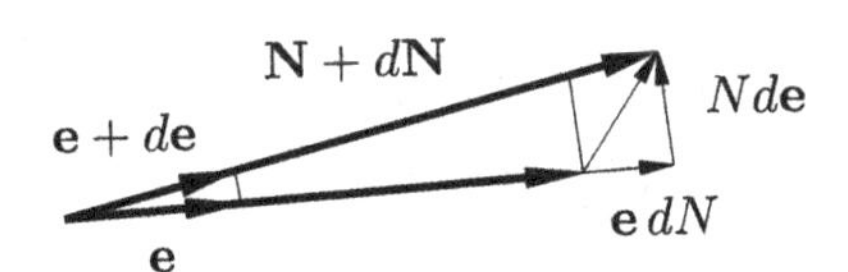

Fig. 2.5. Decomposition of increment into a rotation and a change of length.

In terms of the vector notation (2.22), the tangent stiffness relation (2.30) takes the form

$$d\tilde{\mathbf{q}} = \frac{\partial\tilde{\mathbf{q}}}{\partial\tilde{\mathbf{u}}}\,d\tilde{\mathbf{u}} = \mathbf{K}\,d\tilde{\mathbf{u}}. \tag{2.33}$$

The tangent stiffness matrix $\mathbf{K}$ consists of the following three contributions:

$$\mathbf{K} = \mathbf{K}_0 + \mathbf{K}_u + \mathbf{K}_\sigma, \tag{2.34}$$

with the linear stiffness matrix

$$\mathbf{K}_0 = \frac{EA}{l_0^3}\begin{bmatrix} \mathbf{x}_0\mathbf{x}_0^T & -(\cdots) \\ -(\cdots) & +(\cdots) \end{bmatrix}, \tag{2.35}$$

the initial displacement stiffness matrix

$$\mathbf{K}_u = \frac{EA}{l_0^3}\begin{bmatrix} \mathbf{x}_0\mathbf{u}^T + \mathbf{u}\mathbf{x}_0^T + \mathbf{u}\mathbf{u}^T & -(\cdots) \\ -(\cdots) & +(\cdots) \end{bmatrix}, \tag{2.36}$$

and the initial stress stiffness matrix

$$\mathbf{K}_\sigma = \frac{N}{l_0}\begin{bmatrix} \mathbf{I} & -\mathbf{I} \\ -\mathbf{I} & \mathbf{I} \end{bmatrix}. \tag{2.37}$$

The compact products in terms of $\mathbf{x}$ and $\mathbf{u}$ can be computed by formulae similar to (2.23), but direct computation will be more efficient. In the reference state the initial displacement is zero, and the initial displacement stiffness matrix vanishes.

2.4 Use of shape functions

The bar element is so simple that the full description could be made entirely by reference to the vector $\mathbf{x}$ connecting the end-points A and B. For most other elements the notion of shape functions is needed. Shape functions describe the displacement field in the element, usually in terms of a set of local, or basic, coordinates. In order to illustrate the use of shape functions in a simple case, the equilibrium equations already treated in Section 2.2 are re-derived.

The position vector $\mathbf{r}(\xi)$ of an arbitrary point of the bar AB is defined by interpolation between the end-points $\mathbf{x}_A$ and $\mathbf{x}_B$. In the initial configuration

$$\mathbf{r}_0(\xi) = h_A(\xi)\mathbf{x}_{A_0} + h_B(\xi)\mathbf{x}_{B_0}. \tag{2.38}$$

The shape functions are defined by

$$\begin{aligned} h_A(\xi) &= \tfrac{1}{2}(1-\xi), \quad -1<\xi<1, \\ h_B(\xi) &= \tfrac{1}{2}(1+\xi), \quad -1<\xi<1. \end{aligned} \tag{2.39}$$

The basic coordinate ξ is traditionally normalized to the interval $(-1,1)$. The shape function representation (2.38) can also be written in matrix format:

$$\mathbf{r}_0(\xi) = [\,h_A(\xi)\mathbf{I}\,,\ h_B(\xi)\mathbf{I}\,]\,\tilde{\mathbf{x}}_0. \tag{2.40}$$

In the displaced configuration the same relation holds, but with $\tilde{\mathbf{x}}_0$ replaced by $\tilde{\mathbf{x}} = \tilde{\mathbf{x}}_0 + \tilde{\mathbf{u}}$.

The Green strain at a point defined by ξ is determined by using the definition (2.6) on an infinitesimal vector with initial value $d\mathbf{r}_0(\xi)$ and final value $d\mathbf{r}(\xi)$:

$$\varepsilon_G(\xi) = \frac{d\mathbf{r}^T d\mathbf{r} - d\mathbf{r}_0^T d\mathbf{r}_0}{2\, d\mathbf{r}_0^T d\mathbf{r}_0}. \tag{2.41}$$

The infinitesimal vector $d\mathbf{r}_0$ is found by differentiation of (2.40) as

$$\frac{d\mathbf{r}_0}{d\xi} = \left[\frac{dh_A}{d\xi}\,\mathbf{I}\,,\; \frac{dh_B}{d\xi}\,\mathbf{I} \right] \tilde{\mathbf{x}}_0 = \left[-\tfrac{1}{2}\mathbf{I}\,,\; \tfrac{1}{2}\mathbf{I} \right] \tilde{\mathbf{x}}_0. \tag{2.42}$$

The infinitesimal vector $d\mathbf{r}$ in the displaced configuration follows from a similar formula with $\tilde{\mathbf{x}}_0$ replaced by $\tilde{\mathbf{x}} = \tilde{\mathbf{x}}_0 + \tilde{\mathbf{u}}$. From this the scalar products in the strain definition (2.41) are found as

$$d\mathbf{r}_0^T d\mathbf{r}_0 = \tfrac{1}{4}\tilde{\mathbf{x}}_0^T \begin{bmatrix} \mathbf{I} & -\mathbf{I} \\ -\mathbf{I} & \mathbf{I} \end{bmatrix} \tilde{\mathbf{x}}_0\, d\xi^2 = \tfrac{1}{4}\, l_0^2\, d\xi^2 \tag{2.43}$$

and

$$d\mathbf{r}^T d\mathbf{r} = \tfrac{1}{4}(\tilde{\mathbf{x}}_0 + \tilde{\mathbf{u}})^T \begin{bmatrix} \mathbf{I} & -\mathbf{I} \\ -\mathbf{I} & \mathbf{I} \end{bmatrix} (\tilde{\mathbf{x}}_0 + \tilde{\mathbf{u}})\, d\xi^2 = \tfrac{1}{4}\, l^2\, d\xi^2. \tag{2.44}$$

Substitution of these expressions into the strain definition (2.41) gives

$$\varepsilon_G = \frac{1}{l_0^2}\,(\tilde{\mathbf{x}}_0 + \tfrac{1}{2}\tilde{\mathbf{u}})^T \begin{bmatrix} \mathbf{I} & -\mathbf{I} \\ -\mathbf{I} & \mathbf{I} \end{bmatrix} \tilde{\mathbf{u}} = \frac{1}{l_0^2}\,\tilde{\mathbf{x}}_{1/2}^T \begin{bmatrix} \mathbf{I} & -\mathbf{I} \\ -\mathbf{I} & \mathbf{I} \end{bmatrix} \tilde{\mathbf{u}}. \tag{2.45}$$

This is the formula previously established as (2.24). The strain increment follows by differentiation as in (2.25):

$$\delta\varepsilon_G = \frac{1}{l_0^2}\,\tilde{\mathbf{x}}^T \begin{bmatrix} \mathbf{I} & -\mathbf{I} \\ -\mathbf{I} & \mathbf{I} \end{bmatrix} \delta\tilde{\mathbf{u}}. \tag{2.46}$$

Thus, the strain and strain increment have been established on a pointwise basis.

The equilibrium equations can now be established directly from the principle of virtual work. In the present bar element the only external loads act at the end-points, and therefore the principle of virtual work takes the form

$$\delta V = \int_0^{l_0} \delta\varepsilon_G\, N\, ds - \delta\tilde{\mathbf{u}}^T \tilde{\mathbf{f}} = 0. \tag{2.47}$$

This identity is expressed in matrix form by substitution of the virtual strain from (2.46). In the present case the integration is trivial. For more complicated elements the volume integral must be evaluated by approximate numerical integration in terms of the basic variables $(\xi, \ldots)$.

$$\delta\tilde{\mathbf{u}}^T \left(\frac{N}{l_0} \begin{bmatrix} \mathbf{I} & -\mathbf{I} \\ -\mathbf{I} & \mathbf{I} \end{bmatrix} \tilde{\mathbf{x}} - \tilde{\mathbf{f}} \right) = 0. \tag{2.48}$$

The variation $\delta\tilde{\mathbf{u}}$ of the displacement vector is arbitrary, and thus the scalar identity (2.48) gives the vector equilibrium equation

$$\frac{N}{l_0} \begin{bmatrix} \mathbf{I} & -\mathbf{I} \\ -\mathbf{I} & \mathbf{I} \end{bmatrix} \tilde{\mathbf{x}} = \tilde{\mathbf{f}}. \tag{2.49}$$

The tangent stiffness relation follows by considering increments in the equilibrium equation (2.48). The procedure and results are the same as in Section 2.3.

Example 2.1. Equilibrium and stiffness of two-element truss. The equilibrium equations and the tangent stiffness of the two-element truss shown in Fig. 2.6 are derived and used to describe a lateral bifurcation instability and the associated displacement pattern (Pecknold *et al.*, 1985). The equations are used to illustrate special numerical solution techniques in Chapter 8.

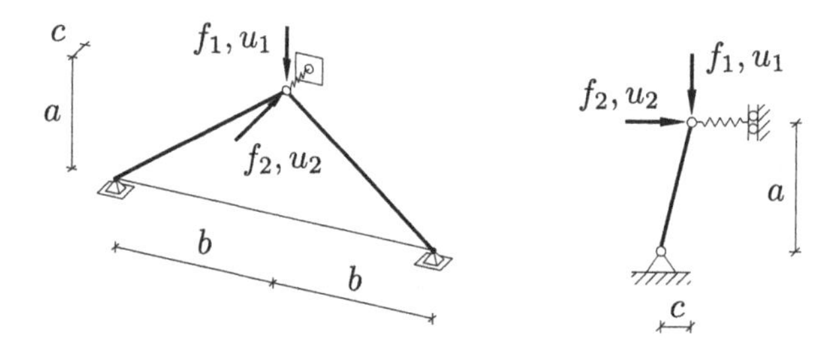

Fig. 2.6. Two-element truss supported by a lateral spring.

Figure 2.6 shows a truss consisting of two identical bar elements with stiffness AE. The initial geometry is specified by the lengths a, b and c, where c indicates the deviation from vertical, e.g. due to a geometric imperfection. The initial length of the elements is denoted l_0. Lateral support is provided by a linear spring with stiffness k. This spring remains horizontal during deformation. The load consists of a vertical force f_1 and a horizontal force f_2 acting at the top of the truss. In the present example the general equilibrium equations and tangent stiffness are derived. The specific results are here limited to the 'perfect' structure, $c = 0$ and $f_2 = 0$.

In the present example the simplest way to obtain the equilibrium equations is by use of the potential energy

$$\Phi = 2\tfrac{1}{2}l_0 EA\varepsilon_G^2 + \tfrac{1}{2}ku_2^2 - f_\alpha u_\alpha,$$

where a repeated Greek subscript implies summation over the terms, i.e. $f_\alpha u_\alpha = \sum_\alpha f_\alpha u_\alpha$. By use of the stiffness AE and the initial length

$$l_0 = \sqrt{a^2 + b^2 + c^2},$$

the following non-dimensional parameters are introduced:

$$\alpha = a/l_0, \qquad \beta = b/l_0, \qquad \gamma = c/l_0,$$

$$v_\alpha = u_\alpha/l_0, \qquad \kappa = kl_0/EA, \qquad p_\alpha = f_\alpha/EA.$$

In terms of these variables the strain ε_G is

$$\varepsilon_G = -v_1(\alpha - \tfrac{1}{2}v_1) + v_2(\gamma + \tfrac{1}{2}v_2)$$

and the potential energy takes the form

$$\Phi = l_0 EA\big(\varepsilon_G^2 + \tfrac{1}{2}\kappa v_2^2 - p_\alpha v_\alpha\big).$$

The equilibrium equations are now established in non-dimensional form from the stationarity condition $\delta\Phi = 0$ using the strain variation

$$\delta\varepsilon_G = -\delta v_1(\alpha - v_1) + \delta v_2(\gamma + v_2) = (\delta v_1, \delta v_2)\begin{bmatrix} -(\alpha - v_1) \\ (\gamma + v_2) \end{bmatrix}.$$

The resulting equilibrium equations are

$$\begin{aligned} -2\varepsilon_G(\alpha - v_1) &= p_1, \\ 2\varepsilon_G(\gamma + v_2) + \kappa v_2 &= p_2. \end{aligned}$$

The tangent stiffness relation follows from the equilibrium equations by differentiation as

$$\begin{bmatrix} 2(\alpha - v_1)^2 + 2\varepsilon_G & -2(\alpha - v_1)(\gamma + v_2) \\ -2(\gamma + v_2)(\alpha - v_1) & 2(\gamma + v_2)^2 + 2\varepsilon_G + \kappa \end{bmatrix} \begin{bmatrix} dv_1 \\ dv_2 \end{bmatrix} = \begin{bmatrix} dp_1 \\ dp_2 \end{bmatrix}.$$

The terms $2\varepsilon_G$ correspond to the initial stress stiffness. These are the two sets of equations used in a non-linear analysis of the two-element truss.

A brief analysis of the ideal, symmetric system with $\gamma = 0$ and $p_2 = 0$ is now given. This analysis establishes the principal behavior of the system and typical load levels to be used in Chapter 8. For the ideal, symmetric truss the transverse equilibrium equation is

$$(2\varepsilon_G + \kappa)\, v_2 \;=\; 0.$$

This equation can be satisfied in two ways: $v_2 = 0$ *or* $2\varepsilon_G + \kappa = 0$. The first solution is symmetric, while the second is non-symmetric. The symmetric branch, $v_2 = 0$, corresponds to the introductory example presented in Section 1.1. In the present notation the load is

$$p_1 \;=\; \left[\alpha^2 - (\alpha - v_1)^2\right](\alpha - v_1) \;=\; 2\alpha^2 v_1 \;-\; 3\alpha v_1^2 + v_1^3,$$

corresponding to (1.6). The maximum load on this branch is attained when $dp_1/dv_1 = 0$, i.e. at

$$v_1 \;=\; \alpha\Big(1 \;-\; \frac{1}{\sqrt{3}}\Big), \qquad p_1 \;=\; \frac{2\alpha^3}{3\sqrt{3}}.$$

The symmetric branch is shown as $ABCD$ in Fig. 2.7. The point C of maximum load is called a *limit point.*

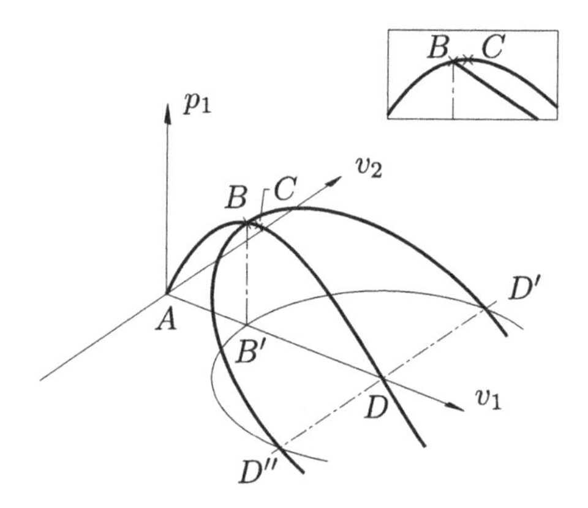

Fig. 2.7. Symmetric and non-symmetric branches for two-element truss.

The non-symmetric solution is only possible when the strain has reached the value $\varepsilon_G = -\frac{1}{2}\kappa$. Initially this corresponds to $v_2 = 0$ *and* $2\varepsilon_G + \kappa = 0$. These combined conditions determine the bifurcation point B shown in Fig. 2.7. At the bifurcation point

$$v_1 \;=\; \alpha \;-\; \sqrt{\alpha^2 - \kappa}, \qquad p_1 \;=\; \kappa\sqrt{\alpha^2 - \kappa}.$$

A comparison between the location of the bifurcation point and the limit point leads to the following conclusions regarding the influence of the spring stiffness κ:

$$\begin{array}{ll} 0 < \kappa < \frac{2}{3}\alpha^2 & \text{bifurcation point before limit point;} \\ \frac{2}{3}\alpha^2 < \kappa < \alpha^2 & \text{bifurcation point after limit point;} \\ \alpha^2 < \kappa & \text{no bifurcation point.} \end{array}$$

For $\kappa < \alpha^2$ the non-symmetric branch satisfies the condition $2\varepsilon_G + \kappa = 0$. Clearly the strain ε_G remains constant, while the displacement components describe a circle with the equation

$$(\alpha - v_1)^2 + v_2^2 = \alpha^2 - \kappa.$$

On this branch the load is determined by

$$p_1 = -2\varepsilon_G(\alpha - v_1) = \kappa(\alpha - v_1)$$

and thus the non-symmetric branches CD' and CD'' form parts of a plane ellipse, while $D'B'D''$ forms a semi-circle.

For a non-symmetric truss the bifurcation and limit loads of the equivalent symmetric truss are never reached. However, for a nearly symmetric truss the load–deformation curves are qualitatively similar to those of Fig. 2.7, and thus the solution for the symmetric truss provides useful information when investigating the non-symmetric situation.

2.5 Assembly of global stiffness and forces

So far the current internal force and the tangent stiffness have been determined for an individual bar element. A structure consists of an assembly of elements, joined at nodes. In the case of bar elements the loads and supports are applied at the nodes. The displacement of the complete structure is described completely in terms of the displacement vector $\mathbf{u}_n$ at each node, and similarly the resulting force at each node can be found by summation of the contributions from all elements attached to that node. Thus, the displacement at a node is shared among elements, while the resulting internal force is accumulated by summation over the elements. In the case of truss structures formed from bar elements this argument is simple and sufficient, but for more general elements, e.g. for shells and solids, a more general approach is needed, as the elements are joined along curves and surfaces and not just at the nodes. The current case of truss structures and bar elements is therefore used to illustrate the general procedure using the principle of virtual work. The use of the principle of virtual work also turns out to play a key role in the formulation of tangent stiffness relations for elements with displacement fields described in terms of translations and rotations, such as beams and shells.

Consider a truss structure formed by joining a number of bar elements at a set of nodes. The principle of virtual work states that equilibrium corresponds to vanishing difference between the internal and the external virtual work. The internal virtual work is the sum of the contributions from each element, and for truss structures, where the loads and supports are applied at the nodes, the external virtual work is the work of the forces at the nodes. Thus, the virtual work of the structure takes the form

$$\delta V = \sum_{\text{elements}} \int \delta\varepsilon \, N \, ds - \sum_{\text{nodes}} \delta\mathbf{u}_n^T \mathbf{f}_n = 0, \tag{2.50}$$

where the virtual strain $\delta\varepsilon$ in the elements is defined by (2.15) in terms of the virtual displacements $\delta\mathbf{u}$ at the element nodes. The incremental virtual strains $\delta\varepsilon$ can be expressed by a derivative of the total strain as

$$\delta\varepsilon = \sum_{\text{nodes}} \frac{\partial\varepsilon}{\partial\mathbf{u}_n}\delta\mathbf{u}_n = \sum_{\text{nodes}} \delta\mathbf{u}_n^T \frac{\partial\varepsilon}{\partial\mathbf{u}_n^T}, \tag{2.51}$$

where the last form brings the displacement factor $\delta\mathbf{u}_n^T$ to the front by transposing and interchanging the order of the two factors in the scalar product. The strain in each element only depends on the displacement vector at the nodes of the element, and thus only these nodes participate in the summation. The virtual work equation (2.50) can now be written as

$$\delta V = \sum_{\text{nodes}} \delta\mathbf{u}_n^T \Big(\sum_{\text{elements}} \int \frac{\partial\varepsilon}{\partial\mathbf{u}_n^T} N \, ds - \mathbf{f}_n \Big) = 0. \tag{2.52}$$

The integrals over the elements are identified as the internal element forces $\mathbf{q}_n$ at the particular node. In fact, it follows from differentiation of the strain definition (2.13) that

$$\mathbf{q}_A = \int \frac{\partial\varepsilon}{\partial\mathbf{u}_A^T} N \, ds = -N\frac{1}{l_0}\mathbf{x}, \quad \mathbf{q}_B = \int \frac{\partial\varepsilon}{\partial\mathbf{u}_B^T} N \, ds = N\frac{1}{l_0}\mathbf{x} \tag{2.53}$$

when the element vector $\mathbf{x}$ is oriented from A to B.

One equilibrium vector equation is recovered for each node from (2.52) by considering one non-vanishing virtual displacement component at a time. When it is realized that the integrals define the internal forces from the individual elements at the appropriate node n, these equilibrium equations take the form

$$\sum_{\text{elements}} \mathbf{q}_n = \mathbf{f}_n, \qquad n = 1, \ldots \tag{2.54}$$

Only the elements connected to the node n participate in the sum relating to the equilibrium of that node. The equations (2.54) express equality between the internal forces, generated within the elements of the structure, and the external loads, applied at the nodes.

In the solution process the relation between an infinitesimal change in the loads and the corresponding change in the displacements is needed. This so-called tangent stiffness relation is obtained by considering the difference between two neighboring states of equilibrium. In each state of equilibrium the virtual work vanishes, and thus the same applies to the increment of the virtual work, symbolically written as

$$d(\delta V) = \lim \big(\delta V_2 - \delta V_1\big). \tag{2.55}$$

The virtual work δV is linear in $\delta\mathbf{u}$ and can be considered as a first variation of a functional V. In this terminology $d(\delta V)$ is the second variation.

In the virtual work equation the displacement variation $\delta\mathbf{u}$ is subject to choice, and if the same displacement variations are selected for both of the neighboring states, the second variation (2.55) takes the form

$$d(\delta V) = \sum_{\text{nodes}} \delta\mathbf{u}_n^T \Big(\sum_{\text{elements}} d\mathbf{q}_n - d\mathbf{f}_n \Big) = 0. \tag{2.56}$$

As before, this relation can be converted into a vector equation for each of the nodes:

$$\sum_{\text{elements}} d\mathbf{q}_n = d\mathbf{f}_n, \qquad n = 1, \ldots, \tag{2.57}$$

this time for the increments of the internal element forces $d\mathbf{q}_n$ and the external loads increment $d\mathbf{f}$. For each element the increment of the internal forces $\tilde{\mathbf{q}}$ and the displacement increments $\tilde{\mathbf{u}}$ of the element nodes are related by the element tangent stiffness matrix,

$$d\tilde{\mathbf{q}} = \frac{\partial\tilde{\mathbf{q}}}{\partial\tilde{\mathbf{u}}}\, d\tilde{\mathbf{u}} = \tilde{\mathbf{K}}\, d\tilde{\mathbf{u}}, \tag{2.58}$$

where the notation $\tilde{\mathbf{K}}$ is introduced to denote the element tangent stiffness matrix. When this expression is substituted into the incremental equilibrium equation (2.57), it takes the symbolic form

$$\Big(\sum_{\text{elements}} \tilde{\mathbf{K}} \Big)\, d\mathbf{u} = d\mathbf{f} \tag{2.59}$$

in which the vectors $d\mathbf{u}$ and $d\mathbf{f}$ contain the components from all the nodes of the structure. The symbolic summation over the elements defines the global tangent stiffness matrix

$$\mathbf{K} = \sum_{\text{elements}} \tilde{\mathbf{K}}. \tag{2.60}$$

The assembly procedure for the global tangent stiffness matrix $\mathbf{K}$ and for the total internal forces $\mathbf{q}$ is illustrated in the following example.

EXAMPLE 2.2. ELEMENT FORMULATION OF TWO-ELEMENT TRUSS. In Example 2.1 the equilibrium equations and the tangent stiffness of the two-element truss shown in Fig. 2.6 were derived directly from the elastic energy. This example shows how the current internal forces and the incremental equilibrium equations of the truss are obtained by assembling the element contributions.

The left support is denoted A, the top node B, and the right support C. The truss then consists of the two elements AB and BC, connected at the node B. The displacement and internal element force vectors of the element AB are

$$\tilde{\mathbf{u}}^T = [\mathbf{u}_A^T, \mathbf{u}_B^T], \quad \tilde{\mathbf{q}}^T = [\mathbf{q}_A^T, \mathbf{q}_B^T],$$

while for the element BC they are

$$\tilde{\mathbf{u}}^T = [\mathbf{u}_B^T, \mathbf{u}_C^T], \quad \tilde{\mathbf{q}}^T = [\mathbf{q}_B^T, \mathbf{q}_C^T].$$

The current internal element forces are given by (2.26),

$$\tilde{\mathbf{q}} = \frac{N}{l_0}\begin{bmatrix} \mathbf{I} & -\mathbf{I} \\ -\mathbf{I} & \mathbf{I} \end{bmatrix}\tilde{\mathbf{x}},$$

with the normal force $N = EA\varepsilon_G$, and the strain ε_G given by (2.24). The vector $\tilde{\mathbf{x}}$ describes the current position of the element nodes. At each node the total internal force is the sum of the contributions from all elements connected to that node. In the present example this can be expressed as

$$\begin{bmatrix} \mathbf{q}_A \\ \mathbf{q}_B \\ \mathbf{q}_C \end{bmatrix} = \begin{bmatrix} \mathbf{q}_A \\ \mathbf{q}_B \\ \; \end{bmatrix}_{AB} + \begin{bmatrix} \; \\ \mathbf{q}_B \\ \mathbf{q}_C \end{bmatrix}_{BC}.$$

The assembly of element forces consists of arranging each element force vector $\tilde{\mathbf{q}}$ in the global numbering scheme, and then adding the contributions.

The element tangent stiffness matrix is given by (2.30) as

$$\tilde{\mathbf{K}}\delta = \frac{EA}{l_0^3}\begin{bmatrix} \mathbf{x}\mathbf{x}^T & -\mathbf{x}\mathbf{x}^T \\ -\mathbf{x}\mathbf{x}^T & \mathbf{x}\mathbf{x}^T \end{bmatrix} + \frac{N}{l_0}\begin{bmatrix} \mathbf{I} & -\mathbf{I} \\ -\mathbf{I} & \mathbf{I} \end{bmatrix},$$

where $\mathbf{x}$ is the vector describing the current position of the element. The element stiffness matrices must be assembled such that the displacement vector at a shared node is common, while the element forces at that node are added. This is accomplished by the format

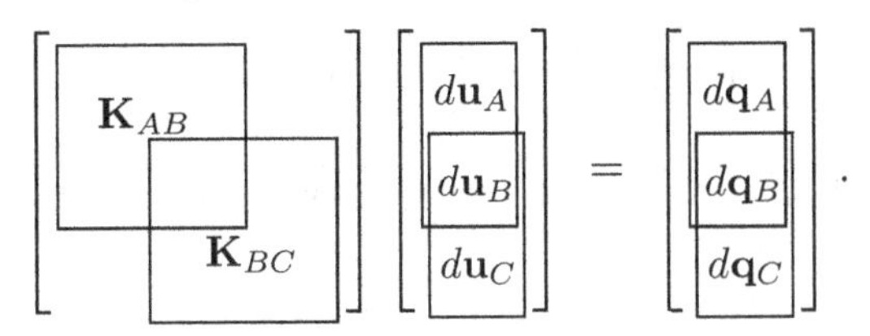

The assembly of the tangent stiffness matrix consists of appropriate positioning of the element stiffness matrices within the global stiffness matrix format and then adding all the element contributions to form the global stiffness matrix.

In the present example the two element stiffness matrices simply overlap with one quarter. In general an element AB is connected to nodes with global numbers i and j. The element stiffness matrix is then divided into four sub-matrices

$$\tilde{\mathbf{K}} = \left[\begin{array}{c|c} \mathbf{k}_{AA} & \mathbf{k}_{AB} \\ \hline \mathbf{k}_{BA} & \mathbf{k}_{BB} \end{array}\right].$$

The contribution from this element to the global tangent stiffness matrix can then be written as

$$\mathbf{K} = \begin{bmatrix} & \vdots & & \vdots & \\ \cdots & \boxed{\mathbf{k}_{AA}} & \cdots & \boxed{\mathbf{k}_{AB}} & \cdots \\ & \vdots & & \vdots & \\ \cdots & \boxed{\mathbf{k}_{BA}} & \cdots & \boxed{\mathbf{k}_{BB}} & \cdots \\ & \vdots & & \vdots & \end{bmatrix} \begin{matrix} \\ i \\ \\ j \\ \\ \end{matrix}.$$
$$\qquad\qquad i \qquad\qquad j$$

It is seen that the key to the assembly of the stiffness matrix is the identification of the element node numbers within the global stiffness matrix.

In the finite element method the element properties are evaluated and assembled for all nodes, and the boundary conditions are introduced subsequently in the global stiffness matrix and load vector. In the two-bar truss the nodes A and C are fully restrained, i.e. $\mathbf{u}_A = \mathbf{0}$ and $\mathbf{u}_C = \mathbf{0}$. This implies that the corresponding internal forces $\mathbf{q}_A$ and $\mathbf{q}_C$ are unknown and have the character of reactions. The fact that $\mathbf{u}_A = \mathbf{0}$ eliminates all contributions from the first block column of the matrix equation. Similarly,

$\mathbf{u}_C = \mathbf{0}$ eliminates all contributions from the last block column. The first and last block rows are not proper equations, as $\mathbf{q}_A$ and $\mathbf{q}_C$ are unknown. This leaves only the central block of the equation system. In the sub-matrix notation the final equations are

$$\left(\mathbf{k}_{BB}^{(AB)} + \mathbf{k}_{AA}^{(BC)}\right) d\mathbf{u}_B = d\mathbf{q}_B.$$

It may be shown that these equations are identical to those of Example 2.1, see Exercise 2.4.

2.6 Total or updated Lagrangian formulation

Finite element formulations that make use of representations of the motion of the material points are called Lagrangian or material. This type of formulation is often used for structures and solids, and is indeed the one used here for truss structures. The alternative formulation, using a representation of the motion of material through a fixed part of space, is called Eulerian or spatial. The Eulerian approach is typically used in fluid mechanics. In recent years mixed Lagrangian–Eulerian methods have been developed, e.g. for modeling very large deformations of solids (Belytschko *et al.*, 2000). The present formulation makes use of an initial configuration $\mathbf{x}_0$ and displacements $\mathbf{u}$ from this configuration. In practice, the solution often involves a number of load steps $\mathbf{f}_1, \ldots, \mathbf{f}_n, \ldots$ and the load step from $\mathbf{f}_{n-1}$ to $\mathbf{f}_n$ may then be formulated in two ways. In the total Lagrangian formulation the initial configuration $\mathbf{x}_0$ is retained as reference during the whole load history, and the total displacements $\mathbf{u}_n$ are used as illustrated in Fig. 2.8. This calls for use of the full tangent stiffness matrix (2.34), including the initial displacement stiffness matrix (2.36), during all stages of the computation.

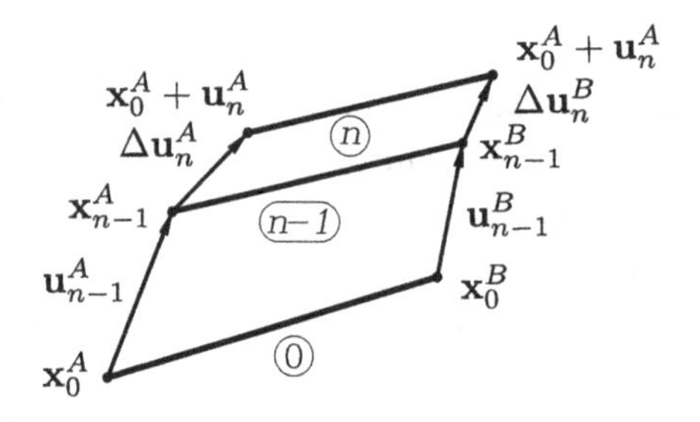

Fig. 2.8. Total Lagrangian formulation.

ALGORITHM 2.1. Total Lagrangian formulation.

Load steps $n = 1, 2, \ldots, n_{\max}$

$$\mathbf{f}_n = \mathbf{f}_{n-1} + \Delta\mathbf{f}_n$$

$$\mathbf{u}_n = \mathbf{u}_{n-1}$$

$$\mathbf{K}_{n-1} = \mathbf{K}_0(\mathbf{x}_0) + \mathbf{K}_u(\mathbf{x}_0, \mathbf{u}_{n-1}) + \mathbf{K}_\sigma(\varepsilon_{n-1})$$

Iterations $i = 1, 2, \ldots, i_{\max}$

$$\varepsilon_n = \frac{1}{l_0^2}\,\tfrac{1}{2}(\tilde{\mathbf{x}}_0 + \tilde{\mathbf{x}}_n)^T \begin{bmatrix} \mathbf{I} & -\mathbf{I} \\ -\mathbf{I} & \mathbf{I} \end{bmatrix} \tilde{\mathbf{u}}_n$$

$$\mathbf{r}_n = \mathbf{f}_n - \sum_{\text{elements}} \frac{EA}{l_0}\,\varepsilon_n \begin{bmatrix} \mathbf{I} & -\mathbf{I} \\ -\mathbf{I} & \mathbf{I} \end{bmatrix} \tilde{\mathbf{x}}_n$$

$$\delta\mathbf{u}_n = (\mathbf{K}_{n-1})^{-1}\mathbf{r}_n$$

$$\mathbf{u}_n = \mathbf{u}_n + \delta\mathbf{u}_n$$

$$\mathbf{x}_n = \mathbf{x}_0 + \mathbf{u}_n$$

Stop iteration when $\|\mathbf{r}_n\| < \epsilon\,\|\mathbf{f}_n\|$

End of load step

The main points of the total Lagrangian formulation are indicated in Algorithm 2.1 using a modified Newton–Raphson method. The formulae in the table are only indications of the main ideas, and no distinction is made between global and element parameters. Each load step starts with incrementing the load, initializing the displacement $\mathbf{u}_n^0$ from the last equilibrium state, and evaluating and factoring the tangent stiffness matrix $\mathbf{K}_{n-1}$ at state $n-1$. This tangent stiffness matrix is retained during the iterations according to the modified Newton–Raphson method. In each iteration the residual force $\mathbf{r}_n^i$ is evaluated as the difference between the external and the internal forces, and the displacement iteration increment $\delta\mathbf{u}_n^i$ is computed by use of the factored tangent stiffness matrix. Then the *total* displacement $\mathbf{u}_n^i$ is updated. The final evaluation of $\mathbf{x}_n^i$ is just a computational convenience and does not change the original reference state $\mathbf{x}_0$. As previously noted the iteration index i does not appear explicitly in the algorithm, as new values overwrite the previous value.

In the updated Lagrangian formulation each load step ends with the definition of an updated configuration $\mathbf{x}_n = \mathbf{x}_{n-1} + \Delta\mathbf{u}_n$, as illustrated in Fig. 2.9 and Algorithm 2.2. Thus, the first evaluation of the tangent stiffness matrix in each load step does not contain any initial displacement contribution. In

ALGORITHM 2.2. Updated Lagrangian formulation.

Load steps $n = 1, 2, \ldots, n_{\max}$

$$\mathbf{f}_n = \mathbf{f}_{n-1} + \Delta\mathbf{f}_n$$

$$\Delta\mathbf{u}_n = \mathbf{0}$$

$$\mathbf{K}_{n-1} = \mathbf{K}_0(\mathbf{x}_{n-1}) + \mathbf{K}_\sigma(\varepsilon_{n-1})$$

Iterations $i = 1, 2, \ldots, i_{\max}$

$$\Delta\varepsilon_n = \frac{1}{l_0^2}\,\tfrac{1}{2}(\tilde{\mathbf{x}}_{n-1} + \tilde{\mathbf{x}}_n)^T \begin{bmatrix} \mathbf{I} & -\mathbf{I} \\ -\mathbf{I} & \mathbf{I} \end{bmatrix} \Delta\tilde{\mathbf{u}}_n$$

$$\varepsilon_n = \varepsilon_{n-1} + \Delta\varepsilon_n$$

$$\mathbf{r}_n = \mathbf{f}_n - \sum_{\text{elements}} \frac{EA}{l_0}\,\varepsilon_n \begin{bmatrix} \mathbf{I} & -\mathbf{I} \\ -\mathbf{I} & \mathbf{I} \end{bmatrix} \tilde{\mathbf{x}}_n$$

$$\delta\mathbf{u}_n = (\mathbf{K}_{n-1})^{-1}\mathbf{r}_n$$

$$\Delta\mathbf{u}_n = \Delta\mathbf{u}_n + \delta\mathbf{u}_n$$

$$\mathbf{x}_n = \mathbf{x}_{n-1} + \Delta\mathbf{u}_n$$

Stop iteration when $\|\mathbf{r}_n\| < \epsilon\,\|\mathbf{f}_n\|$

End of load step

---

each iteration the residual force $\mathbf{r}_n^i$ is evaluated and the displacement iteration increment $\delta\mathbf{u}_n^i$ is computed by use of the factored tangent stiffness matrix as in the total Lagrangian formulation. The difference appears in the updates. In the updated Lagrangian formulation $\delta\mathbf{u}_n^i$ is used to update the displacement $\Delta\mathbf{u}_n^i$ from the last equilibrium state, and the strain increment $\Delta\varepsilon_n^i$ also is relative to state $n-1$.

The change of reference configuration in a Lagrangian formulation requires some care, because the strain definition (2.6) and the internal force definition

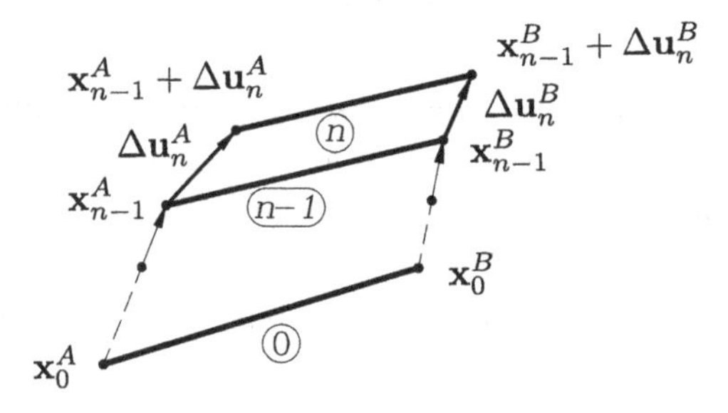

Fig. 2.9. Updated Lagrangian formulation.

(2.19) make use of a reference length from this state. If the current length l_{n-2} was used as reference length in load step $n-1$, the actual normal force in the bar at the end of load step $n-1$ is $N_{n-1}\, l_{n-1}/l_{n-2}$. In the following load step the new reference length would be l_{n-1} and N should therefore be multiplied by the factor l_{n-1}/l_{n-2} when initiating load step n. For small strains this correction factor is without importance, and it is usually omitted. In Algorithm 2.2 the reference length l_0 has been used throughout, whereby the updated method is just a reformulation of the total method. Differences arising in the total and updated Lagrangian formulations due to the use of different reference lengths are discussed by Yang and Leu (1991).

2.7 Summing up the principles

The results obtained in this chapter for bar elements are particularly simple because the deformation is one-dimensional and homogeneously distributed within each bar element. However, the results are of general nature, and it is therefore worthwhile to restate the principal results in general terms, applicable also to other geometrically non-linear elements. The discussion makes use of the current coordinate vector $\mathbf{x}$, the displacement vector $\mathbf{u}$ and the load vector $\mathbf{f}$, and no explicit distinction is made between the element and the global level.

The basic relation (2.50) is a statement of equilibrium formulated in terms of virtual work in the form

$$\delta V = \int \delta\varepsilon\, N\, ds - \delta\mathbf{u}^T\mathbf{f} = 0, \tag{2.61}$$

where the integral covers all elements, and the vectors $\mathbf{u}$ and $\mathbf{f}$ include all nodal displacements and the corresponding loads. The state of the structure is expressed in terms of the current geometry described by $\mathbf{x}$ and $\mathbf{u}$, the internal state of stress described by the normal force N in each element, and the external load $\mathbf{f}$. $\delta\varepsilon$ is the *virtual* strain corresponding to the virtual nodal displacements $\delta\mathbf{u}$, i.e. $\delta\varepsilon$ is the strain increment corresponding to a virtual change of the nodal positions from $\mathbf{x}$ to $\mathbf{x}+\delta\mathbf{u}$. The increments $\delta\varepsilon$ and $\delta\mathbf{u}$ are considered as virtual – i.e. corresponding to a hypothetical change of displacement of the structure. Thus, they do not change the actual state of the structure, and the internal stresses and external loads remain unchanged.

In the statement (2.61) of the virtual work principle the strain ε was deliberately stated without reference to the particular choice of strain measure. The important point is that the strain measure must be independent of any rigid body displacement. For the bar this will be the case for any strain measures based only on the length of the bar. In Section 2.1 three different strain measures were introduced, all meeting this requirement: the engineering strain ε_E, the Green strain ε_G, and the logarithmic strain ε_L. When formulated in two- or three-dimensional space, all of these strain measures are non-linear functions of the displacements. Indeed, any strain measure that is invariant to (large) rotations must be a non-linear function of the displacements. Also, the curve length parameter s was left unspecified in (2.61). The choice of curve parameter depends on the choice of strain. For engineering and Green strain the initial curve length s_0 is used, while logarithmic strain corresponds to the use of current curve length s. Similar considerations hold for the internal force N, the interpretation of which also depends on the chosen strain measure.

The strain increment is found by differentiation of the total strain ε,

$$\delta\varepsilon = \frac{\partial\varepsilon}{\partial\mathbf{u}}\,\delta\mathbf{u} = \delta\mathbf{u}^T\frac{\partial\varepsilon}{\partial\mathbf{u}^T}. \tag{2.62}$$

For bars with Green strain this relation corresponds to (2.46). For a non-linear strain definition the derivative $\partial\varepsilon/\partial\mathbf{x}$ depends on the current geometry $\mathbf{x}$. In the virtual work equation the internal work is the product of the stresses and strains, and thus the stress measure depends on the selected strain measure, or conversely. This relation is explored further in relation to rotations in Chapters 3 and 4 and in relation to solids in Chapter 6.

When the virtual strain $\delta\varepsilon$ in the principle of virtual work is expressed in terms of the virtual nodal displacements $\delta\mathbf{u}$, the statement of equilibrium takes the form

$$\delta\mathbf{u}^T\Big(\underbrace{\int\frac{\partial\varepsilon}{\partial\mathbf{u}^T}N\,ds}_{\text{internal force}} - \mathbf{f}\Big) = 0. \tag{2.63}$$

The components of the virtual displacement vector $\delta\mathbf{u}$ can be associated with arbitrary values, and therefore the internal and external force vectors inside the parentheses must be equal. Thus, the scalar equation (2.63) is equivalent to the system of equilibrium equations

$$\int\frac{\partial\varepsilon}{\partial\mathbf{u}^T}N\,ds = \mathbf{f} \tag{2.64}$$

corresponding to (2.54). These equilibrium equations only involve the current state of the structure, with $\partial\varepsilon/\partial\mathbf{u}^T$ appearing as a weight function on the internal state of stress N.

In non-linear analysis the equilibrium equations derived from virtual work are used to check equilibrium and to evaluate the corresponding residual, which is an expression for the unbalanced loads. When equilibrium has not been attained, it is necessary to consider a change of an equilibrium configuration $\mathbf{x}$, N with load $\mathbf{f}$ to a neighboring equilibrium configuration $\mathbf{x}+d\mathbf{u}$, $N+dN$ with load $\mathbf{f}+d\mathbf{f}$. In contrast to the virtual increments considered above, the increments $d\mathbf{u}$, dN and $d\mathbf{f}$ correspond to actual changes, and thus the corresponding changes in the state of deformation $d\varepsilon$ lead to changes dN in the internal state of stress.

Equilibrium in the new state is expressed in terms of the principle of virtual work. The change in virtual work from the original equilibrium state to the neighboring state is found by differentiation of δV, and when the virtual displacements $\delta\mathbf{u}$ are kept constant, the resulting second variation is found to be

$$d(\delta V) = \int \Big(\underbrace{\delta\varepsilon\, dN}_{\substack{\text{constitutive}\\ \text{change}}} + \underbrace{d(\delta\varepsilon)\, N}_{\substack{\text{geometry}\\ \text{change}}} \Big) ds - \delta\mathbf{u}^T d\mathbf{f} = 0. \tag{2.65}$$

Note that the load term has been obtained from $d(\delta\mathbf{u}^T\mathbf{f}) = \delta\mathbf{u}^T d\mathbf{f}$, because there is no change of the virtual nodal displacement, i.e. $d(\delta\mathbf{u}) = \mathbf{0}$. It is intuitively clear that the same virtual translation $\delta\mathbf{u}$ can be used for both neighboring configurations. However, when the nodal displacements also contain finite rotations $\boldsymbol{\varphi}$, as in the case of beams and shells, independence can no longer be assumed, and $d(\delta\boldsymbol{\varphi}) \neq \mathbf{0}$. This problem and its solution are considered in detail in Chapter 3 dealing with finite rotations.

The first part of the integrand in (2.65) contains the effect of the change of the internal state of stress dN. The change of the state of stress is due to the change in the state of strain $d\varepsilon$, and the constitutive relation will usually provide the incremental stiffness factor in the relation

$$dN = \frac{dN}{d\varepsilon}\, d\varepsilon. \tag{2.66}$$

In the case of linear elasticity the stiffness of the bar is $dN/d\varepsilon = AE$, but non-linear elasticity or plasticity could also provide the incremental stiffness needed in this format.

The second part of the integrand represents the effect of the change of geometry on the equilibrium of the internal state of stress. The normal force in the bar changes direction due to the incremental displacement $d\mathbf{u}$. Note that this effect is not related to deformation as such, but is a consequence of the rotation of the internal stress state. The virtual strain increment $\delta\varepsilon$ is given by (2.15). In a geometrically non-linear theory the virtual strain increment $\delta\varepsilon$ depends on the current geometry, and thus the factor $\partial\varepsilon/\partial\mathbf{u}$ will generally be a function of the current node positions $\mathbf{x}$. The incremental change is found by differentiation,

$$d(\delta\varepsilon) = \frac{\partial}{\partial\mathbf{u}}\Big(\frac{\partial\varepsilon}{\partial\mathbf{u}}\,\delta\mathbf{u}\Big)d\mathbf{u} = \delta\mathbf{u}^T\frac{\partial^2\varepsilon}{\partial\mathbf{u}^T\partial\mathbf{u}}\,d\mathbf{u}. \tag{2.67}$$

Note that also here it is used that the virtual displacement $\delta\mathbf{u}$ remains unaffected by the change of state $d(\;)$.

When (2.66) and (2.67) are substituted into the increment of the virtual work as given by (2.65), the following form is obtained:

$$d(\delta V) = \int\Big(\delta\varepsilon\,\frac{dN}{d\varepsilon}\,d\varepsilon + \delta\mathbf{u}^T\Big(N\,\frac{\partial^2\varepsilon}{\partial\mathbf{u}^T\partial\mathbf{u}}\Big)d\mathbf{u}\Big)\,ds - \delta\mathbf{u}^T\,d\mathbf{f} = 0. \tag{2.68}$$

The first term can be expressed in terms of the nodal displacement increments by use of (2.62) for $\delta\varepsilon$ and the similar formula for $d\varepsilon$, whereby

$$\delta\mathbf{u}^T\Big\{\Big[\int\Big(\frac{\partial\varepsilon}{\partial\mathbf{u}^T}\,\frac{dN}{d\varepsilon}\,\frac{\partial\varepsilon}{\partial\mathbf{u}} + N\,\frac{\partial^2\varepsilon}{\partial\mathbf{u}^T\partial\mathbf{u}}\Big)ds\Big]d\mathbf{u} - d\mathbf{f}\Big\} = 0. \tag{2.69}$$

The variation $\delta\mathbf{u}$ is arbitrary, and thus the scalar equation (2.69) is equivalent to the *incremental stiffness relation*

$$\Big[\int\Big(\frac{\partial\varepsilon}{\partial\mathbf{u}^T}\,\frac{dN}{d\varepsilon}\,\frac{\partial\varepsilon}{\partial\mathbf{u}} + N\,\frac{\partial^2\varepsilon}{\partial\mathbf{u}^T\partial\mathbf{u}}\Big)ds\Big]d\mathbf{u} = d\mathbf{f}. \tag{2.70}$$

The square brackets contain the tangent stiffness $\mathbf{K}$. It is seen that the constitutive contribution to the tangent stiffness depends on the first derivatives of the strain with respect to the coordinates, while the initial stress contribution depends on the second derivative. Thus, the geometric effect of rotating the internal stress state can only be represented by use of a non-linear strain measure, and this non-linearity leads to state-dependent factors in the constitutive term. Note that the form of the incremental stiffness in (2.70) leads to a symmetric matrix – also for non-linear material behavior. In the general continuum theory with more than one strain component at a point, symmetry depends on the symmetry properties of the constitutive matrix. This is dealt with in Chapter 7 on plasticity.

For the elastic bar element the tangent stiffness relation was given in (2.30). The first and second derivatives of the axial Green strain follow from the incremental strain relation (2.25), and it is easily verified that (2.30) follows as a special case of (2.70).

In the special case of an elastic material, there exists a strain energy density function, which is an expression for the accumulated internal work,

$$\varphi(\varepsilon) = \int_0^\varepsilon N\, d\varepsilon. \tag{2.71}$$

It follows from this definition that the internal stress N is the derivative of the strain energy density,

$$N = \frac{d\varphi(\varepsilon)}{d\varepsilon}. \tag{2.72}$$

For elastic bodies the virtual work can be considered as the first variation of a potential energy function. Thus, the potential energy corresponding to (2.61) is

$$\Phi(\mathbf{u}) = \int \varphi(\varepsilon)\, ds - \mathbf{u}^T \mathbf{f} = 0. \tag{2.73}$$

The first variation of the potential energy gives the equilibrium equations in the form

$$\delta\Phi(\mathbf{u}) = \int \delta\varepsilon\, \frac{d\varphi}{d\varepsilon}\, ds - \delta\mathbf{u}^T \mathbf{f} = 0. \tag{2.74}$$

The incremental stiffness relation follows by differentiation as above, leading to the equations

$$\left[\int \left(\frac{\partial\varepsilon}{\partial\mathbf{u}^T} \frac{d^2\varphi}{d\varepsilon d\varepsilon} \frac{\partial\varepsilon}{\partial\mathbf{u}} + \frac{d\varphi}{d\varepsilon} \frac{\partial^2\varepsilon}{\partial\mathbf{u}^T\partial\mathbf{u}} \right) ds \right] d\mathbf{u} = d\mathbf{f}. \tag{2.75}$$

This form of the incremental stiffness equations is symmetric, even for the general case of multi-dimensional stresses due to the existence of an energy potential $\varphi(\varepsilon)$.

The results of this chapter can be generalized in two directions: by introducing rotations at the nodes in addition to the translation displacements, leading to beams and shells, and by extending the notion of strain to two or three dimensions, leading to continuum models for solids. The first route is followed in Chapters 3–5 dealing with finite rotations and beam elements. The second is taken up in Chapter 6 on the non-linear relations of solid mechanics and the following chapter on plasticity models for solids.

2.8 Exercises

Exercise 2.1 Consider the bar element AB in Fig. 2.9. In the initial state the end-points have the coordinates $\mathbf{x}_0^A$ and $\mathbf{x}_0^B$. In the state $n-1$ the coordinates are $\mathbf{x}_{n-1}^A$ and $\mathbf{x}_{n-1}^B$. In the following state n the coordinates are $\mathbf{x}_n^A = \mathbf{x}_{n-1}^A + \Delta\mathbf{u}_n^A$ and $\mathbf{x}_n^B = \mathbf{x}_{n-1}^B + \Delta\mathbf{u}_n^B$.

(a) Derive the mean position formula for the increment in Green strain used in the updated Lagrangian formulation in Algorithm 2.2:

$$\Delta\varepsilon_n = \varepsilon_n - \varepsilon_{n-1} = \frac{1}{l_0^2}\,\tfrac{1}{2}(\mathbf{x}_{n-1} + \mathbf{x}_n)^T\,\Delta\mathbf{u}_n.$$

Note that the common reference length is l_0.

Exercise 2.2 The virtual work equation (2.16) was used to find the precise definition of the internal force N corresponding to a given strain measure.

(a) Find $\delta\varepsilon_E$ corresponding to the engineering strain $\varepsilon_E = (l - l_0)/l_0$, i.e. the formula similar to (2.15).
(b) Use the variation $\delta\varepsilon_E$ and the initial length l_0 in the integral of the virtual work equation (2.16) to find $\mathbf{f}_A$ and $\mathbf{f}_B$.
(c) What is the physical meaning of the internal force N, when used together with engineering strain.

Exercise 2.3 The general framework of geometrically non-linear bar elements outlined in Section 2.7 does not depend on the particular strain measure used in the theory.

(a) Use the results of Section 2.6 to establish the equilibrium equations of the current state (2.64) and the incremental stiffness relation (2.70) in explicit matrix form, when formulated in terms of engineering strain ε_E.
(b) Obtain the explicit matrix expressions when the theory is formulated in logarithmic strain ε_L (using current length measures ds and l).

Exercise 2.4 Example 2.2 describes the assembly of the total internal forces and the global tangent stiffness matrix with specific reference to the two-bar truss treated in Example 2.1. In the notation introduced in Example 2.1 the current coordinates of the three nodes of the truss are

$$A = (0, 0, -b), \quad B = (-c + u_1, u_2, u_3), \quad C = (0, 0, b).$$

Consider the symmetric case, in which $u_3 = 0$. The following computations can then be based on the element AB alone.

(a) Find the total strain ε_G and the strain increment $\delta\varepsilon_G$ in the element AB.
(b) Form the current position vector $\mathbf{x}$ for the element AB, and find the internal element forces $\mathbf{q}_A$ and $\mathbf{q}_B$ in terms of ε_G.
(c) Determine the tangent stiffness matrix of the element AB, and find the part of the global tangent stiffness matrix that enters into the equilibrium equations after elimination of the support conditions.
(d) Explain how the spring shown in Fig. 2.6 enters into the stiffness matrix.

Exercise 2.5* Set up a finite element model of the two-bar truss shown in Fig. 2.6. Subroutines from FemFiles can be used. The model has 2 bar elements, connecting 3 nodes, of which 2 are fully constrained. The horizontal linear spring is represented by a direct entry of the spring stiffness in the corresponding diagonal term.

(a) Write a data script file truss01 for a non-linear finite element analysis of the truss structure. The file must define the following data arrays:
  - X = [x,y,z;...], containing node coordinates;
  - T = [n1,n2,matl;...], for topology and material identification;
  - D = [A,E;...], defining material property table;
  - F = [n1,Fx,Fy,Fz;...], defining the load increment vector;
  - C = [n1,dof,u0;...], defining the boundary constraints.

  In addition, the file must provide some algorithmic parameters: number of load steps n_max, max number of iterations i_max, and tolerance of force residual unbalance TOL.

(b) Write subroutines kbar(K,T,X,D,u) and gbar(g,T,X,D,u) for the global tangent stiffness matrix and global internal force vector, respectively.

(c) Write a Newton–Raphson driver routine nrtruss that calls the data script file truss01 and makes use of the functions kbar(K,T,X,D,u) and gbar(g,T,X,D,u).

(d) Use truss01 and nrtruss to study the equilibrium path of the truss from Fig. 2.6 for ideal symmetry and for a small inclination of the truss with vertical. Illustrate the results with plots of the displacement components u_1, u_2 and the load f_1.

Exercise 2.6* The shallow dome shown in Fig. 8.10 is made of elastic bars. The dimensions of the dome are specified by the reference length $b_1 = 1.0$ and $b_2 = 2.0b_1$, $h_1 = 0.25b_1$ and $h_2 = 1.3h_1$. The dome is loaded by a concentrated downward force $-f$ at nodes 3 and 4 and an upward force f at nodes 6 and 7.

(a) Write a data script file dome01 for non-linear finite element analysis of the dome. The maximum load may be estimated around $f/EA \simeq 10^{-3}$, corresponding to downward displacement of node 3 around $w_3/b_1 \simeq 0.1$. The file must define the structure by the arrays listed in question (a) of Exercise 2.5. In addition, it should give data for a f–w plot.

(b) Write a Newton–Raphson driver routine nrtruss that calls the data script file dome01 and makes use of the functions kbar(K,T,X,D,u) and gbar(g,T,X,D,u) from Exercise 2.5 for the global tangent stiffness matrix and global internal force vector.

(c) Use `dome01` and `nrtruss` to calculate the deformation of the dome and make load–displacement plots for nodes 3 and 6.

(d) Is the displacement field anti-symmetric? Can the full load–displacement path be traced using the Newton–Raphson method with a set of vertical springs?

3

Finite rotations

Beams and shells can be considered as three-dimensional solids, in which two or one of the dimensions are small compared to the remaining dimensions. This has given rise to beam and shell models in which the representation of the transverse displacement components is simplified. These models are formulated in terms of the translation and rotation of a reference curve or surface in the case of beams and shells, respectively. As it will appear, there are fundamental differences between the representation of translations and rotations. It is important to account for the special characteristics of rotations in the formulation and analysis of kinematically non-linear theories for beams and shells, and this chapter provides a concise presentation of the special properties of rotations needed for the development of general non-linear beam and shell theories. A detailed discussion of rotations and their various representations has been given by Argyris (1982) and by Géradin and Cardona (2001).

The translation of a point is described by a vector $\mathbf{u}$, and the result of a sequence of translations $\Delta\mathbf{u}_1, \Delta\mathbf{u}_2, \ldots$ is simply the sum of the individual translation vectors. The result is independent of the order of the individual terms. Finite rotations cannot be described in this simple manner, as may be seen by considering two rotations of 90° about orthogonal axes, and comparing the result with that of the same rotations taken in the opposite order (Goldstein, 1980). The top sequence of Fig. 3.1 shows a box which is first rotated 90° around the x_1-axis and then −90° around the x_2-axis. This leaves the box standing on its end. The bottom sequence shows the box when rotated in the opposite order, first −90° around the x_2-axis and then 90° around the x_1-axis. This leaves the box lying on its side. The two results are distinctly different, and there is a clear need for a representation of rotations that reflects their non-commuting character.

The basic representation of finite rotations is considered in Sections 3.1

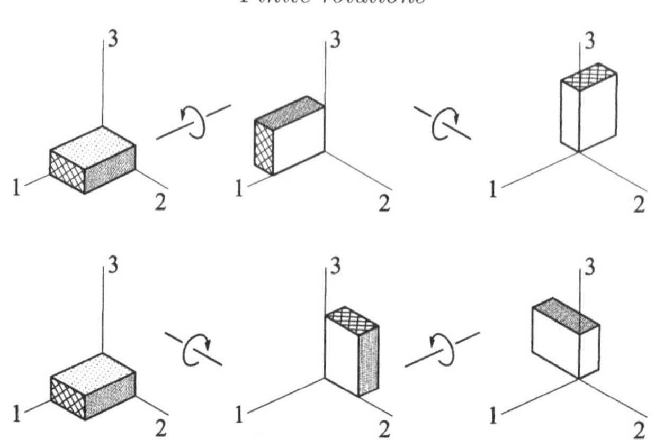

Fig. 3.1. Rotation of box around x_1- and x_2-axes.

and 3.2. It is demonstrated that finite rotations can be expressed in the form of a rotation tensor, the components of which form a 3×3 matrix. The components of the rotation tensor can be expressed in terms of a unit vector, defining the direction of the rotation axis, and the angle of rotation. This indicates that a rotation can be described in a vector-like format $\boldsymbol{\varphi}$ in which the direction of the vector defines the rotation axis, while the length of the vector defines the magnitude of the rotation. In structural models it is convenient to represent the rotations in this three-component pseudo-vector format.

In structural theories involving rotations, small rotation increments occur in two contexts: as a virtual rotation $\delta\boldsymbol{\varphi}$ used in the principle of virtual work δV, and as a change of rotation between neighboring points $d\boldsymbol{\varphi}/ds$, expressing curvature or deformation of the structure. The role of rotation increments in the principle of virtual work is dealt with in Section 3.3. In particular, the comparison of virtual work in neighboring states of equilibrium, required to establish the tangent stiffness, leads to the important result that the virtual rotation $\delta\boldsymbol{\varphi}$ cannot be considered identical in the two states, and therefore an extra term appears in the tangent stiffness relation. The second role, in which the derivatives $d\boldsymbol{\varphi}/ds$ are needed, is discussed in Section 3.4.

In kinematically non-linear structural models rotations need repeatedly to be updated from computed rotation increments. It is therefore of interest to find the most efficient way to combine two rotations. This is accomplished by an extended pseudo-vector representation of the rotations, called quaternion parameters. The quaternion parameter representation of rotations and the associated product rule that adds the corresponding rotations are derived in Section 3.5 from concepts of vector analysis. Finally, Section 3.6 gives a simple geometric derivation of a slightly different pseudo-vector representation of rotations, and a direct transformation of this vector to quaternion parameters.

3.1 The rotation tensor

The basic properties of finite rotations are important for the development of kinematically non-linear theories of beams and shells, where the rotation is used as an independent representation of displacement. Let $\mathbf{a}_0$ be a vector that is rotated into the position $\mathbf{a}$ as shown in Fig. 3.2(a). This can be described by a relation of the form

$$\mathbf{a} = \mathbf{R}\,\mathbf{a}_0, \tag{3.1}$$

where $\mathbf{R}$ is the rotation tensor. In this chapter it is convenient to use the notation of matrix analysis. A vector is represented by its Cartesian components in column format. This implies that the scalar product of two vectors is written as $\mathbf{a}^T\mathbf{b}$ and the exterior product as $\mathbf{a}\mathbf{b}^T$.

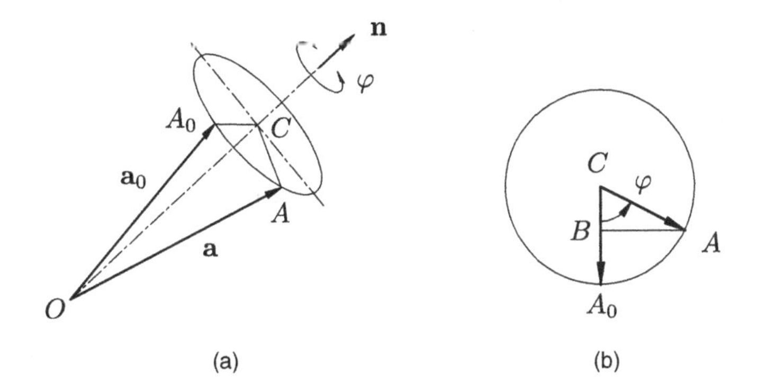

Fig. 3.2. Rotation of vector $\mathbf{a}_0$ by the angle φ around the direction $\mathbf{n}$.

The components of the rotation tensor constitute a 3×3 matrix. However, it is seen from the figure that a rotation can be specified by a direction $\mathbf{n}$ and an angle of rotation φ. It is therefore clear that not all the components of the rotation tensor $\mathbf{R}$ are independent. The relations between the components are found by using the fact that the length of any vector as well as the mutual orientation of any two vectors must be conserved by the rotation. Let $\mathbf{a}_0$ and $\mathbf{b}_0$ be two vectors. They are now rotated by the common rotation tensor $\mathbf{R}$ into the new vectors $\mathbf{a}$ and $\mathbf{b}$. The scalar product of the vectors is left unchanged by the rotation, and therefore

$$\mathbf{a}^T\mathbf{b} = \mathbf{a}_0^T\,\mathbf{R}^T\mathbf{R}\mathbf{b}_0 = \mathbf{a}_0^T\,\mathbf{I}\,\mathbf{b}_0 = \mathbf{a}_0^T\mathbf{b}_0, \tag{3.2}$$

where $\mathbf{I}$ is the unit tensor. Thus $\mathbf{R}$ must satisfy the relations

$$\mathbf{R}^T\mathbf{R} = \mathbf{R}\,\mathbf{R}^T = \mathbf{I}, \quad \det(\mathbf{R}) = +1. \tag{3.3}$$

The condition of positive determinant ensures that a right-hand vector triple retains its right-hand orientation after the rotation. These relations impose six constraints on the components of the tensor $\mathbf{R}$, thus leaving three independent components to specify the magnitude and direction of the rotation. The relations (3.3) define the rotation tensor $\mathbf{R}$ as a proper orthogonal tensor.

Figure 3.2(a) shows the rotation of a vector $\mathbf{a}_0$ by the angle φ around the direction $\mathbf{n}$ into the new position $\mathbf{a}$. The rotation parameters may be combined into the pseudo-vector

$$\boldsymbol{\varphi} = \varphi\,\mathbf{n}, \tag{3.4}$$

often called the rotation vector. The corresponding rotation tensor can be obtained directly in terms of $\mathbf{n}$ and φ by the following geometric argument, illustrated in Fig. 3.2. First the original vector is decomposed into its projection on the rotation axis $\overrightarrow{OC} = \mathbf{n}\,(\mathbf{n}^T\mathbf{a}_0)$ and the remainder $\overrightarrow{CA_0} = [\mathbf{a}_0 - \mathbf{n}(\mathbf{n}^T\mathbf{a}_0)]$. The vector $\overrightarrow{CA_0}$ is rotated by the angle φ in the plane orthogonal to $\mathbf{n}$. Hereby the component $\overrightarrow{CA_0}$ in the initial direction is reduced to $\overrightarrow{CB} = \cos\varphi\,\overrightarrow{CA_0}$, while a new component $\overrightarrow{BA} = \sin\varphi\,(\mathbf{n}\times\mathbf{a}_0)$ is generated, using the vector product denoted by $\times$. This gives the rotated vector in the form

$$\mathbf{a} = \mathbf{n}\,(\mathbf{n}^T\mathbf{a}_0) + \cos\varphi\,[\,\mathbf{a}_0 - \mathbf{n}\,(\mathbf{n}^T\mathbf{a}_0)\,] + \sin\varphi\,(\mathbf{n}\times\mathbf{a}_0), \tag{3.5}$$

i.e. in terms of the vectors $\mathbf{a}_0$ and $\mathbf{n}$, and trigonometric functions of the angle φ. This relation is often termed Rodrigues' formula, although it was in fact derived earlier by Euler, see e.g. Cheng and Gupta (1989).

The vector $\mathbf{a}_0$ can be extracted as a factor by introducing the notation

$$\mathbf{n}\times\mathbf{a}_0 = \hat{\mathbf{n}}\,\mathbf{a}_0, \tag{3.6}$$

where $\hat{\mathbf{n}}$ is a skew-symmetric two-dimensional tensor with the component matrix

$$\underbrace{(\hat{\mathbf{n}})_{ij} = -\varepsilon_{ijk}n_k} = \begin{bmatrix} 0 & -n_3 & n_2 \\ n_3 & 0 & -n_1 \\ -n_2 & n_1 & 0 \end{bmatrix}. \tag{3.7}$$

The skew-symmetric component matrix is also indicated in compact form by use of the summation convention and the permutation symbol ε_{ijk}. The summation convention implies that a subscript that appears with the same symbol twice corresponds to a summation, and thus (3.7) is a sum of three terms corresponding to $k = 1, 2, 3$, respectively. The permutation symbol ε_{ijk} equals $+1$ or -1 when ijk is an even or an odd permutation of the numbers $1, 2, 3$, and equals zero if any index value occurs more than once. With the vector product notation, the rotated vector can be expressed as

$$\mathbf{a} = \Big[\cos\varphi\, \mathbf{I} + \sin\varphi\, \hat{\mathbf{n}} + (1 - \cos\varphi)\, \mathbf{n}\,\mathbf{n}^T \Big] \mathbf{a}_0. \tag{3.8}$$

In this expression the rotation tensor corresponds to the terms in the square brackets, and thus

$$\mathbf{R} = \cos\varphi\, \mathbf{I} + \sin\varphi\, \hat{\mathbf{n}} + (1 - \cos\varphi)\, \mathbf{n}\,\mathbf{n}^T. \tag{3.9}$$

The component form can be written using Kronecker's delta δ_{ij} for the unit matrix and the permutation symbol ε_{ijk},

$$R_{ij} = \cos\varphi\, \delta_{ij} - \sin\varphi\, \varepsilon_{ijk} n_k + (1 - \cos\varphi)\, n_i n_j. \tag{3.10}$$

It is noted that the first and last terms are symmetric, while the middle term is skew-symmetric.

In the relations (3.8) and (3.9) the rotation tensor is expressed in terms of the direction $\mathbf{n}$ and the angle of rotation φ. For a given rotation tensor the rotation angle and the direction can be extracted in the following way. Properties in vector and tensor analysis that do not change if the coordinate system is changed are called scalars or invariants. It is known from tensor analysis that the trace of a tensor, defined as the sum of the diagonal terms, is invariant, see e.g. Malvern (1969) or Holzapfel (2000). It is clear from the definition of the angle of rotation φ that it is independent of any choice of coordinate system, while the components $[n_1, n_2, n_3]$ of the direction of the rotation axis refer to the chosen coordinate system. It is therefore to be expected that the angle of rotation can be extracted from the trace of the rotation tensor. This is confirmed by direct calculation from (3.10), whereby the trace of the rotation tensor is

$$\mathrm{tr}(\mathbf{R}) = R_{ii} = 1 + 2\cos\varphi. \tag{3.11}$$

Once the rotation angle has been obtained from (3.11), the direction components can be extracted from the skew-symmetric part of R_{ij}, giving the second term in (3.10). The principles used to extract the rotation angle φ and the components $[n_1, n_2, n_3]$ of the axis of rotation constitute a simple example of the use of invariance and symmetry. In Section 3.5 similar principles are used to derive a compact addition formula for rotations.

EXAMPLE 3.1. SOME SIMPLE ROTATIONS. For given direction $\mathbf{n}$ and rotation angle φ the components of the rotation tensor $\mathbf{R}$ follow from (3.10). The rotation in Fig. 3.1 around the x_1-axis is given by

$$\begin{aligned} \varphi_1 &= \tfrac{1}{2}\pi \\ \mathbf{n}_1 &= [1,0,0]^T \end{aligned} \qquad \mathbf{R}_1 = \begin{bmatrix} 1 & 0 & 0 \\ 0 & 0 & -1 \\ 0 & 1 & 0 \end{bmatrix},$$

while the negative rotation around the x_2-axis is given by

$$\begin{aligned} \varphi_2 &= -\tfrac{1}{2}\pi \\ \mathbf{n}_2 &= [0,1,0]^T \end{aligned} \qquad \mathbf{R}_2 = \begin{bmatrix} 0 & 0 & -1 \\ 0 & 1 & 0 \\ 1 & 0 & 0 \end{bmatrix}.$$

The rotation tensor $\mathbf{R}_{12}$ corresponding to first rotating by $\mathbf{R}_1$ and then rotating by $\mathbf{R}_2$, as illustrated in the top sequence of Fig. 3.1, follows from the product rule (3.1) as

$$\mathbf{R}_{12} = \mathbf{R}_2\,\mathbf{R}_1 = \begin{bmatrix} 0 & -1 & 0 \\ 0 & 0 & -1 \\ 1 & 0 & 0 \end{bmatrix}.$$

It is noted that the first rotation appears to the right in the product.

The angle φ_{12} of the combined rotation is found from the trace of the combined rotation tensor by (3.11),

$$\mathrm{tr}(\mathbf{R}_{12}) = 1 + 2\cos(\varphi_{12}) = 0 \qquad \Rightarrow \qquad \varphi_{12} = \pm\tfrac{2}{3}\pi.$$

The different signs correspond to opposite choices of the positive direction of the axis of rotation. Here the positive angle $\varphi_{12} = \frac{2}{3}\pi$ is chosen. The direction vector $\mathbf{n}_{12}$ of the rotation axis is extracted from the skew-symmetric part of the rotation tensor $\mathbf{R}_{12}$, given by (3.9):

$$\hat{\mathbf{n}}_{12} = \frac{1}{2\sin\varphi_{12}}\left(\mathbf{R}_{12} - \mathbf{R}_{12}^T\right) = \frac{1}{\sqrt{3}}\begin{bmatrix} 0 & -1 & -1 \\ 1 & 0 & -1 \\ 1 & 1 & 0 \end{bmatrix},$$

where it has been used that $\sin\varphi_{12} = \frac{\sqrt{3}}{2}$. This identifies the direction vector of the rotation axis of the combined rotation as

$$\mathbf{n}_{12} = \tfrac{1}{\sqrt{3}}[1,-1,1]^T.$$

When the order of the rotations is changed, as illustrated in the bottom sequence of Fig. 3.1, the rotation tensor components are

$$\mathbf{R}_{21} = \mathbf{R}_1\,\mathbf{R}_2 = \begin{bmatrix} 0 & 0 & -1 \\ -1 & 0 & 0 \\ 0 & 1 & 0 \end{bmatrix}.$$

The rotation angle and the direction of the rotation axis are extracted by the process described above, yielding

$$\varphi_{21} = \tfrac{2}{3}\pi, \quad \mathbf{n}_{21} = \tfrac{1}{\sqrt{3}}[1, -1, -1]^T.$$

3.2 Rotation of a vector into a specified direction

The rotation tensor $\mathbf{R}$ has been expressed above in terms of the direction vector $\mathbf{n}$ of the rotation axis and the magnitude φ of the rotation. In applications the rotation may be specified as a rotation of a unit vector $\mathbf{e}_0$ into a unit vector $\mathbf{e}_1$ with rotation axis normal to the plane containing the two vectors, as illustrated in Fig. 3.3. This is the smallest rotation that accomplishes the mapping of $\mathbf{e}_0$ on $\mathbf{e}_1$. In this case the rotation tensor can be expressed directly in terms of the two unit vectors $\mathbf{e}_0$ and $\mathbf{e}_1$.

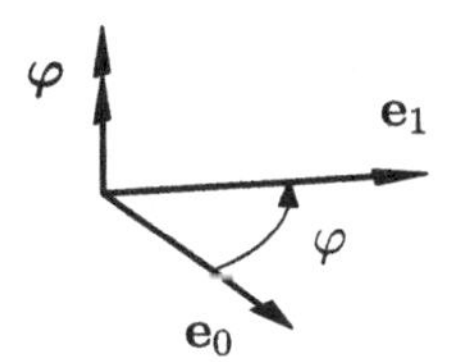

Fig. 3.3. Rotation of unit vector $\mathbf{e}_0$ into specified direction $\mathbf{e}_1$.

First, the rotation tensor (3.9) is written in the form

$$\mathbf{R} = \mathbf{I} + \sin\varphi\,\hat{\mathbf{n}} - (1 - \cos\varphi)(\mathbf{I} - \mathbf{n}\,\mathbf{n}^T) \tag{3.12}$$

and the two last terms are then expressed in terms of the vectors $\mathbf{e}_0$ and $\mathbf{e}_1$. The second term is rewritten by introducing the rotation axis via the vector product, $\mathbf{n} = \mathbf{e}_0 \times \mathbf{e}_1 / \sin\varphi$. This gives, by the well-known double vector product formula,

$$\begin{aligned} \sin\varphi\,\hat{\mathbf{n}}\,\mathbf{a} &= \sin\varphi\,(\mathbf{n} \times \mathbf{a}) = (\mathbf{e}_0 \times \mathbf{e}_1) \times \mathbf{a} \\ &= \mathbf{e}_1(\mathbf{e}_0^T\mathbf{a}) - \mathbf{e}_0(\mathbf{e}_1^T\mathbf{a}) = \left(\mathbf{e}_1\mathbf{e}_0^T - \mathbf{e}_0\mathbf{e}_1^T\right)\mathbf{a}. \end{aligned} \tag{3.13}$$

In the last term of (3.12) the tensor $(\mathbf{I} - \mathbf{n}\,\mathbf{n}^T)$ represents a projection on a plane normal to the direction $\mathbf{n}$ of the rotation axis. This plane contains the two vectors $\mathbf{e}_0$ and $\mathbf{e}_1$. They can be combined into the two orthogonal vectors $\mathbf{e}_0 + \mathbf{e}_1$ and $\mathbf{e}_0 - \mathbf{e}_1$. When these vectors are normalized the projection operator can be expressed as the sum of the projection on each of these vectors,

$$\mathbf{I} - \mathbf{n}\,\mathbf{n}^T = \frac{(\mathbf{e}_0 + \mathbf{e}_1)(\mathbf{e}_0 + \mathbf{e}_1)^T}{(\mathbf{e}_0 + \mathbf{e}_1)^T(\mathbf{e}_0 + \mathbf{e}_1)} + \frac{(\mathbf{e}_0 - \mathbf{e}_1)(\mathbf{e}_0 - \mathbf{e}_1)^T}{(\mathbf{e}_0 - \mathbf{e}_1)^T(\mathbf{e}_0 - \mathbf{e}_1)}. \tag{3.14}$$

Note the exterior vector products as numerators, and the scalar products as denominators. It is now used that $\mathbf{e}_0$ and $\mathbf{e}_1$ are unit vectors, and that $\cos\varphi = \mathbf{e}_0^T\mathbf{e}_1$, whereby $|\mathbf{e}_0 \pm \mathbf{e}_1|^2 = 2(1 \pm \cos\varphi)$. The rotation tensor (3.12) can then be expressed in the form

$$\mathbf{R} = \mathbf{I} + 2\,\mathbf{e}_1\mathbf{e}_0^T - 2\frac{(\mathbf{e}_0 + \mathbf{e}_1)(\mathbf{e}_0 + \mathbf{e}_1)^T}{|\mathbf{e}_0 + \mathbf{e}_1|^2}. \tag{3.15}$$

This constitutes the expression for the rotation tensor directly in terms of the two unit vectors $\mathbf{e}_0$ and $\mathbf{e}_1$. It is easily verified that $\mathbf{e}_1 = \mathbf{R}\mathbf{e}_0$, as required.

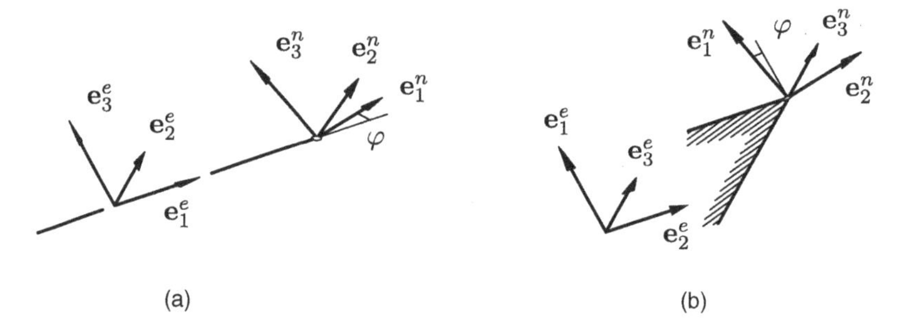

Fig. 3.4. Unit vectors $\mathbf{e}_j^n$ at node and $\mathbf{e}_j^e$ in element. (a) Beam, (b) shell.

In the formulation of geometrically non-linear theories for beams, plates and shells, the nodes may be associated with a unit vector triple $\mathbf{e}_j^n = [\mathbf{e}_1^n, \mathbf{e}_2^n, \mathbf{e}_3^n]$ as shown in Fig. 3.4. In the co-rotational formulation, discussed in Chapter 5, it is of interest to know the rotation of the vector triple $\mathbf{e}_j^n$ relative to a coordinate system $\mathbf{e}_j^e$ moving with the element. This involves rotating the triple $\mathbf{e}_j^n$ such that one of the vectors, e.g. $\mathbf{e}_1^n$, is brought into a direction $\mathbf{e}_1^e$, defined by the element-based coordinate system. The rotation is defined as a rotation of the vector $\mathbf{e}_1^n$ into the new vector $\mathbf{e}_1^e$ about an axis orthogonal to the plane of these two vectors, i.e. by the smallest rotation of the vector $\mathbf{e}_1^n$ into $\mathbf{e}_1^e$. Thus, the vector triple is rotated into the new unit vector triple $\mathbf{e}_j^e = [\mathbf{e}_1^e, \mathbf{e}_2^e, \mathbf{e}_3^e]$. The two rotated vectors $\mathbf{e}_2^e$ and $\mathbf{e}_3^e$ follow directly from the rotation formula (3.15). The two initial vectors $\mathbf{e}_2^n$ and $\mathbf{e}_3^n$ are orthogonal to the vector $\mathbf{e}_1^n$, and therefore the second term in (3.15) does not contribute. Thus, the transformation formula for the original base vectors $\mathbf{e}_j^n$ can be written in the simple symmetric form

$$\mathbf{e}_j^e = (\mathbf{I} - 2\bar{\mathbf{e}}\bar{\mathbf{e}}^T)\,\mathbf{e}_j^n, \quad j = 2,3 \qquad \text{with} \qquad \bar{\mathbf{e}} = \frac{\mathbf{e}_1^n + \mathbf{e}_1^e}{|\,\mathbf{e}_1^n + \mathbf{e}_1^e|}. \tag{3.16}$$

It is seen that the transformation of the two vectors $\mathbf{e}_j^n$ is described completely by the unit vector $\bar{\mathbf{e}}$. This transformation corresponds to a reflection in a plane normal to the unit vector $\bar{\mathbf{e}}$, see Exercise 3.2. This type of transformation is used in numerical analysis under the name of a Householder transformation (Press *et al.*, 1986).

In the case of a co-rotating beam element, treated in Chapter 5, the unit vectors $\mathbf{e}_\alpha^n$ define the beam cross-section or shell surface and the rotation defines the bending of the element.

3.3 The increment of the rotation variation

In the formulation of theories of structures with explicit representation of rotations, a need arises for comparing a state of rotation described by the rotation pseudo-vector $\boldsymbol{\varphi} = (\varphi_1, \varphi_2, \varphi_3)$ with a neighboring state with rotation pseudo-vector $\boldsymbol{\varphi} + \delta\boldsymbol{\varphi} = (\varphi_1 + \delta\varphi_1, \varphi_2 + \delta\varphi_2, \varphi_3 + \delta\varphi_3)$. According to Fig. 3.1 and the findings in Example 3.1, the components of the total rotation pseudo-vector will in general not be the algebraic sum of the components of the constituent rotations. Thus, the components $(\delta\varphi_1, \delta\varphi_2, \delta\varphi_3)$ will in general not describe an incremental rotation. Here the notation $\delta\bar{\boldsymbol{\varphi}} = (\delta\bar{\varphi}_1, \delta\bar{\varphi}_2, \delta\bar{\varphi}_3)$ will be used for the incremental rotation leading from the state $\boldsymbol{\varphi}$ to the state $\boldsymbol{\varphi} + \delta\boldsymbol{\varphi}$. The relation between the three pseudo-vectors $\boldsymbol{\varphi}$, $\delta\boldsymbol{\varphi}$ and $\delta\bar{\boldsymbol{\varphi}}$ is illustrated in Fig. 3.5.

The principle of virtual work is concerned with the change from the cur-

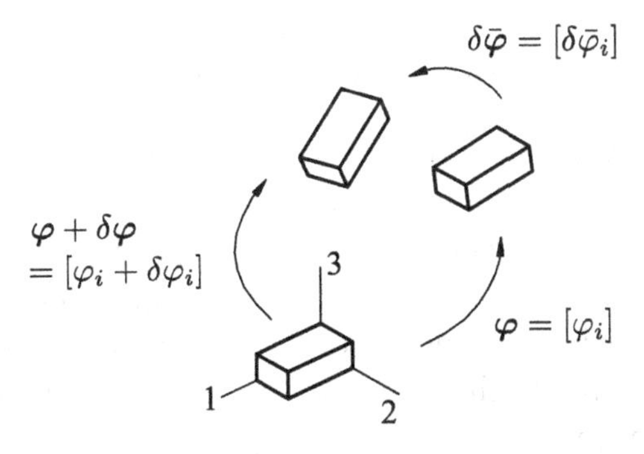

Fig. 3.5. Total rotation $\boldsymbol{\varphi} + \delta\boldsymbol{\varphi}$ as rotation $\boldsymbol{\varphi}$ followed by incremental rotation $\delta\bar{\boldsymbol{\varphi}}$.

rent state to a virtual neighboring state. Thus, the principle of virtual work including rotations is formulated in terms of the incremental rotation $\delta\bar{\boldsymbol{\varphi}}$. This has a profound influence on the tangent stiffness relations, where the change $d(\delta\bar{\boldsymbol{\varphi}})$ has to be considered. The questions relating to the use of the incremental rotation $\delta\bar{\boldsymbol{\varphi}}$ in the principle of virtual work and its higher-order increment $d(\delta\bar{\boldsymbol{\varphi}})$ in stiffness relations are treated in this section, while the general relation between an incremental rotation $d\bar{\boldsymbol{\varphi}}$ and the corresponding parameter increments $d\boldsymbol{\varphi}$, needed e.g. for describing curvature, is derived in the following section.

Let $\mathbf{a}_0$ be a vector that is rotated into the new position $\mathbf{a}$ as shown in Fig. 3.2. The rotation is described by the rotation tensor $\mathbf{R}(\boldsymbol{\varphi})$, defined as a function of the rotation pseudo-vector components $\boldsymbol{\varphi}$. Now consider a small variation $\delta\mathbf{a}$ generated by an infinitesimal rotation $\delta\mathbf{R}(\boldsymbol{\varphi})$ from the rotated position $\mathbf{a}$. It follows from (3.1) and (3.2) that

$$\delta\mathbf{a} = \delta\mathbf{R}\,\mathbf{a}_0 = \delta\mathbf{R}\,\mathbf{R}^T\mathbf{a}. \tag{3.17}$$

It follows from differentiation of (3.3a) that the tensor $\delta\mathbf{R}\mathbf{R}^T$ is skew-symmetric,

$$\delta(\mathbf{R}\,\mathbf{R}^T) = \delta\mathbf{R}\,\mathbf{R}^T + \mathbf{R}\,\delta\mathbf{R}^T = \delta\mathbf{R}\,\mathbf{R}^T + (\delta\mathbf{R}\,\mathbf{R}^T)^T = \mathbf{0}. \tag{3.18}$$

An incremental rotation of the vector $\mathbf{a}$ by the infinitesimal vector $\delta\bar{\boldsymbol{\varphi}}$ is given by the vector product

$$\delta\mathbf{a} = \delta\bar{\boldsymbol{\varphi}} \times \mathbf{a} = \widehat{\delta\bar{\boldsymbol{\varphi}}}\,\mathbf{a}, \tag{3.19}$$

where the symbol $\widehat{\ }$ was introduced in (3.7) for the skew-symmetric matrix associated with a three-dimensional vector and representing a vector product.

From this it follows that the skew-symmetric matrix $\widehat{\delta\bar{\boldsymbol{\varphi}}}$ corresponding to the incremental rotation vector $\delta\bar{\boldsymbol{\varphi}}$ is expressed in terms of the rotation tensor as

$$\delta\bar{\boldsymbol{\varphi}}\times = \widehat{\delta\bar{\boldsymbol{\varphi}}} = \delta\mathbf{R}\,\mathbf{R}^T. \tag{3.20}$$

The explicit non-linear relation between the incremental rotation vector $\delta\bar{\boldsymbol{\varphi}}$ and the increment of the pseudo-vector components $\delta\boldsymbol{\varphi}$ for an arbitrary value of the rotation parameter vector $\boldsymbol{\varphi}$ is derived in Section 3.4. However, many of the physical implications of the non-commuting property of rotations can be identified by considering the simpler case of small rotations around the state $\boldsymbol{\varphi} \simeq \mathbf{0}$, and therefore this special case is considered independently first.

First, small rotations relative to a configuration in which $\boldsymbol{\varphi} = \mathbf{0}$ are considered with the objective of studying the changes in the variation $\delta\bar{\boldsymbol{\varphi}}$ for an infinitesimal change $d\boldsymbol{\varphi} = [d\varphi_1, d\varphi_2, d\varphi_3]^T$ in the pseudo-vector components. For this purpose it is sufficient to consider expansions of $\delta\mathbf{R}$ and $\mathbf{R}^T$ including only constant and linear terms in the rotation pseudo-vector $\boldsymbol{\varphi}$. While the general relation (3.9) between the rotation tensor $\mathbf{R}$ and the rotation pseudo-vector $\boldsymbol{\varphi}$ is rather indirect, because it relies entirely upon a split into the direction $\mathbf{n}$ and the magnitude of the rotation φ, the second-order expansion of $\mathbf{R}$ can be expressed directly in terms of the pseudo-vector $\boldsymbol{\varphi}$ as

$$\mathbf{R} = \mathbf{I} + (\boldsymbol{\varphi}\times) - \tfrac{1}{2}(\boldsymbol{\varphi}^T\boldsymbol{\varphi})\,\mathbf{I} + \tfrac{1}{2}\,\boldsymbol{\varphi}\,\boldsymbol{\varphi}^T + O(\varphi^3). \tag{3.21}$$

From this the following expansions are obtained:

$$\delta\mathbf{R} = (\delta\boldsymbol{\varphi}\times) - (\delta\boldsymbol{\varphi}^T\boldsymbol{\varphi})\,\mathbf{I} + \tfrac{1}{2}(\delta\boldsymbol{\varphi}\,\boldsymbol{\varphi}^T + \boldsymbol{\varphi}\,\delta\boldsymbol{\varphi}^T) + O(\varphi^2) \tag{3.22}$$

and

$$\mathbf{R}^T = \mathbf{I} - (\boldsymbol{\varphi}\times) + O(\varphi^2). \tag{3.23}$$

In these expansions only linear terms in $\boldsymbol{\varphi}$ have been retained. The expansion of the product $\delta\mathbf{R}\,\mathbf{R}^T$, including linear terms in $\boldsymbol{\varphi}$, is then

$$\begin{aligned}\delta\mathbf{R}\,\mathbf{R}^T &= (\delta\boldsymbol{\varphi}\times) - \delta\boldsymbol{\varphi}\times(\boldsymbol{\varphi}\times) - (\delta\boldsymbol{\varphi}^T\boldsymbol{\varphi})\,\mathbf{I} \\ &\quad + \tfrac{1}{2}(\delta\boldsymbol{\varphi}\,\boldsymbol{\varphi}^T + \boldsymbol{\varphi}\,\delta\boldsymbol{\varphi}^T) + O(\varphi^2).\end{aligned} \tag{3.24}$$

Finally, this expansion is reformulated by rewriting the double vector product in terms of scalar products using the well-known vector relation

$$\mathbf{a}\times(\mathbf{b}\times\mathbf{c}) = (\mathbf{a}^T\mathbf{c})\,\mathbf{b} - (\mathbf{a}^T\mathbf{b})\,\mathbf{c} = \left[\,\mathbf{b}\,\mathbf{c}^T - \mathbf{c}\,\mathbf{b}^T\,\right]\mathbf{a} \tag{3.25}$$

and then reassembling the resulting scalar products into vector product form again. The result of this rearrangement of the terms is

$$\delta\mathbf{R}\,\mathbf{R}^T = \left[\,\delta\boldsymbol{\varphi} - \tfrac{1}{2}\,\delta\boldsymbol{\varphi}\times\boldsymbol{\varphi} + O(\varphi^2)\,\right]\times \tag{3.26}$$

It then follows from (3.20) that the incremental rotation $\delta\bar{\boldsymbol{\varphi}}$ is related to the pseudo-vector $\boldsymbol{\varphi}$ and its increment $\delta\boldsymbol{\varphi}$ by

$$\delta\bar{\boldsymbol{\varphi}} = \delta\boldsymbol{\varphi} - \tfrac{1}{2}\,\delta\boldsymbol{\varphi}\times\boldsymbol{\varphi} + O(\varphi^2). \tag{3.27}$$

It follows from this result that in a configuration with $\boldsymbol{\varphi} = \mathbf{0}$ the two first-order increments are equal, i.e.

$$\delta\bar{\boldsymbol{\varphi}} = \delta\boldsymbol{\varphi} \quad \text{for} \quad \boldsymbol{\varphi} = \mathbf{0}. \tag{3.28}$$

Thus, the infinitesimal parameter increments $\delta\boldsymbol{\varphi} = [\delta\varphi_1, \delta\varphi_2, \delta\varphi_3]^T$ do represent the components of an infinitesimal rotation vector $\delta\bar{\boldsymbol{\varphi}}$. However, this result is limited to the linear part of the generally non-linear relation between the rotation parameters and the incremental rotation.

The second variation $d(\delta\bar{\varphi})$ describes the change of the incremental rotation $\delta\bar{\varphi}$, when the rotation components are changed by $d\varphi$ and the variation $\delta\varphi = [\delta\varphi_1, \delta\varphi_2, \delta\varphi_3]^T$ of the components remains constant. It follows from differentiation of (3.27), where the linear term in φ gives rise to the second variation

$$d(\delta\bar{\varphi}) \;=\; -\,\tfrac{1}{2}\,\delta\varphi \times d\varphi \quad \text{for} \quad \varphi = \mathbf{0}. \tag{3.29}$$

In the second variation the increments may be replaced by their corresponding 'physical' representations in terms of incremental rotations $\delta\bar{\varphi}$ and $d\bar{\varphi}$ on account of the 'tangency condition' (3.28). Hereby any reference to the parameter representation in terms of φ disappears, and the second variation takes the form

$$d(\delta\bar{\varphi}) \;=\; -\,\tfrac{1}{2}\,\delta\bar{\varphi} \times d\bar{\varphi}. \tag{3.30}$$

Note that the sign of the second variation is determined by the order of the operations: first a rotation variation $\delta\bar{\varphi}$ is introduced, and then a change of configuration $d\bar{\varphi}$ is considered. This effect is purely three-dimensional; in plane problems the two rotation increments are co-linear, and their vector product vanishes. The implications of this new term are illustrated in the following two examples.

EXAMPLE 3.2. CHANGES IN EXTERNAL MOMENTS. In finite element analysis the effect of an external moment $\mathbf{M}$ is introduced via the contribution to external virtual work,

$$\delta V_M \;=\; \delta\bar{\varphi}^T\mathbf{M}.$$

The contribution to the tangent stiffness relation is evaluated from the second variation,

$$d(\delta V_M) \;=\; d(\delta\bar{\varphi}^T\mathbf{M}) \;=\; \delta\bar{\varphi}^T d\mathbf{M} \,+\, d(\delta\bar{\varphi})^T\mathbf{M}.$$

The first term is the well-known contribution from the change of the moment, while the second term is the change in the virtual rotation because of the non-linearity of the rotation parametrization. When the second variation $d(\delta\bar{\varphi})$ is substituted from (3.30), and the factors in the resulting triple product are rearranged,

$$d(\delta V_M) \;=\; \delta\bar{\varphi}^T d\mathbf{M} \,-\, \tfrac{1}{2}(\delta\bar{\varphi} \times d\bar{\varphi})^T\mathbf{M} \;=\; \delta\bar{\varphi}^T\big(\,d\mathbf{M} \,-\, \tfrac{1}{2}\,d\bar{\varphi} \times \mathbf{M}\,\big).$$

This relation can be written in the form

$$d(\delta V_M) \;=\; \delta\bar{\varphi}^T d_*\mathbf{M},$$

where $d\mathbf{M}_*$ is the effective change of the external moment, defined as

$$d_*\mathbf{M} = d\mathbf{M} - \tfrac{1}{2}\, d\bar{\boldsymbol{\varphi}} \times \mathbf{M}.$$

Thus, an external moment for which $d_*\mathbf{M} = \mathbf{0}$ will appear as a constant moment in tangent stiffness relations. A similar conclusion was obtained by Christoffersen (1989) using the reciprocity relation for conservative systems.

A case of particular interest is the motion of a beam element with an initial distribution of forces and moments. In this case the force and moment at the nodes of the beam element are rotated with the element. If the incremental rotation of the element is $d\bar{\boldsymbol{\varphi}}$, the corresponding moment increment is $d\mathbf{M} = d\bar{\boldsymbol{\varphi}} \times \mathbf{M}$. However, the apparent change of the moment $d_*\mathbf{M}$ as seen via the change in external virtual work is

$$d_*\mathbf{M} = d\mathbf{M} - \tfrac{1}{2}\, d\bar{\boldsymbol{\varphi}} \times \mathbf{M} = \tfrac{1}{2}\, d\bar{\boldsymbol{\varphi}} \times \mathbf{M},$$

i.e. *half* the physical change due to the rotation of the element. This is a crucial point in the development of geometric stiffness of beams and shells as discussed in Chapters 4 and 5.

EXAMPLE 3.3. SEMI-TANGENTIAL MOMENT. Figure 3.6(a) shows the end of a beam loaded by a moment $\mathbf{M}$. The tangent of the beam is described by the unit vector $\mathbf{n}$, and a rigid circular disk of radius a is mounted on the end of the beam, normal to $\mathbf{n}$. The moment $\mathbf{M} = M\mathbf{n}$ is generated via four long strings wound around the circular disk. Each string has the tension force P, thus producing the moment $M = 4aP$. When the forces are kept constant in magnitude and direction, the moment is conservative.

Now, consider a small rotation $d\bar{\boldsymbol{\varphi}}$ of the end of the beam around the direction of one pair of forces as shown in Fig. 3.6(b). The contribution to

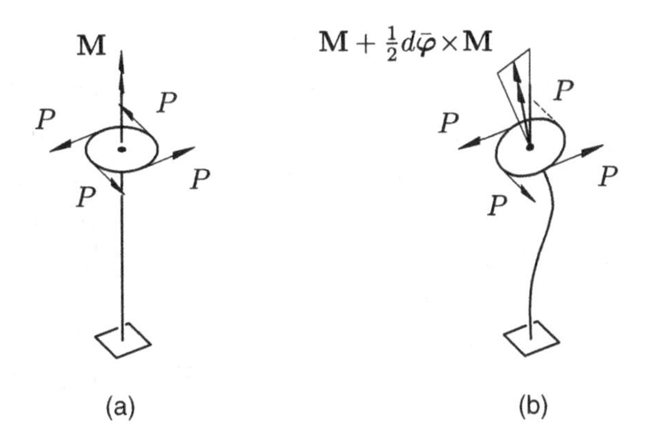

Fig. 3.6. Semi-tangential moment.

the moment from the pair of forces parallel to the rotation vector is rotated with the disk, while the moment contribution from the other pair of forces remains unchanged. The total effect is a change of the moment

$$d\mathbf{M} = \tfrac{1}{2}\, d\bar{\boldsymbol{\varphi}} \times \mathbf{M}.$$

Thus, the moment rotates with *half* the rotation increment $d\bar{\boldsymbol{\varphi}}$. This type of moment is called semi-tangential and was introduced by Ziegler (1977) in the context of follower forces and moments, and their influence on stability. Note that the expression for $d\mathbf{M}$ remains valid also if the rotation increment $d\bar{\boldsymbol{\varphi}}$ has a component along the tangent $\mathbf{n}$, although the intuitive interpretation of the moment rotating through half the rotation increment becomes more indirect.

For a semi-tangential moment the effective moment increment, introduced in Example 3.2, is

$$d_*\mathbf{M} = d\mathbf{M} - \tfrac{1}{2}\, d\bar{\boldsymbol{\varphi}} \times \mathbf{M} = \mathbf{0}.$$

Thus, the rotation $d\bar{\boldsymbol{\varphi}}$ of the point of application produces a change of the moment vector $d\mathbf{M}$, but the non-linearity of the rotation parametrization compensates for this change in the incremental form of the principle of virtual work, making a semi-tangential moment *appear* as constant during the change of configuration. Several authors, e.g. Argyris *et al.* (1979) and Yang and Kuo (1994), have made use of the concept of semi-tangential moments in the development of non-linear beam theories. However, the properties of semi-tangential moments rely on the constant direction of the forces generating the moment. In a solid body the stresses are convected with the material, and thus the essential problem of non-commutative rotations remains in the description of the stress components.

3.4 Parameter representation of an incremental rotation

In Section 3.3 the implications of the non-linear parametrization of rotations were investigated for small rotations around the state $\boldsymbol{\varphi} = \mathbf{0}$. In beam theory the rotations vary along the beam, and it is therefore necessary to obtain the general relation between the rotation increment vector $d\bar{\boldsymbol{\varphi}}$ and the analogous pseudo-vector rotation component increment $d\boldsymbol{\varphi}$. The relation plays a role in the parametrization of curved beams and shells, and has been derived by several authors, e.g. Simo (1985), Simo and Vu-Quoc (1986), Atluri and Cazzani (1995).

In this section an explicit relation is obtained between the incremental rotation $d\bar{\boldsymbol{\varphi}}$ and the increment of the rotation pseudo-vector components $d\boldsymbol{\varphi}$ from the basic relation (3.20),

$$d\bar{\boldsymbol{\varphi}}\times \;=\; \widehat{d\bar{\boldsymbol{\varphi}}} \;=\; d\mathbf{R}\,\mathbf{R}^T. \tag{3.31}$$

The rotation tensor was given in terms of the rotation pseudo-vector $\boldsymbol{\varphi}$ in (3.9) as

$$\mathbf{R} \;=\; \cos\varphi\,\mathbf{I} \,+\, \sin\varphi\,(\mathbf{n}\times) \,+\, (1-\cos\varphi)\,\mathbf{n}\,\mathbf{n}^T. \tag{3.32}$$

The dependence on the pseudo-vector $\boldsymbol{\varphi}$ is through its magnitude φ and the direction $\mathbf{n}$, and it is therefore convenient first to express the increment of $\mathbf{R}$ in terms of the increments $d\varphi$ and $d\mathbf{n}$:

$$\begin{aligned} d\mathbf{R} \;&=\; \big[\,-\sin\varphi\,\mathbf{I} + \cos\varphi\,(\mathbf{n}\times) + \sin\varphi\,\mathbf{n}\,\mathbf{n}^T\big]\,d\varphi \\ &\quad + \sin\varphi\,(d\mathbf{n}\times) + (1-\cos\varphi)(d\mathbf{n}\,\mathbf{n}^T + \mathbf{n}\,d\mathbf{n}^T). \end{aligned} \tag{3.33}$$

The contribution to $d\mathbf{R}\mathbf{R}^T$ from the increment in magnitude $d\varphi$ is deduced using the relations

$$\mathbf{n}^T(\mathbf{n}\times) \;=\; \mathbf{n}^T\,\hat{\mathbf{n}} \;=\; \mathbf{0}, \quad (\mathbf{n}\times)^2 \;=\; \hat{\mathbf{n}}\,\hat{\mathbf{n}} \;=\; (\,\mathbf{n}\,\mathbf{n}^T \,-\, \mathbf{I}\,). \tag{3.34}$$

This leads to the simple relation

$$d\mathbf{R}_{d\varphi}\,\mathbf{R}^T \;=\; d\varphi\,(\mathbf{n}\times). \tag{3.35}$$

This implies that for a rotation increment $d\boldsymbol{\varphi}$ proportional to the current rotation pseudo-vector $\boldsymbol{\varphi}$, the incremental rotation $d\bar{\boldsymbol{\varphi}}$ is identical to the increment of the rotation parameter vector $d\boldsymbol{\varphi}$. This is the case in plane problems, where the rotation vector is always orthogonal to the plane, and thus this class of problems can be formulated without the distinction between the two types of rotation increments.

In the general case also the direction changes. The contribution to $d\mathbf{R}\,\mathbf{R}^T$ from this increment can be reduced to the form

$$\begin{aligned} d\mathbf{R}_{d\mathbf{n}}\,\mathbf{R}^T \;&=\; \sin\varphi\cos\varphi\,(d\mathbf{n}\times) + (1-\cos\varphi)\,\big[d\mathbf{n}\,\mathbf{n}^T - \mathbf{n}\,d\mathbf{n}^T\big] \\ &\quad + \sin\varphi\,(1-\cos\varphi)\,\big[(d\mathbf{n}\times\mathbf{n})\,\mathbf{n}^T \,-\, \mathbf{n}\,(d\mathbf{n}\times\mathbf{n})^T\big]. \end{aligned} \tag{3.36}$$

In the last two terms the brackets contain exterior vector products of the form $\mathbf{a}\mathbf{n} - \mathbf{n}\mathbf{a}$ with $\mathbf{a} = d\mathbf{n}$ and $\mathbf{a} = d\mathbf{n}\times\mathbf{n}$, respectively. This form is equivalent to a double vector product as expressed by a reformulation of the vector product relation (3.25),

$$(\mathbf{n}\times\mathbf{a})\times \;=\; \mathbf{a}\,\mathbf{n}^T \,-\, \mathbf{n}\,\mathbf{a}^T. \tag{3.37}$$

When this relation is used to reduce the last two terms,

$$\begin{aligned} d\mathbf{R}_{dn}\,\mathbf{R}^T &= \sin\varphi\cos\varphi\,(d\mathbf{n}\times) + (1-\cos\varphi)\,[\,(\mathbf{n}\times d\mathbf{n})\times\,] \\ &\quad + \sin\varphi\,(1-\cos\varphi)\,[\,(\mathbf{n}\times(d\mathbf{n}\times\mathbf{n}))\times\,]. \end{aligned} \tag{3.38}$$

The vector $\mathbf{n}$ is a unit vector, and therefore $d\mathbf{n}$ is orthogonal to $\mathbf{n}$. This in turn implies that $\mathbf{n}\times(d\mathbf{n}\times\mathbf{n}) = d\mathbf{n}$, whereby the last term is simplified. After combination of the first and last term, the final formula for the increment from the change of direction $d\mathbf{n}$ is

$$d\mathbf{R}_{dn}\mathbf{R}^T = \sin\varphi\,(d\mathbf{n}\times) + (1-\cos\varphi)\,[\,(\mathbf{n}\times d\mathbf{n})\times\,]. \tag{3.39}$$

Addition of the contributions from $d\varphi$ and $d\mathbf{n}$ then gives the total increment in the form

$$d\mathbf{R}\,\mathbf{R}^T = d\varphi\,(\mathbf{n}\times) + \sin\varphi\,(d\mathbf{n}\times) + (1-\cos\varphi)\,[\,(\mathbf{n}\times d\mathbf{n})\times\,]. \tag{3.40}$$

This expression consists of three terms, each corresponding to a vector product. The incremental rotation vector $d\bar{\boldsymbol{\varphi}}$ can therefore be identified directly from (3.31) as the sum of the corresponding three vectors,

$$d\bar{\boldsymbol{\varphi}} = \mathbf{n}\,d\varphi + \sin\varphi\,d\mathbf{n} + (1-\cos\varphi)\,(\mathbf{n}\times d\mathbf{n}). \tag{3.41}$$

The three terms in this expression give the components of $d\bar{\boldsymbol{\varphi}}$ along each of the three orthogonal directions $\mathbf{n}, d\mathbf{n}, \mathbf{n}\times d\mathbf{n}$.

The increments $d\varphi$ and $d\mathbf{n}$ are now expressed in terms of the parameter pseudo-vector $d\boldsymbol{\varphi}$. Differentiation of the defining relations $\varphi^2 = \boldsymbol{\varphi}^T\boldsymbol{\varphi}$ and $\mathbf{n} = \boldsymbol{\varphi}/\varphi$ gives

$$d\varphi = \mathbf{n}^T d\boldsymbol{\varphi}, \quad d\mathbf{n} = \frac{1}{\varphi}(\mathbf{I} - \mathbf{n}\,\mathbf{n}^T)\,d\boldsymbol{\varphi}. \tag{3.42}$$

The final relation between the incremental rotation vector $d\bar{\boldsymbol{\varphi}}$ and the increment $d\boldsymbol{\varphi}$ of the rotation pseudo-vector then follows by substitution of these expressions into (3.41):

$$d\bar{\boldsymbol{\varphi}} = \Big[\,\mathbf{n}\,\mathbf{n}^T + \frac{\sin\varphi}{\varphi}(\mathbf{I} - \mathbf{n}\,\mathbf{n}^T) + \frac{1-\cos\varphi}{\varphi}(\mathbf{n}\times)\Big]\,d\boldsymbol{\varphi}. \tag{3.43}$$

This relation is written as

$$d\bar{\boldsymbol{\varphi}} = \mathbf{T}(\boldsymbol{\varphi})\,d\boldsymbol{\varphi}, \tag{3.44}$$

with the tensor $\mathbf{T}(\boldsymbol{\varphi})$ defined by

$$\mathbf{T}(\boldsymbol{\varphi}) = \mathbf{n}\,\mathbf{n}^T + \frac{\sin\varphi}{\varphi}(\mathbf{I} - \mathbf{n}\,\mathbf{n}^T) + \frac{1-\cos\varphi}{\varphi}\,\hat{\mathbf{n}}. \tag{3.45}$$

Note that while the transformation tensor $\mathbf{T}(\boldsymbol{\varphi})$ is formed as the sum of three terms of the same structure as the rotation tensor $\mathbf{R}(\boldsymbol{\varphi})$, the scalar factors are different, and $\mathbf{T}(\boldsymbol{\varphi})$ is not an orthogonal tensor.

In the formulation of beam and shell theories, it is convenient to represent the incremental rotation vector $d\bar{\boldsymbol{\varphi}}$ directly, and to obtain the parameter increments $d\boldsymbol{\varphi}$ subsequently from $d\bar{\boldsymbol{\varphi}}$. This requires the inverse of the transformation tensor $\mathbf{T}(\boldsymbol{\varphi})$. The inverse $\mathbf{T}(\boldsymbol{\varphi})^{-1}$ is most easily obtained from an assumed vector representation in the form

$$\mathbf{T}(\boldsymbol{\varphi})^{-1} = \alpha\, \mathbf{n}\,\mathbf{n}^T + \beta\,(\mathbf{I} - \mathbf{n}\,\mathbf{n}^T) + \gamma\,\hat{\mathbf{n}} \tag{3.46}$$

similar to the form of $\mathbf{T}(\boldsymbol{\varphi})$, where each term represents an independent vector component in space. The parameters α, β, γ are determined from the fundamental relation $\mathbf{T}\,\mathbf{T}^{-1} = \mathbf{I}$, giving the inverse as

$$\mathbf{T}(\boldsymbol{\varphi})^{-1} = \mathbf{n}\,\mathbf{n}^T + \tfrac{1}{2}\varphi\cot(\tfrac{1}{2}\varphi)\,(\mathbf{I} - \mathbf{n}\,\mathbf{n}^T) - \tfrac{1}{2}\varphi\,\hat{\mathbf{n}} \tag{3.47}$$

and thus confirming the assumed form of the inverse.

In some applications the tensor $\mathbf{T}(\boldsymbol{\varphi})$ or its inverse $\mathbf{T}(\boldsymbol{\varphi})^{-1}$ are needed for very small angles φ. In that case it may be preferable with respect to numerical accuracy to use a representation of the argument directly in terms of the rotation pseudo-vector $\boldsymbol{\varphi}$ to avoid extraction of the normalized direction $\mathbf{n}$, see Exercise 3.3.

3.5 Quaternion parameter representation

It is seen from the form (3.9) of the rotation tensor that it only depends on the rotation angle φ through trigonometric functions. This has led to several pseudo-vector formats in terms of trigonometric functions. Of these the quaternion parameter format deserves special attention, because it is non-singular for all angles, and it leads to a simple algebraic formula for the non-commutative addition of finite rotations. The concept of quaternions as four-dimensional vectors with a special product rule was originally introduced by Hamilton, and the addition formula developed by Cayley. A description of this theory has been given by e.g. Corben and Stehle (1974), and Géradin and Cardona (2001) present a survey of different three- and four-parameter representations of finite rotations. In this presentation the quaternion parameters will simply be used as a convenient means of representing the rotation tensor and the addition of rotations, and the necessary formulae will be developed directly in vector format, using techniques from vector analysis.

3.5.1 Representation of the rotation tensor

The quaternion parameter representation of a rotation φ about the axis $\mathbf{n}$ consists of a scalar r_0 and a vector $\mathbf{r}$, defined by

$$r_0 = \cos(\tfrac{1}{2}\varphi), \quad \mathbf{r} = \sin(\tfrac{1}{2}\varphi)\mathbf{n}. \tag{3.48}$$

It turns out to be convenient to retain the four-parameter format, although the quaternion parameters satisfy the identity

$$r_0^2 + \mathbf{r}^T\mathbf{r} = r_0^2 + r_1^2 + r_2^2 + r_3^2 = 1. \tag{3.49}$$

In terms of the four quaternion parameters $r_0, \mathbf{r}$ the rotation tensor (3.9) takes the homogeneous quadratic form

$$\mathbf{R} = (r_0^2 - \mathbf{r}^T\mathbf{r})\,\mathbf{I} + 2\,r_0\,\hat{\mathbf{r}} + 2\,\mathbf{r}\,\mathbf{r}^T. \tag{3.50}$$

The corresponding component representation is

$$R_{ij} = (r_0^2 - r_k r_k)\,\delta_{ij} - 2\varepsilon_{ijk}\,r_0 r_k + 2\,r_i r_j \tag{3.51}$$

or, in full matrix form,

$$\mathbf{R} = \begin{bmatrix} r_0^2 + r_1^2 - r_2^2 - r_3^2 & 2(r_1r_2 - r_0r_3) & 2(r_1r_3 + r_0r_2) \\ 2(r_2r_1 + r_0r_3) & r_0^2 - r_1^2 + r_2^2 - r_3^2 & 2(r_2r_3 - r_0r_1) \\ 2(r_3r_1 - r_0r_2) & 2(r_3r_2 + r_0r_1) & r_0^2 - r_1^2 - r_2^2 + r_3^2 \end{bmatrix}. \tag{3.52}$$

The scalar quaternion parameter r_0^2 is determined directly by the trace of the rotation matrix. From (3.52) and the normalization condition (3.49),

$$\mathrm{tr}(\mathbf{R}) = R_{ii} = 4\,r_0^2 - 1. \tag{3.53}$$

The quaternion vector components r_l can be extracted from the skew-symmetric part of R_{ij}. Multiplication of R_{ij} in (3.51) by the permutation symbol ε_{lij} gives

$$\varepsilon_{lij}R_{ij} = -2\varepsilon_{lij}\varepsilon_{ijk}\,r_0 r_k = -4\,r_0 r_l, \tag{3.54}$$

where the identity $\varepsilon_{lij}\varepsilon_{ijk} = 2\delta_{lk}$ has been used.

Example 3.4. Extraction of quaternion parameters. While the formulae (3.53) and (3.54) illustrate the principle behind extraction of the quaternion parameters from the rotation tensor components, they become singular at rotations of 180°, and lose numerical accuracy for angles around this value. A scheme that is numerically stable for all angles was proposed by Spurrier (1978) and given in simplified form by Géradin and Cardona (2001). The current status is available in Markley (2008). In the extraction process it is necessary to divide by one of the quaternion components $[r_0, r_1, r_2, r_3]$, and it is numerically advantageous to select the largest. This is conveniently done by forming the quaternion component product matrix

$$\mathbf{S} = 4\begin{bmatrix} r_0^2 & r_0 r_1 & r_0 r_2 & r_0 r_3 \\ r_0 r_1 & r_1^2 & r_1 r_2 & r_1 r_3 \\ r_0 r_2 & r_2 r_1 & r_2^2 & r_2 r_3 \\ r_0 r_3 & r_3 r_1 & r_3 r_2 & r_3^2 \end{bmatrix}$$

in terms of the rotation tensor components R_{ij} from (3.52),

$$\mathbf{R} = \begin{bmatrix} 1+R_{11}+R_{22}+R_{33} & R_{32}-R_{23} & R_{13}-R_{31} & R_{21}-R_{12} \\ R_{32}-R_{32} & 1+R_{11}-R_{22}-R_{33} & R_{12}+R_{21} & R_{13}+R_{31} \\ R_{13}-R_{31} & R_{21}+R_{12} & 1-R_{11}+R_{22}-R_{33} & R_{23}+R_{32} \\ R_{21}-R_{12} & R_{13}+R_{31} & R_{23}+R_{32} & 1-R_{11}-R_{22}+R_{33} \end{bmatrix}.$$

The largest component r_i is identified from the diagonal element as $4\,r_i^2 = \max_j S_{jj}$, without summation of the repeated subscript. Thus, $r_i = \frac{1}{2}\sqrt{S_{ii}}$, while the full set of components follows from the corresponding row of $\mathbf{S}$ as $r_j = \frac{1}{4}S_{ij}/r_i$, $j = 1, \ldots, 4$, again without summation.

3.5.2 Addition of two rotations

Let two rotations be represented by the rotation tensors $\mathbf{P}$ and $\mathbf{Q}$. If a vector $\mathbf{a}_0$ is first rotated into $\mathbf{a}_1$ by $\mathbf{Q}$ and then into $\mathbf{a}_2$ by $\mathbf{P}$, the total rotation is given by the product

$$\mathbf{a}_2 = \mathbf{P}\,\mathbf{a}_1 = \mathbf{P}\,\mathbf{Q}\,\mathbf{a}_0. \tag{3.55}$$

Thus, the combined rotation matrix is

$$\mathbf{R} = \mathbf{P}\,\mathbf{Q} \tag{3.56}$$

or, in component form,

$$R_{ij} = P_{ik}\,Q_{kj}. \tag{3.57}$$

It is desirable to have formulae for this matrix product directly in terms of the corresponding pseudo-vector representations, in order to compute all rotation changes directly in terms of the corresponding pseudo-vectors. The combination rule for rotations is particularly simple when expressed in terms of the quaternion parameters.

Let the rotation tensors $\mathbf{P}$ and $\mathbf{Q}$ have the quaternion representations $p_0, \mathbf{p}$ and $q_0, \mathbf{q}$, respectively. The objective then is to determine the quaternion parameters $r_0, \mathbf{r}$ of the combined rotation. The scalar quaternion parameter

r_0 is determined by the trace R_{ii} as shown in (3.53). Substitution of the subscript form (3.51) of the quaternion representations of P_{ik} and Q_{ki} into the trace of (3.57) gives

$$R_{ii} = [(2p_0^2 - 1)\delta_{ik} + 2p_ip_k - 2\varepsilon_{ikl}p_0p_l][(2q_0^2 - 1)\delta_{ki} + 2q_kq_i - 2\varepsilon_{kim}q_0q_m]. \tag{3.58}$$

Products of symmetric and skew-symmetric terms cancel, and the result is therefore easily reduced to the form

$$R_{ii} = 4\,(p_0q_0 - p_iq_i)^2 - 1 \tag{3.59}$$

and thus, by (3.53), the scalar quaternion parameter of the combined rotation is

$$r_0 = p_0\,q_0 - p_i\,q_i. \tag{3.60}$$

The vector part of the quaternion for the combined rotation follows from the skew-symmetric part of R_{ij} by substitution of the product $P_{ik}Q_{kj}$ into (3.54). After reduction of the permutation symbol products, the following result is obtained:

$$\varepsilon_{rij}R_{ij} = -4\,(p_0q_0 - p_iq_i)(p_0q_r + q_0p_r + \varepsilon_{rij}p_iq_j). \tag{3.61}$$

This identifies the vector part as

$$r_r = p_0q_r + q_0p_r + \varepsilon_{rij}p_iq_j. \tag{3.62}$$

In vector form the formulae for the combined rotation $\mathbf{q}$ followed by $\mathbf{p}$ are

$$r_0 = p_0\,q_0 - \mathbf{p}^T\mathbf{q}, \qquad \mathbf{r} = p_0\,\mathbf{q} + q_0\,\mathbf{p} + \mathbf{p}\times\mathbf{q}. \tag{3.63}$$

The last term takes care of the effect of different directions of the two rotations $\mathbf{p}$ and $\mathbf{q}$. If the direction is the same, (3.63) reduces to the trigonometric addition formulae for the two angles.

EXAMPLE 3.5. ADDITION OF ROTATIONS BY QUATERNIONS. This example demonstrates the representation and addition of the rotations shown in Fig. 3.1 by use of quaternions. The first is a 90° rotation around the x_1-axis. The quaternion representation follows directly from the defining relations (3.48) with $\cos(\frac{1}{4}\pi) = \sin(\frac{1}{4}\pi) = \frac{\sqrt{2}}{2}$,

$$r_0^{(1)} = \tfrac{\sqrt{2}}{2}, \quad \mathbf{r}^{(1)} = [\tfrac{\sqrt{2}}{2}, 0, 0]^T.$$

The other is a 90° rotation around the negative x_2-axis, corresponding to the quaternion representation

$$r_0^{(2)} = \tfrac{\sqrt{2}}{2}, \quad \mathbf{r}^{(2)} = [0, -\tfrac{\sqrt{2}}{2}, 0]^T.$$

The quaternion of the combined rotation from first rotating around the x_1-axis and then around the x_2-axis follows from (3.63) as

$$r_0^{(12)} = \tfrac{1}{2}, \quad \mathbf{r}^{(12)} = [\tfrac{1}{2}, -\tfrac{1}{2}, \tfrac{1}{2}]^T.$$

The angle of rotation is then obtained from

$$\cos(\tfrac{1}{2}\varphi_{12}) = \tfrac{1}{2}, \quad \sin(\tfrac{1}{2}\varphi_{12}) = \tfrac{\sqrt{3}}{2} \qquad \Rightarrow \qquad \varphi_{12} = \tfrac{2}{3}\pi$$

and the direction vector follows from $\mathbf{r}^{(12)}$ as $\mathbf{n}_{12} = \frac{1}{\sqrt{3}}[1, -1, 1]^T$.

Changing the order of the rotations just changes the sign of the vector product in (3.61), whereby

$$r_0^{(21)} = \tfrac{1}{2}, \quad \mathbf{r}^{(21)} = [\tfrac{1}{2}, -\tfrac{1}{2}, -\tfrac{1}{2}]^T.$$

The angle and rotation axis direction are then found to be $\varphi_{21} = \frac{2}{3}\pi$ and $\mathbf{n}_{21} = \frac{1}{\sqrt{3}}[1, -1, -1]^T$. As expected, the same results are obtained as in Example 3.1, but somewhat more directly.

3.5.3 Incremental rotation from quaternion parameters

The problem treated in Section 3.4 of expressing an incremental rotation $d\bar{\boldsymbol{\varphi}}$ in terms of the increments of a suitable set of parameters describing an initial state of rotation can be solved in a compact form, when the rotation is represented in terms of quaternion parameters. The angular velocity $\boldsymbol{\omega}$ is defined as the incremental rotation $d\bar{\boldsymbol{\varphi}}$ normalized with the corresponding time increment dt, i.e. by the relation $\boldsymbol{\omega} = d\bar{\boldsymbol{\varphi}}/dt$, and the relation between the angular velocity and the time derivative of the quaternion parameters plays an important role in multibody dynamics, see e.g. Géradin and Cardona (2001).

Let a rotation state be described by the quaternion parameters $(r_0, \mathbf{r})$. From this state an infinitesimal rotation $d\bar{\boldsymbol{\varphi}}$ is applied, leading to a final state described by the quaternion parameters $(r_0 + dr_0, \mathbf{r} + d\mathbf{r})$. The problem is to express the incremental rotation $d\bar{\boldsymbol{\varphi}}$ in terms of the quaternion parameter increments $(dr_0, d\mathbf{r})$.

It follows from the quaternion parameter definition (3.48) that the quaternion representation of an infinitesimal rotation $d\bar{\boldsymbol{\varphi}}$ is $(1, \frac{1}{2}d\bar{\boldsymbol{\varphi}})$ plus second- and higher-order terms. Use of the quaternion addition formulae (3.63) now gives the quaternion parameter representation of the final rotation,

$$r_0 + dr_0 \simeq 1\, r_0 - \tfrac{1}{2}d\bar{\boldsymbol{\varphi}}^T\mathbf{r}, \quad \mathbf{r} + d\mathbf{r} \simeq 1\,\mathbf{r} + r_0\,\tfrac{1}{2}d\bar{\boldsymbol{\varphi}} + \tfrac{1}{2}d\bar{\boldsymbol{\varphi}}\times\mathbf{r}. \quad (3.64)$$

After cancellation of the initial rotation parameters this formula gives the quaternion parameter increments in terms of the original quaternion parameters and the incremental rotation vector,

$$2dr_0 = -\mathbf{r}^T d\bar{\boldsymbol{\varphi}}, \qquad 2d\mathbf{r} = r_0\, d\bar{\boldsymbol{\varphi}} - \mathbf{r} \times d\bar{\boldsymbol{\varphi}}. \tag{3.65}$$

The inverse relation can be obtained by multiplication of the vector part by $\mathbf{r}\times$, yielding

$$2\mathbf{r} \times d\mathbf{r} = r_0\, \mathbf{r} \times d\bar{\boldsymbol{\varphi}} - \mathbf{r} \times (\mathbf{r} \times d\bar{\boldsymbol{\varphi}}). \tag{3.66}$$

This formula is reduced by writing the double vector product in the form

$$\mathbf{r} \times (\mathbf{r} \times d\bar{\boldsymbol{\varphi}}) = -(\,|\mathbf{r}|^2\, \mathbf{I} - \mathbf{r}\,\mathbf{r}^T)\, d\bar{\boldsymbol{\varphi}} \tag{3.67}$$

and substituting the product $\mathbf{r}\times d\boldsymbol{\varphi}$ from (3.65b). The normalization relation (3.49) for the quaternion parameters implies that $|\mathbf{r}|^2 = 1 - r_0^2$, whereby the inverse relation is obtained from (3.66) in the form

$$\tfrac{1}{2} d\bar{\boldsymbol{\varphi}} = -\mathbf{r}\, dr_0 + r_0\, d\mathbf{r} + \mathbf{r} \times d\mathbf{r}. \tag{3.68}$$

This is the desired relation, giving the incremental rotation $d\bar{\boldsymbol{\varphi}}$ in terms of the current quaternion parameters and their increments. It plays the same role for the quaternion representation as (3.44) plays for the rotation pseudo-vector representation. The relation also gives the angular velocity $\boldsymbol{\omega}$, when the quaternion parameter increment $(dr_0, d\mathbf{r})$ is replaced by its time derivative $(dr_0/dt, d\mathbf{r}/dt)$.

3.5.4 Mean and difference of two rotations

In the formulation of simple elements the rotation may be known at two states A and B, and the questions then arise, what is the mean rotation at some mean state C, and what is the rotation increment from A to B? In the case of displacements these questions are answered simply by the algebraic mean and difference of the corresponding displacement vector components. However, in the case of rotations that are not coaxial a slightly more elaborate formulation of the questions and answers is needed. It turns out to be convenient to pose the problem in terms of quaternion parameters. Let the rotation at A be described by the quaternion $(p_A^0, \mathbf{p}_A)$ and the rotation at B similarly by $(p_B^0, \mathbf{p}_B)$. The mean rotation, with quaternion representation $(r_0, \mathbf{r})$, is now defined such that the rotation from C to A is the reverse of the rotation from C to B. Let the rotation from C to B be represented by the quaternion $(s_0, \mathbf{s})$. The reverse rotation from C to A is then represented by $(s_0, -\mathbf{s})$, and the mean value property of C can be expressed by the quaternion addition formulae (3.63) as

$$\begin{aligned} p_A^0 &= s_0 r_0 + \mathbf{s}^T\mathbf{r}, & \mathbf{p}_A &= s_0\mathbf{r} - r_0\mathbf{s} - \mathbf{s}\times\mathbf{r}, \\ p_B^0 &= s_0 r_0 - \mathbf{s}^T\mathbf{r}, & \mathbf{p}_B &= s_0\mathbf{r} + r_0\mathbf{s} + \mathbf{s}\times\mathbf{r}. \end{aligned} \tag{3.69}$$

The explicit extraction of the mean rotation $(r_0, \mathbf{r})$ and the increment from the mean to the original values $(s_0, \mathbf{s})$ now proceeds as follows. The quaternion of the mean rotation can be expressed directly by the sum of the scalar and vector relations in the form

$$s_0\, r_0 = \tfrac{1}{2}(p_A^0 + p_B^0), \qquad s_0\,\mathbf{r} = \tfrac{1}{2}(\mathbf{p}_A + \mathbf{p}_B), \tag{3.70}$$

where the scalar factor s_0 is determined from the quaternion normalization relation (3.49),

$$s_0^2 = \tfrac{1}{4}(p_A^0 + p_B^0)^2 + \tfrac{1}{4}|\mathbf{p}_A + \mathbf{p}_B|^2. \tag{3.71}$$

The vector part $\mathbf{s}$ is determined from the difference between the two sets of relations in (3.69),

$$\mathbf{s}^T\mathbf{r} = \tfrac{1}{2}(p_A^0 - p_B^0), \qquad r_0\mathbf{s} - \mathbf{r}\times\mathbf{s} = \tfrac{1}{2}(\mathbf{p}_B - \mathbf{p}_A). \tag{3.72}$$

Pre-multiplication of the last of these equations by $\mathbf{r}\times$ gives

$$r_0\,\mathbf{r}\times\mathbf{s} - \mathbf{r}\times(\mathbf{r}\times\mathbf{s}) = \mathbf{r}\times\tfrac{1}{2}(\mathbf{p}_B - \mathbf{p}_A). \tag{3.73}$$

In this equation $\mathbf{r}\times\mathbf{s}$ is substituted from (3.72b), the double vector product is expressed by scalar products, and $\mathbf{r}$ is substituted from (3.70b),

$$r_0[\,\tfrac{1}{2}(\mathbf{p}_A - \mathbf{p}_B) + r_0\,\mathbf{s}\,] + |\mathbf{r}|^2\mathbf{s} - \mathbf{r}(\mathbf{r}^T\mathbf{s}) = \tfrac{1}{4}(\mathbf{p}_B + \mathbf{p}_A)\times(\mathbf{p}_B - \mathbf{p}_A)\frac{1}{s_0}. \tag{3.74}$$

The final result is obtained from this relation by introduction of $|\mathbf{r}|^2 = 1 - r_0^2$, substitution of $\mathbf{r}^T\mathbf{s}$ from (3.72a), and multiplication by $2s_0$,

$$2s_0\,\mathbf{s} = p_A^0\mathbf{p}_B - p_B^0\mathbf{p}_A + \mathbf{p}_A\times\mathbf{p}_B. \tag{3.75}$$

In the case of coaxial rotations at A and B the vector product vanishes, and the formula reduces to the trigonometric formula for the difference between the angles $\frac{1}{2}\varphi_B$ and $\frac{1}{2}\varphi_A$.

3.6 Alternative representation of the rotation tensor

An alternative pseudo-vector representation of the rotation tensor can be obtained by a simple geometric argument illustrated in Fig. 3.7. The figure shows the rotation of a vector $\mathbf{a}_0$ by the rotation pseudo-vector $\boldsymbol{\varphi} = \varphi\mathbf{n}$ into

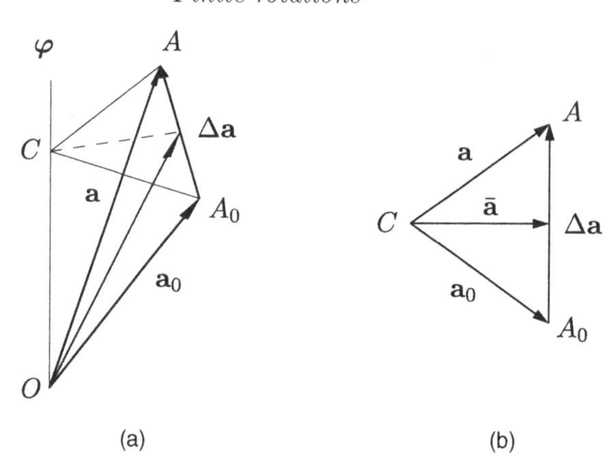

Fig. 3.7. Vector increment $\Delta\mathbf{a}$ in terms of mean vector $\bar{\mathbf{a}}$.

its final position $\mathbf{a}$. This rotation can be characterized in simple terms by using the vector increment $\Delta\mathbf{a} = \mathbf{a} - \mathbf{a}_0$ and the mean vector $\bar{\mathbf{a}} = \frac{1}{2}(\mathbf{a} + \mathbf{a}_0)$. It is seen from Fig. 3.7(a) that the vector increment $\Delta\mathbf{a}$ is orthogonal to both the mean vector $\bar{\mathbf{a}}$ and the rotation pseudo-vector $\boldsymbol{\varphi}$. It can therefore be expressed as a scalar factor times the vector product $\mathbf{n} \times \bar{\mathbf{a}}$. The two angles at C in Fig. 3.7 are equal to $\frac{1}{2}\varphi$, and the geometry of the triangle CA_0A then implies that

$$\tfrac{1}{2}(\mathbf{a} - \mathbf{a}_0) \;=\; \tan(\tfrac{1}{2}\varphi)\,\mathbf{n} \times \tfrac{1}{2}(\mathbf{a} + \mathbf{a}_0). \tag{3.76}$$

It is seen from this formulation that the rotation only appears in the form of the pseudo-vector,

$$\boldsymbol{\psi} \;=\; \tan(\tfrac{1}{2}\varphi)\,\mathbf{n} \;=\; \frac{1}{r_0}\,\mathbf{r}. \tag{3.77}$$

Thus, the rotation can be introduced via $\boldsymbol{\psi}$ instead of the pseudo-vector $\boldsymbol{\varphi} = \varphi\,\mathbf{n}$ with length equal to the rotation angle φ, or the quaternion pseudo-vector $\mathbf{r} = \sin(\frac{1}{2}\varphi)\mathbf{n}$ with length $\sin(\frac{1}{2}\varphi)$. The pseudo-vector $\boldsymbol{\psi}$ leads to an algebraic representation of the rotation tensor in terms of three components, but has a singularity for $\varphi = \pm\pi, \pm 3\pi, \ldots$ The pseudo-vector $\boldsymbol{\psi}$, sometimes called the Rodrigues vector, has been used extensively by e.g. Argyris (1982), while here the rotation vector $\boldsymbol{\varphi}$ and the quaternion representation $(r_0, \mathbf{r})$ are used as the basic variables.

The relation (3.76) can be written in terms of the skew-symmetric matrix $\widehat{\boldsymbol{\psi}} = (\boldsymbol{\psi}\times)$ as

$$(\,\mathbf{I} \,-\, \widehat{\boldsymbol{\psi}}\,)\,\mathbf{a} \;=\; (\,\mathbf{I} \,+\, \widehat{\boldsymbol{\psi}}\,)\,\mathbf{a}_0. \tag{3.78}$$

The rotation tensor $\mathbf{R}$ can now be determined by solving this equation for $\mathbf{a}$ in terms of $\mathbf{a}_0$. The solution has the form

$$\mathbf{a} = (\mathbf{I} - \widehat{\boldsymbol{\psi}})^{-1}(\mathbf{I} + \widehat{\boldsymbol{\psi}})\,\mathbf{a}_0, \tag{3.79}$$

whereby the rotation tensor is identified as

$$\mathbf{R} = (\mathbf{I} - \widehat{\boldsymbol{\psi}})^{-1}(\mathbf{I} + \widehat{\boldsymbol{\psi}}). \tag{3.80}$$

This form, called the Cayley representation, gives the rotation tensor $\mathbf{R}$ as the product of the inverse of the tensor $(\mathbf{I} - \widehat{\boldsymbol{\psi}})$ with the tensor $(\mathbf{I} + \widehat{\boldsymbol{\psi}})$. This format can be used e.g. in the integration of the equations of rigid body dynamics, where the equations can be simplified by using the parameters $\boldsymbol{\psi}$ directly – see, for example, Simo and Wong (1991) and Krenk (2007b) for the basic algorithms. It has also been used with the vector $\boldsymbol{\psi}$ replaced by $\frac{1}{2}\boldsymbol{\varphi}$ as an approximation of the rotation tensor. The format gives an orthogonal tensor, irrespective of the specific choice of the vector denoted $\boldsymbol{\psi}$.

An explicit form of the rotation tensor can be obtained from (3.80) by introducing the identity tensor in the form $(\mathbf{I} + \widehat{\boldsymbol{\psi}})^{-1}(\mathbf{I} + \widehat{\boldsymbol{\psi}})$ between the factors. It then follows that

$$\mathbf{R} = [(\mathbf{I} + \widehat{\boldsymbol{\psi}})\,(\mathbf{I} - \widehat{\boldsymbol{\psi}})]^{-1}(\mathbf{I}+\widehat{\boldsymbol{\psi}})^2 = (\mathbf{I} - \widehat{\boldsymbol{\psi}}^2)^{-1}(\mathbf{I}+\widehat{\boldsymbol{\psi}})^2, \tag{3.81}$$

where the linear terms in the inverse have cancelled. It follows from the expansion of vector triple products (3.25) that

$$\widehat{\boldsymbol{\psi}}\,\widehat{\boldsymbol{\psi}}\,\mathbf{a} = \boldsymbol{\psi}\,(\boldsymbol{\psi}^T\mathbf{a}) - (\boldsymbol{\psi}^T\boldsymbol{\psi})\,\mathbf{a} = (\,\boldsymbol{\psi}\boldsymbol{\psi}^T - \psi^2\mathbf{I}\,)\,\mathbf{a}. \tag{3.82}$$

When this is introduced into the first factor of (3.81), the formula takes the form

$$\mathbf{R} = [\,(1+\psi^2)\,\mathbf{I} - \boldsymbol{\psi}\boldsymbol{\psi}^T\,]^{-1}\,(\mathbf{I}+\widehat{\boldsymbol{\psi}})^2. \tag{3.83}$$

Now, the pseudo-vector $\boldsymbol{\psi}$ is expressed in terms of the quaternion parameters, $\boldsymbol{\psi} = \mathbf{r}/r_0$, and the scalar factor $(1 + \psi^2) = \cos^{-2}(\frac{1}{2}\varphi) = r_0^{-2}$ is moved to the second factor. The formula then takes the simplified form

$$\mathbf{R} = (\mathbf{I} - \mathbf{r}\,\mathbf{r}^T)^{-1}\,(\,r_0\,\mathbf{I} + \hat{\mathbf{r}}\,)^2. \tag{3.84}$$

The first factor is evaluated by observing that it is the inverse of the identity tensor minus an exterior vector product. The inverse then has a similar format, and it is easily verified by direct multiplication that

$$(\mathbf{I} - \mathbf{r}\,\mathbf{r}^T)^{-1} = \mathbf{I} + \frac{1}{r_0^2}\,\mathbf{r}\,\mathbf{r}^T. \tag{3.85}$$

Finally, substitution of this gives the rotation tensor in the explicit form

$$\mathbf{R} = (r_0\mathbf{I} + \hat{\mathbf{r}})^2 + \mathbf{r}\mathbf{r}^T = (r_0^2 - \mathbf{r}^T\mathbf{r})\mathbf{I} + 2r_0\hat{\mathbf{r}} + 2\mathbf{r}\mathbf{r}^T. \quad (3.86)$$

This establishes the representation (3.50) of the rotation tensor in terms of quaternion parameters by an alternative procedure.

3.7 Summary of rotations and their virtual work

Rotations occur in two different contexts in the formulation of theories for structures and solids: as a finite rotation describing a configuration at a particular instant, and as a 'small' change of the state of rotation typically encountered when setting up conditions of equilibrium, and in the calculation of the tangent stiffness relations of models described in terms of rotations. Rotations are principally different from translations in the fact that they cannot be reduced to linear operations, and that they are non-commuting in three-dimensional space. Thus, there are two aspects of rotations in relation to mechanical theories for structures and solids. The first is the representation of the rotation associated with a particular state, and the change of this state by a finite additional rotation. Due to the non-linear character of motion associated with rotations, these operations are described by rotation tensors and the addition of consecutive rotations by tensor products. The basic formulae were derived in Section 3.1 in terms of the rotation pseudo-vector $\boldsymbol{\varphi}$. An alternative representation in terms of a vector before and after rotation was presented in Section 3.2. This form is particularly relevant for determining the rotation of element nodes relative to element axes.

The special character of rotations has important implications for the equilibrium equations and tangent stiffness of solid bodies described by use of rotations. The tangent stiffness is an expression for forces and moments arising from an incremental change of a state of equilibrium. When dealing with continuous bodies these relations are generally expressed via the principle of virtual work. This is very important when using rotations to describe the motion. The equilibrium equation will then contain virtual work contributions of the form $\delta\bar{\boldsymbol{\varphi}}^T\mathbf{M}$, where $\delta\bar{\boldsymbol{\varphi}}$ is a virtual rotation and $\mathbf{M}$ is an internal or external moment. In the analysis of problems described in terms of translations the variation $\delta\mathbf{u}$ can be kept constant, when considering changes around the state of equilibrium. However, in the case of rotations a change of state also implies a change of the variation of the form $d(\delta\bar{\boldsymbol{\varphi}})$. This 'second variation' was derived in Section 3.3 and the specific form given in (3.30). The implications were indicated briefly in Examples 3.2 and 3.3.

This led to the interesting conclusion that, when treated in a virtual work context, rotation of a moment will only contribute $\frac{1}{2}d\bar{\boldsymbol{\varphi}} \times \mathbf{M}$, corresponding to an apparent rotation of only half the actual incremental angle. This has important implications for theories involving rotations, and the consequences in terms of symmetry and geometric stiffness are demonstrated and discussed in the following chapters.

Finally, a number of special issues have been discussed, notably the representation of rotations in terms of quaternion parameters presented in Section 3.5. These parameters describe a rotation in a homogeneous four component format in terms of trigonometric functions of half the rotation angle. This homogeneous format of order two enables a compact combination rule for rotations, expressed directly in the parameters in vector format, thereby avoiding the use of matrix multiplications.

3.8 Exercises

Exercise 3.1 The finite rotation tensor formula (3.8) can also be derived from a vector differential equation. Let $\mathbf{n}$ be a constant unit vector, and let φ be the finite angle of rotation as shown in Fig. 3.2. For the fixed direction $\mathbf{n}$, the incremental rotation can be written as

$$d\mathbf{a} = d\varphi\, \mathbf{n} \times \mathbf{a} = d\varphi\, \hat{\mathbf{n}}\, \mathbf{a}$$

or as the differential equation

$$\frac{d\mathbf{a}}{d\varphi} = \hat{\mathbf{n}}\, \mathbf{a}.$$

The formal integral to this differential equation is

$$\mathbf{a} = \mathrm{e}^{\hat{\mathbf{n}}\varphi}\, \mathbf{a}_0,$$

where the exponential function of a matrix is defined via its Taylor series expansion

$$\mathrm{e}^{\mathbf{A}} = \mathbf{I} + \mathbf{A} + \tfrac{1}{2!}\mathbf{A}^2 + \tfrac{1}{3!}\mathbf{A}^3 + \cdots$$

(a) Demonstrate by substitution that the series expansion of the matrix exponential function is a solution of the differential equation.

(b) Establish the vector product identity

$$\mathbf{n} \times \mathbf{n} \times \mathbf{n} \times \mathbf{a} = -\,\mathbf{n} \times \mathbf{a} \qquad \text{or} \qquad \hat{\mathbf{n}}^3\, \mathbf{a} = -\,\hat{\mathbf{n}}\, \mathbf{a}.$$

(*Hint*: $\mathbf{n} \times \mathbf{a}$ is orthogonal to $\mathbf{n}$.)

(c) Use the result from (b) to sum the even and odd power terms in the Taylor expansion of the solution to get the rotation tensor

$$\mathbf{R} = \mathrm{e}^{\hat{\mathbf{n}}\varphi} = \mathbf{I} + \sin\varphi\,\hat{\mathbf{n}} + (1-\cos\varphi)\,\hat{\mathbf{n}}\,\hat{\mathbf{n}}.$$

Exercise 3.2 Let $\mathbf{n}_j$ be a set of orthonormal base vectors, and let $\mathbf{n}'_j$ be the orthonormal basis obtained by rotating the vector $\mathbf{n}_1$ into $\mathbf{n}'_1$. It was shown in Section 3.2 that the remaining two base vectors transform according to

$$\mathbf{n}'_\alpha = (\mathbf{I} - 2\bar{\mathbf{n}}\,\bar{\mathbf{n}}^T)\,\mathbf{n}_\alpha, \qquad \bar{\mathbf{n}} = \frac{\mathbf{n}_1 + \mathbf{n}'_1}{|\mathbf{n}_1 + \mathbf{n}'_1|}.$$

This transformation is symmetric and defined solely by the unit vector $\bar{\mathbf{n}}$.

(a) Find the inverse transformation, giving the vectors $\mathbf{n}_2$ and $\mathbf{n}_3$ in terms of the vectors $\mathbf{n}'_2$ and $\mathbf{n}'_3$.
(b) Show that the transformation corresponds to reflection of the vectors $\mathbf{n}_2$ and $\mathbf{n}_3$ in a plane orthogonal to $\bar{\mathbf{n}}$, and illustrate this by a sketch.
(c) Show that the magnitude of the rotation angle of each of the vectors $\mathbf{n}_j$ can be determined by

$$\sin(\tfrac{1}{2}\varphi_j) = \tfrac{1}{2}(\mathbf{n}_j - \mathbf{n}'_j)^T\bar{\mathbf{n}} = \mathbf{n}_j^T\,\bar{\mathbf{n}}.$$

Exercise 3.3 The formula (3.9) for the rotation tensor $\mathbf{R}(\boldsymbol{\varphi})$, and (3.45) and (3.47) for the transformation tensor $\mathbf{T}(\boldsymbol{\varphi})$ and its inverse all contain potential numerical singularities for $\varphi \to 0$. The problem is that in the limit the rotation direction vector $\mathbf{n}$ is not well defined.

Rewrite the expressions for $\mathbf{R}(\boldsymbol{\varphi})$, $\mathbf{T}(\boldsymbol{\varphi})$ and $\mathbf{T}^{-1}(\boldsymbol{\varphi})$ in terms of the rotation pseudo-vector $\boldsymbol{\varphi}$ and suitable scalar factors containing powers and trigonometric functions of φ or $\frac{1}{2}\varphi$ with a well-defined limit for $\varphi \to 0$. Indicate a two- to three-term Taylor series expansion of the scalar factors for small φ.

Exercise 3.4 Quaternions can be considered as four-component numbers in a similar way as complex numbers are two-component numbers. Like the complex numbers the addition rule is trivial, while the special properties are associated with the product rule. This exercise presents some of the basic properties of quaternions.

(a) Let the quaternions associated with the rotation tensors $\mathbf{P}$ and $\mathbf{Q}$ be denoted $(p_0, \mathbf{p})$ and $(q_0, \mathbf{q})$, respectively. The product, here denoted by the asterisk $*$, is then defined by the rules from (3.63),

$$(p_0, \mathbf{p}) * (q_0, \mathbf{q}) = (\,p_0q_0 - \mathbf{p}^T\mathbf{q}\,,\; p_0\mathbf{q} + q_0\mathbf{p} + \mathbf{p}\times\mathbf{q}\,).$$

(b) Ordinary vectors $\mathbf{a}$ and $\mathbf{b}$ are associated with quaternions with zero scalar part, $(0, \mathbf{a})$ and $(0, \mathbf{b})$. Find the corresponding quaternion product $(0, \mathbf{a}) * (0, \mathbf{b})$.

(c) The adjoint quaternion is defined as $(r_0, \mathbf{r})' = (r_0, -\mathbf{r})$. Let the quaternion $(r_0, \mathbf{r})$ correspond to the rotation tensor $\mathbf{R}$. Show that the rotated vector $\mathbf{a}_1 = \mathbf{R}\,\mathbf{a}$ can be evaluated by the two-sided quaternion product

$$(0, \mathbf{a}_1) \;=\; (r_0, \mathbf{r}) * (0, \mathbf{a}) * (r_0, \mathbf{r})',$$

where each factor is linear in the quaternion parameters r_0 and $\mathbf{r}$.

(d) Explain how the rotation formula in (c) leads to the bi-linear product formula (3.63) for addition of rotations.

Exercise 3.5* Let $\boldsymbol{\varphi}$ denote the rotation pseudo-vector $\boldsymbol{\varphi} = \varphi\mathbf{n}$ introduced in (3.4) and $\mathbf{R}$ the corresponding rotation tensor.

(a) Use the quaternion addition formula (3.63) to write a MATLAB function `phi = addrot(phi1,phi2)` giving the rotation pseudo-vector $\boldsymbol{\varphi}$ corresponding to the rotation $\boldsymbol{\varphi}_1$ *followed* by the rotation $\boldsymbol{\varphi}_2$.

(b) Write a MATLAB function `R = rotmat(phi)` giving the rotation matrix $\mathbf{R}$ corresponding to the rotation $\boldsymbol{\varphi}$.

(c) Introduce the three rotations $\boldsymbol{\varphi}_1 = [\frac{1}{2}\pi, 0, 0]^T$, $\boldsymbol{\varphi}_2 = [0, \frac{1}{2}\pi, 0]^T$ and $\boldsymbol{\varphi}_3 = [0, 0, \frac{1}{2}\pi]^T$. Obtain the rotation pseudo-vector and rotation matrix corresponding to the three cases: (i) $\boldsymbol{\varphi}_1$ followed by $\boldsymbol{\varphi}_2$, (ii) $\boldsymbol{\varphi}_2$ followed by $\boldsymbol{\varphi}_1$ and (iii) $\boldsymbol{\varphi}_1$ followed by $\boldsymbol{\varphi}_2$ followed by $\boldsymbol{\varphi}_3$.

Exercise 3.6 Consider two coaxial rotations with pseudo-vectors $\boldsymbol{\varphi}_A = \varphi_A\mathbf{n}$ and $\boldsymbol{\varphi}_B = \varphi_B\mathbf{n}$. Use the relations of Section 3.5.4 to find the mean rotation and the rotation increment from the mean to B.

Exercise 3.7 Consider the rotation tensors $\mathbf{R}_A$ and $\mathbf{R}_B$ associated with two points A and B, e.g. the end-points of a beam element. Let $\mathbf{R}_C$ be the mean rotation tensor. This implies that there exists a rotation tensor $\mathbf{S}$ representing the rotation from A to C as well as from C to B. By the product rule of rotation tensors this implies that

$$\mathbf{R}_C \;=\; \mathbf{S}\,\mathbf{R}_A \quad \text{and} \quad \mathbf{R}_B \;=\; \mathbf{S}\,\mathbf{R}_C.$$

The rotations represented by the tensors $\mathbf{R}_C$ and $\mathbf{S}$ were derived in Section 3.5.4. A more direct procedure for $\mathbf{S}$ is indicated here.

(a) Eliminate $\mathbf{R}_C$ to obtain an expression for $\mathbf{S}^2$ in terms of $\mathbf{R}_A$ and $\mathbf{R}_B$.

(b) Introduce the quaternion representation $(s_0, \mathbf{s})$ for the rotation $\mathbf{S}$, and find the corresponding quaternion representation for $\mathbf{S}^2$.

(c) Use the results from (a) and (b) to establish the formula (3.75) for $(s_0, \mathbf{s})$.

4

Finite rotation beam theory

Beams are slender structural elements, usually defined by their axis and cross-sections. This permits two basically different ways of modeling beams, one based on the translation and rotation of the points on the beam axis and the connected cross-sections (Fig. 4.1), the other treating the beam as a special example of a fully three-dimensional continuum. This chapter and the next develop fully non-linear beam models of the first type. In this chapter the fully non-linear theory of a beam represented as a curve with elastic properties – a so-called elastica – is developed, and its finite element implementation discussed. This theory is in principle unique, and can therefore be called *the* theory of the elastica. It describes the deformation of an elastic space curve in terms of total translation and rotation, and relies heavily on the properties of finite rotations, presented in Chapter 3. An alternative formulation of non-linear beams is presented in Chapter 5. In that description the motion is decomposed into a local deformation described in a frame of reference following the beam, and a motion of the local frame of reference. The local frame of reference moves with each beam element, and this type of formulation is therefore often called co-rotating. Simple beam theories can be used within a co-rotating formulation, where finite rotation contributions are added. The application-oriented reader may want to go directly to the co-rotational formulation in Chapter 5.

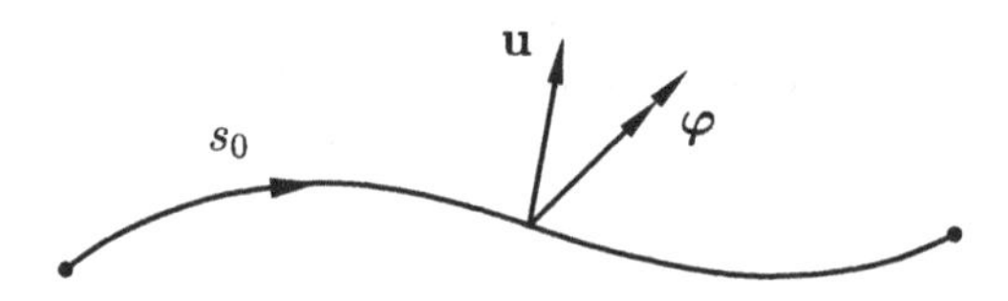

Fig. 4.1. Displacement and rotation vectors in beam element.

4.1 Equilibrium equations

In this chapter a non-linear beam theory is developed by generalizing the virtual work approach of Chapter 2 to include rotations, using the results established in Chapter 3. The beam is considered as a curve segment $\mathbf{x}(s_0)$, where each point is associated with a cross-section. The orientation of each cross-section is determined by two orthogonal unit vectors $\mathbf{n}_2, \mathbf{n}_3$. A third unit vector, normal to the cross-section, is defined by $\mathbf{n}_1 = \mathbf{n}_2 \times \mathbf{n}_3$. Thus, the triple $\mathbf{n}_j = [\mathbf{n}_1, \mathbf{n}_2, \mathbf{n}_3]$ constitutes a local orthonormal basis, that rotates with the cross-section, see Fig. 4.2. This type of formulation, originally introduced by Dupuis (1969) and Reissner (1981), has been pursued more recently in connection with non-linear finite element analysis by Simo (1985), Simo and Vu-Quoc (1986, 1988), Cardona and Géradin (1988), Lo (1992), Mathiesen (1993), Ibrahimbegović (1995) and Géradin and Cardona (2001).

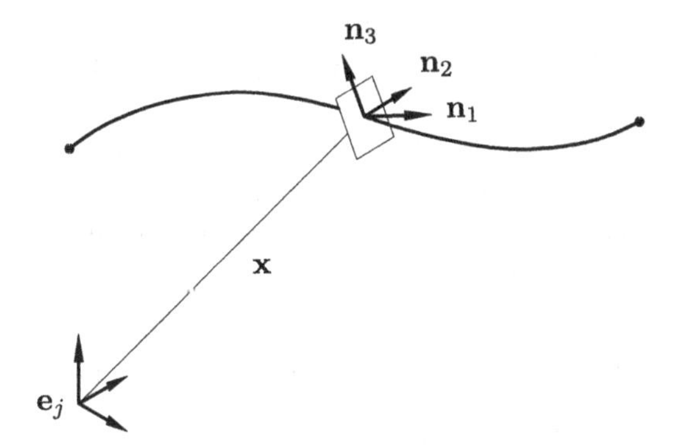

Fig. 4.2. Beam representation as a curve with orthogonal directors.

The theory will here be developed for an elastica – i.e. a flexible curve – described in terms of an initial arc-length parameter s_0. The curve is loaded by distributed forces $\mathbf{p}(s_0)$ and moments $\mathbf{m}(s_0)$ per unit initial length s_0. The section force and moment are $\mathbf{N}(s_0)$ and $\mathbf{M}(s_0)$, respectively. Consideration of a small length of the beam in its current, deformed, configuration gives the force equilibrium equation

$$\frac{d\mathbf{N}}{ds_0} + \mathbf{p} = \mathbf{0} \tag{4.1}$$

and the moment equilibrium equation

$$\frac{d\mathbf{M}}{ds_0} + \frac{d\mathbf{x}}{ds_0} \times \mathbf{N} + \mathbf{m} = \mathbf{0}. \tag{4.2}$$

Note that $d\mathbf{x}/ds_0$ is a tangent vector, of unit length in the initial but usually not in the deformed geometry.

4.2 Virtual work, strain and curvature

The equation of virtual work is now formed by introducing suitable displacement and rotation variation fields $\delta\mathbf{u}(s_0)$ and $\delta\bar{\boldsymbol{\varphi}}(s_0)$. Here, $\delta\bar{\boldsymbol{\varphi}}(s_0)$ is a small incremental rotation as discussed in Section 3.3, and

$$\int_0^{l_0} \left\{ \delta\mathbf{u}^T \Big(\frac{d\mathbf{N}}{ds_0} + \mathbf{p} \Big) + \delta\bar{\boldsymbol{\varphi}}^T \Big(\frac{d\mathbf{M}}{ds_0} + \frac{d\mathbf{x}}{ds_0} \times \mathbf{N} + \mathbf{m} \Big) \right\} ds_0 = 0. \quad (4.3)$$

The equation is reformulated via integration by parts, whereby

$$\begin{aligned} &\int_0^{l_0} \left\{ \Big(\frac{d(\delta\mathbf{u})}{ds_0} - \delta\bar{\boldsymbol{\varphi}} \times \frac{d\mathbf{x}}{ds_0} \Big)^T \mathbf{N} + \frac{d}{ds_0}(\delta\bar{\boldsymbol{\varphi}})^T \mathbf{M} \right\} ds_0 \\ &- \Big[\delta\mathbf{u}^T\mathbf{N} + \delta\bar{\boldsymbol{\varphi}}^T\mathbf{M} \Big]_0^{l_0} - \int_0^{l_0} \Big(\delta\mathbf{u}^T\mathbf{p} + \delta\bar{\boldsymbol{\varphi}}^T\mathbf{m} \Big) ds_0 = 0. \end{aligned} \quad (4.4)$$

This is a statement of the principle of virtual work in the form that the internal virtual work, expressed in terms of internal forces and their conjugate deformation measures, minus the external virtual work equals zero for arbitrary admissible variations $\delta\mathbf{u}(s_0), \delta\bar{\boldsymbol{\varphi}}(s_0)$.

The constitutive relations of the beam provide the components of the section force and moment in the local director basis, i.e. the components

$$\mathbf{N} = N_j\,\mathbf{n}_j, \qquad \mathbf{M} = M_j\,\mathbf{n}_j. \quad (4.5)$$

The virtual work equation (4.4) then takes the form

$$\begin{aligned} \delta V = &\int_0^{l_0} \Big(\delta\varepsilon_j N_j + \delta\kappa_j M_j \Big) ds_0 - \Big[\delta\mathbf{u}^T\mathbf{N} + \delta\bar{\boldsymbol{\varphi}}^T\mathbf{M} \Big]_0^{l_0} \\ &- \int_0^{l_0} \Big(\delta\mathbf{u}^T\mathbf{p} + \delta\bar{\boldsymbol{\varphi}}^T\mathbf{m} \Big) ds_0 = 0. \end{aligned} \quad (4.6)$$

In this equation $\delta\varepsilon_j$ and $\delta\kappa_j$ are the variational strain and curvature components in the local director basis, defined via the virtual work equation. The use of local components in the representation of the internal work leads to a natural separation of each term into a factor representing a kinematic variation and a factor determined from the constitutive relations. In the further derivation of the incremental stiffness relation these factors generate the geometric and constitutive stiffness contributions, respectively.

In the treatment of the bar element in Chapter 2, the strain definition was introduced *a priori* and the conjugate internal force was then determined from the virtual work equation. In the case of the elastica beam theory the internal forces are introduced via the equilibrium equation, and the virtual work equation then determines the appropriate virtual strain components. Thus, in general appropriate generalized stresses and their conjugate generalized virtual strains are determined by the virtual work equation, when either one of the two sets is given.

The virtual strain components follow from (4.4)–(4.6) as

$$\delta\varepsilon_j = \left(\frac{d(\delta\mathbf{u})}{ds_0} - \delta\bar{\varphi}\times\frac{d\mathbf{x}}{ds_0}\right)^T\mathbf{n}_j. \tag{4.7}$$

It is easily verified that this is the variation of the total strain components

$$\varepsilon_j = \left(\frac{d\mathbf{x}}{ds_0} - \mathbf{n}_1\right)^T\mathbf{n}_j \tag{4.8}$$

when use is made of the fact that $\delta(\mathbf{n}_1^T\mathbf{n}_j) = 0$, because the directors remain orthonormal. The corresponding vector form of the total strain is

$$\boldsymbol{\varepsilon} = \left(\frac{d\mathbf{x}}{ds_0} - \mathbf{n}_1\right). \tag{4.9}$$

This definition of the total strain components leads to representation of the tangent vector as

$$\frac{d\mathbf{x}}{ds_0} = \mathbf{n}_1 + \varepsilon_j\mathbf{n}_j = \mathbf{n}_1 + \boldsymbol{\varepsilon}. \tag{4.10}$$

The local components $[1+\varepsilon_1, \varepsilon_2, \varepsilon_3]$ of the tangent vector $d\mathbf{x}/ds_0$ are illustrated in Fig. 4.3.

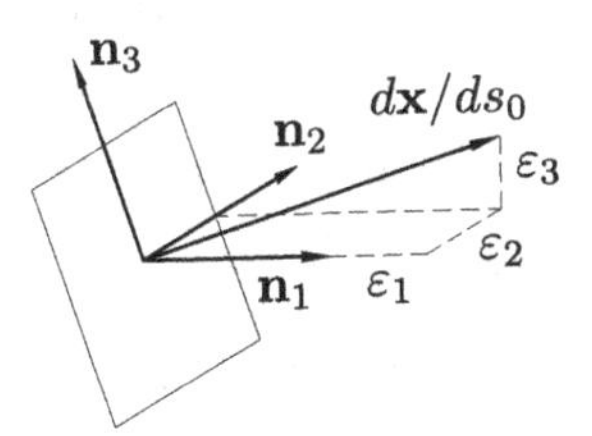

Fig. 4.3. Local components of tangent vector in terms of total strain.

The components $\delta\kappa_j$ of the variation of curvature components also follow from (4.4)–(4.6) as

$$\delta\kappa_j = \frac{d(\delta\bar{\varphi})}{ds_0}^T\mathbf{n}_j. \tag{4.11}$$

The variations $\delta\kappa_j$ correspond to total curvature components κ_j defined by the derivatives of the rotation vector along the reference curve of the beam,

$$\kappa_j = \frac{d\bar{\varphi}}{ds_0}^T \mathbf{n}_j. \tag{4.12}$$

The fact that $\delta\kappa_j$ are the variations of κ_j is demonstrated by differentiation of (4.12),

$$\delta\kappa_j = \delta\Big(\frac{d\bar{\varphi}}{ds_0}\Big)^T \mathbf{n}_j + \frac{d\bar{\varphi}}{ds_0}^T \delta\mathbf{n}_j. \tag{4.13}$$

The order of the operators δ and d in the second-order increment $\delta(d\bar{\varphi})$ is now interchanged by use of (3.30), and the variation of the unit vector $\delta\mathbf{n}_j$ is expressed as a vector product,

$$\delta\kappa_j = \frac{d(\delta\bar{\varphi})}{ds_0}^T \mathbf{n}_j + \Big(\delta\bar{\varphi} \times \frac{d\bar{\varphi}}{ds_0}\Big)^T \mathbf{n}_j + \frac{d\bar{\varphi}}{ds_0}^T (\delta\bar{\varphi} \times \mathbf{n}_j). \tag{4.14}$$

The factors in the triple vector products in the two last terms are identical, but occur in opposite order. These terms therefore cancel, demonstrating that $\delta\kappa_j$ defined by (4.11) is indeed the variation of the total curvature components κ_j defined by (4.12). The vector form of the curvature corresponding to (4.12) is

$$\boldsymbol{\kappa} = \frac{d\bar{\varphi}}{ds_0} = \mathbf{T}(\boldsymbol{\varphi}) \frac{d\boldsymbol{\varphi}}{ds_0}, \tag{4.15}$$

where the tensor $\mathbf{T}(\boldsymbol{\varphi})$ given in (3.45) relates the incremental rotation $d\bar{\varphi}$ to the increment of the rotation parameters $d\boldsymbol{\varphi}$.

REMARK: It should be noted that $\delta\varepsilon_j$ and $\delta\kappa_j$ are the variations of the local *components* ε_j and κ_j of strain and curvature. The variations do not include contributions from rotation of the directors, and thus they are not the components of the variations of the corresponding vectors $\boldsymbol{\varepsilon}$ and $\boldsymbol{\kappa}$. The vectorial form of the variations can be expressed as

$$\begin{aligned} \delta\boldsymbol{\varepsilon} &= \delta\varepsilon_j\,\mathbf{n}_j = \mathbf{R}\,(\delta\varepsilon_j\,\mathbf{R}^T\mathbf{n}_j) = \mathbf{R}\,\delta(\mathbf{R}^T\boldsymbol{\varepsilon}),\\ \delta\boldsymbol{\kappa} &= \delta\kappa_j\,\mathbf{n}_j = \mathbf{R}\,(\delta\kappa_j\,\mathbf{R}^T\mathbf{n}_j) = \mathbf{R}\,\delta(\mathbf{R}^T\boldsymbol{\kappa}), \end{aligned}$$

where the role of $\mathbf{R}^T$ is to rotate from the local basis $\mathbf{n}_j$ back to the fixed global basis $\mathbf{e}_j = \mathbf{R}^T\mathbf{n}_j$. When the component variation has been taken, the result is then rotated back into the local basis $\mathbf{n}_j$ by $\mathbf{R}$. These operations are termed pull-back and push-forward by Marsden and Hughes (1983). However, the correct component form arises naturally from the principle of virtual work, and the present remark merely serves to identify the physical nature of $\delta\varepsilon_j$ and $\delta\kappa_j$ and to explain a terminology sometimes used in the computational mechanics literature.

4.3 Increment of the virtual work equation

The virtual work equation (4.6) with the definitions (4.7) and (4.11) of the virtual strain and curvature components permits a check of equilibrium by evaluation of residual forces and moments. In order to prescribe a desired change of the configuration of the beam, an incremental change of configuration must be introduced in the virtual work equation as discussed in Section 2.7. The change of configuration is described by the incremental displacements and rotations $d\mathbf{u}$ and $d\bar{\boldsymbol{\varphi}}$, the changes dN_j and dM_j in internal forces and moments, and the changes $d\mathbf{p}, d\mathbf{m}$ and $d\mathbf{N}, d\mathbf{M}$ in distributed and end-point forces and moments. In the derivation of the incremental relations it is important to note that the rotation increment $d\bar{\boldsymbol{\varphi}}$ is in the form of a small rotation and is not generally equal to the increment $d\boldsymbol{\varphi}$ of the components in the rotation pseudo-vector $\boldsymbol{\varphi}$.

The change of the internal virtual work is

$$
\begin{aligned}
d(\delta V_{\text{int}}) &= d \int_0^{l_0} \Big(\delta\varepsilon_j N_j \,+\, \delta\kappa_j M_j \Big) ds_0 \\
&= \int_0^{l_0} \Big(\delta\varepsilon_j dN_j \,+\, \delta\kappa_j dM_j \,+\, d(\delta\varepsilon_j) N_j \,+\, d(\delta\kappa_j) M_j \Big) ds_0 .
\end{aligned}
\tag{4.16}
$$

The first two terms of the integrand come from constitutive changes, while the two last represent the effect of the change of geometry on the current stress state. For simplicity of notation it is assumed that the constitutive relations for the section forces N_j are uncoupled from the constitutive relations for the moments M_j. More general formulations may include the coupling between torsion and extension found in twisted beams, e.g. Krenk (1983a,b). By the assumed independence, the incremental constitutive relations are of the form

$$
dN_j \;=\; \frac{\partial N_j}{\partial \varepsilon_k}\, d\varepsilon_k, \qquad dM_j \;=\; \frac{\partial M_j}{\partial \kappa_k}\, d\kappa_k . \tag{4.17}
$$

Substitution into the incremental form (4.16) of the internal virtual work then gives

$$
\begin{aligned}
&d(\delta V_{\text{int}}) \;= \\
&\int_0^{l_0} \Big(\delta\varepsilon_j \frac{\partial N_j}{\partial \varepsilon_k} d\varepsilon_k \,+\, \delta\kappa_j \frac{\partial M_j}{\partial \kappa_k} d\kappa_k \,+\, d(\delta\varepsilon_j) N_j \,+\, d(\delta\kappa_j) M_j \Big) ds_0 .
\end{aligned}
\tag{4.18}
$$

For symmetric constitutive relations (4.17) the first two terms are seen to be symmetric in the virtual and actual increments $(\delta\varepsilon_j, \delta\kappa_j)$ and $(d\varepsilon_k, d\kappa_k)$. In particular, symmetric stiffness terms would follow from an internal energy density function.

4.3.1 Constitutive stiffness

The incremental constitutive relations are given in terms of the local tensor components

$$C^{\varepsilon}_{jk} = \frac{\partial N_j}{\partial \varepsilon_k}, \qquad C^{\kappa}_{jk} = \frac{\partial M_j}{\partial \kappa_k}, \tag{4.19}$$

where the components C^{ε}_{jk} and C^{κ}_{jk} refer to the local base vectors $\mathbf{n}_j$ following the beam. A typical example is the linear elastic relation expressed in principal axes,

$$C^{\varepsilon}_{jk} = \begin{bmatrix} EA & & \\ & GA_2 & \\ & & GA_3 \end{bmatrix}, \qquad C^{\kappa}_{jk} = \begin{bmatrix} GJ & & \\ & EI_2 & \\ & & EI_3 \end{bmatrix}. \tag{4.20}$$

If the displacement and rotation fields are represented by global components, the constitutive tensors must be introduced via their global form

$$\begin{aligned} \mathbf{C}^{\varepsilon}_{*} &= \mathbf{n}_j C^{\varepsilon}_{jk} \mathbf{n}_k = \mathbf{R}\,\mathbf{C}^{\varepsilon}\,\mathbf{R}^T, \\ \mathbf{C}^{\kappa}_{*} &= \mathbf{n}_j C^{\kappa}_{jk} \mathbf{n}_k = \mathbf{R}\,\mathbf{C}^{\kappa}\,\mathbf{R}^T, \end{aligned} \tag{4.21}$$

where the rotation tensor $\mathbf{R}$ represents the current orientation of the axes, and the subscript $*$ indicates the global form of the constitutive parameters.

The global vector form of the virtual strain increment follows from (4.7) as

$$\delta\boldsymbol{\varepsilon} = \frac{d(\delta\mathbf{u})}{ds_0} - \delta\bar{\boldsymbol{\varphi}} \times \frac{d\mathbf{x}}{ds_0} = \frac{d(\delta\mathbf{u})}{ds_0} + \frac{d\mathbf{x}}{ds_0} \times \delta\bar{\boldsymbol{\varphi}}. \tag{4.22}$$

Similarly, the virtual curvature increment vector follows from (4.11) as

$$\delta\boldsymbol{\kappa} = \frac{d(\delta\bar{\boldsymbol{\varphi}})}{ds_0}. \tag{4.23}$$

With these expressions and the notation $(\ )' = d(\ )/ds_0$ the constitutive contribution to the incremental stiffness can be expressed in the following block matrix format:

$$\begin{gathered} \delta\varepsilon_j\, C^{\varepsilon}_{jk}\, d\varepsilon_k + \delta\kappa_j\, C^{\kappa}_{jk}\, d\kappa_k = \\ [\,\delta\mathbf{u}'\ \delta\bar{\boldsymbol{\varphi}}'\ \delta\bar{\boldsymbol{\varphi}}\,] \begin{bmatrix} \mathbf{C}^{\varepsilon}_{*} & \mathbf{0} & \mathbf{C}^{\varepsilon}_{*}\hat{\mathbf{x}}' \\ \mathbf{0} & \mathbf{C}^{\kappa}_{*} & \mathbf{0} \\ \hat{\mathbf{x}}'^T\mathbf{C}^{\varepsilon}_{*} & \mathbf{0} & \hat{\mathbf{x}}'^T\mathbf{C}^{\varepsilon}_{*}\hat{\mathbf{x}}' \end{bmatrix} \begin{bmatrix} d\mathbf{u}' \\ d\bar{\boldsymbol{\varphi}}' \\ d\bar{\boldsymbol{\varphi}} \end{bmatrix}, \end{gathered} \tag{4.24}$$

where the notation $\hat{\mathbf{x}} = (\mathbf{x}\times)$ from (3.7) has been used for the skew-symmetric matrix equivalent of a vector. The constitutive stiffness contribution to a non-linear beam element follows by substitution of a suitable representation of the variations $\delta\mathbf{u}', \delta\bar{\boldsymbol{\varphi}}', \delta\bar{\boldsymbol{\varphi}}$ and the increments $d\mathbf{u}', d\bar{\boldsymbol{\varphi}}', d\bar{\boldsymbol{\varphi}}$ in terms of shape functions, as discussed in Section 4.4.

4.3.2 Geometric stiffness

The geometric stiffness is represented by the last two terms of the integrand in (4.18). The second variations $d(\delta\kappa_j)$ and $d(\delta\varepsilon_j)$ are now evaluated using the results of Section 3.3 on second-order rotation increments. It follows directly from the definition (4.11) of the curvature component variations that

$$d(\delta\kappa_j) = \frac{d(d\delta\bar{\varphi})^T}{ds_0}\mathbf{n}_j + \frac{d(\delta\bar{\varphi})^T}{ds_0}(d\bar{\varphi}\times\mathbf{n}_j). \tag{4.25}$$

As explained in Section 3.3, the *parameter variation* $\delta\varphi$ is kept constant when changing the configuration. The non-linear form of the rotation parametrization then implies that the virtual rotation $\delta\bar{\varphi}$ must have the second-order variation given by (3.30). Substitution of the second-order variation into the first term of (4.25) gives

$$d(\delta\kappa_j) = \frac{d}{ds_0}\big(-\tfrac{1}{2}\,\delta\bar{\varphi}\times d\bar{\varphi}\big)^T\mathbf{n}_j + \frac{d(\delta\bar{\varphi})^T}{ds_0}(d\bar{\varphi}\times\mathbf{n}_j). \tag{4.26}$$

Differentiation of the first factor in the first term leads to cancellation of half of the last term, leaving

$$d(\delta\kappa_j) = -\tfrac{1}{2}\Big(\delta\bar{\varphi}\times\frac{d(d\bar{\varphi})}{ds_0}\Big)^T\mathbf{n}_j + \tfrac{1}{2}\frac{d(\delta\bar{\varphi})^T}{ds_0}(d\bar{\varphi}\times\mathbf{n}_j). \tag{4.27}$$

Upon rearrangement of the order of the triple products this result can be written in the tensor format

$$d(\delta\kappa_j) = \delta\bar{\varphi}^T\big(\tfrac{1}{2}\,\mathbf{n}_j\times\big)\frac{d(d\bar{\varphi})}{ds_0} - \frac{d(\delta\bar{\varphi})^T}{ds_0}\big(\tfrac{1}{2}\,\mathbf{n}_j\times\big)d\bar{\varphi}. \tag{4.28}$$

The tensor in the parentheses is skew-symmetric, and the second variation $d(\delta\kappa_j)$ therefore is a symmetric function of the increments $\delta\bar{\varphi}$ and $d\bar{\varphi}$.

The contribution $d(\delta\kappa_j)M_j$ to the internal virtual work can be given in tensor form, by observing that the internal moment vector is $\mathbf{M} = M_j\mathbf{n}_j$, and it then follows from (4.28) that

$$d(\delta\kappa_j)M_j = \delta\bar{\varphi}^T\big(\tfrac{1}{2}\mathbf{M}\times\big)\frac{d(d\bar{\varphi})}{ds_0} - \frac{d(\delta\bar{\varphi})^T}{ds_0}\big(\tfrac{1}{2}\mathbf{M}\times\big)d\bar{\varphi}. \tag{4.29}$$

The invariant form of (4.29) in terms of tensors makes it valid for computations in global, local or element system components, provided the differentiation of the base vectors is included. The component format of the skew-symmetric tensor was given in (3.6).

The first step in the evaluation of the second variation of the strain components follows from (4.7) by taking differentials of all factors, and recalling that the variation of the displacement is selected such that $d(\delta\mathbf{u})$ vanishes,

$$\begin{aligned} d(\delta\varepsilon_j) &= \frac{d(\delta\mathbf{u})}{ds_0}^T (d\bar{\varphi}\times\mathbf{n}_j) - \Big(\delta\bar{\varphi}\times\frac{d(d\mathbf{u})}{ds_0}\Big)^T\mathbf{n}_j \\ &\quad - \Big(d(\delta\bar{\varphi})\times\frac{d\mathbf{x}}{ds_0}\Big)^T\mathbf{n}_j - \Big(\delta\bar{\varphi}\times\frac{d\mathbf{x}}{ds_0}\Big)^T(d\bar{\varphi}\times\mathbf{n}_j). \end{aligned} \tag{4.30}$$

The first two terms are rearranged into a form similar to (4.28), and the second variation $d(\delta\bar{\varphi})$ is substituted from (3.30), whereby

$$\begin{aligned} d(\delta\varepsilon_j) &= \delta\bar{\varphi}^T(\mathbf{n}_j\times)\frac{d(d\mathbf{u})}{ds_0} - \frac{d(\delta\mathbf{u})}{ds_0}^T(\mathbf{n}_j\times)\,d\bar{\varphi} \\ &+ \Big(\tfrac{1}{2}(\delta\bar{\varphi}\times d\bar{\varphi})\times\frac{d\mathbf{x}}{ds_0}\Big)^T\mathbf{n}_j - \Big(\delta\bar{\varphi}\times\frac{d\mathbf{x}}{ds_0}\Big)^T(d\bar{\varphi}\times\mathbf{n}_j). \end{aligned} \tag{4.31}$$

The last two terms are reduced by use of the vector product identities

$$(\mathbf{a}\times\mathbf{b})\cdot(\mathbf{c}\times\mathbf{d}) = \mathbf{a}\cdot[\mathbf{b}\times(\mathbf{c}\times\mathbf{d})] = (\mathbf{a}\cdot\mathbf{c})(\mathbf{b}\cdot\mathbf{d}) - (\mathbf{a}\cdot\mathbf{d})(\mathbf{b}\cdot\mathbf{c}). \tag{4.32}$$

One of the terms with factor $\frac{1}{2}$ cancels half of a similar term, resulting in the fully symmetric form

$$\begin{aligned} d(\delta\varepsilon_j) &= \delta\bar{\varphi}^T(\mathbf{n}_j\times)\frac{d(d\mathbf{u})}{ds_0} - \frac{d(\delta\mathbf{u})}{ds_0}^T(\mathbf{n}_j\times)\,d\bar{\varphi} \\ &+ \delta\bar{\varphi}^T\Big(\tfrac{1}{2}\mathbf{n}_j\frac{d\mathbf{x}}{ds_0}^T + \tfrac{1}{2}\frac{d\mathbf{x}}{ds_0}\mathbf{n}_j^T\Big)\,d\bar{\varphi} - (\delta\bar{\varphi}^T d\bar{\varphi})\Big(\mathbf{n}_j^T\frac{d\mathbf{x}}{ds_0}\Big), \end{aligned} \tag{4.33}$$

where the terms with factor $\frac{1}{2}$ are exterior products.

Also in this case the introduction of the vector form of the internal force $\mathbf{N} = N_j\mathbf{n}_j$ reduces the result to the convenient tensor format

$$\begin{aligned} d(\delta\varepsilon_j)\,N_j &= \delta\bar{\varphi}^T(\mathbf{N}\times)\frac{d(d\mathbf{u})}{ds_0} - \frac{d(\delta\mathbf{u})}{ds_0}^T(\mathbf{N}\times)\,d\bar{\varphi} \\ &+ \delta\bar{\varphi}^T\Big(\tfrac{1}{2}\mathbf{N}\frac{d\mathbf{x}}{ds_0}^T + \tfrac{1}{2}\frac{d\mathbf{x}}{ds_0}\mathbf{N}^T - (\mathbf{N}^T\frac{d\mathbf{x}}{ds_0})\,\mathbf{I}\Big)\,d\bar{\varphi}. \end{aligned} \tag{4.34}$$

The format of this contribution is similar to (4.29) from the moments.

The initial stress contributions (4.34) and (4.29) from section force and moment can be combined in block matrix format, where the notation $(\ )' = d(\ )/ds_0$ is used,

$$d(\delta\varepsilon_j)\,N_j + d(\delta\kappa_j)\,M_j \;=\; \begin{bmatrix}\delta\mathbf{u}'^T & \delta\bar{\boldsymbol{\varphi}}'^T & \delta\bar{\boldsymbol{\varphi}}^T\end{bmatrix} \begin{bmatrix} \mathbf{0} & \mathbf{0} & \hat{\mathbf{N}}^T \\ \mathbf{0} & \mathbf{0} & \frac{1}{2}\hat{\mathbf{M}}^T \\ \hat{\mathbf{N}} & \frac{1}{2}\hat{\mathbf{M}} & \begin{matrix}\frac{1}{2}(\mathbf{N}\,\mathbf{x}'^T + \mathbf{x}'\,\mathbf{N}^T) \\ -(\mathbf{N}^T\mathbf{x}')\,\mathbf{I}\end{matrix} \end{bmatrix} \begin{bmatrix} d\mathbf{u}' \\ d\bar{\boldsymbol{\varphi}}' \\ d\bar{\boldsymbol{\varphi}} \end{bmatrix}. \tag{4.35}$$

This formula provides the initial stress, or geometric stiffness, contribution to a beam element, when the variations $\delta\mathbf{u}', \delta\bar{\boldsymbol{\varphi}}', \delta\bar{\boldsymbol{\varphi}}$ and the increments $d\mathbf{u}', d\bar{\boldsymbol{\varphi}}', d\bar{\boldsymbol{\varphi}}$ are represented by suitable shape functions. In particular, if identical shape functions are used for each of the three vector components, the formula enables an efficient global implementation of a general element for large deformation of curved beams as discussed in the following section.

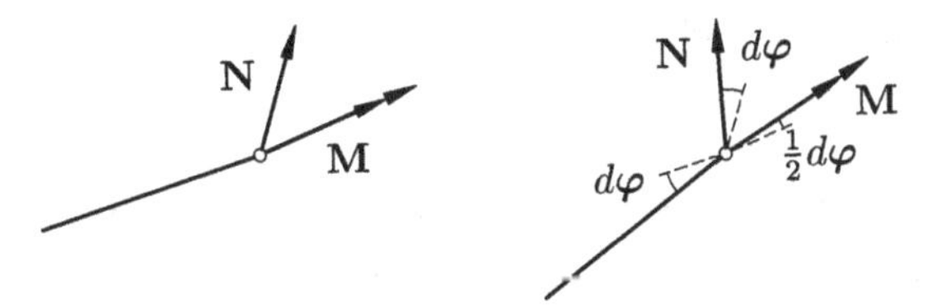

Fig. 4.4. Incremental rotation of force and moment at rotating node.

There is an interesting difference between the effect of rotating the internal force $\mathbf{N}$ and rotating the internal moment $\mathbf{M}$. While rotation of the internal force gives a contribution of the form $d\bar{\boldsymbol{\varphi}} \times \mathbf{N}$ corresponding to the classic result for infinitesimal rotation of a vector, the similar contribution from the moment is $\frac{1}{2}d\bar{\boldsymbol{\varphi}} \times \mathbf{M}$, i.e. reduced to precisely half. The reason for the difference lies in a reduction introduced via the second variation of the rotation $d(\delta\boldsymbol{\varphi})$ as discussed in Examples 3.2 and 3.3. In connection with convected internal forces in structures, it *appears* as if the moment is only rotated by half the angle as illustrated in Fig. 4.4. Similar effects may occur in relation to applied moment load, and these should be included via their contribution to the virtual work.

4.3.3 The load increments

The external virtual work serves to define the external loads. In the case of the 'elastica' beam theory the external virtual work is the last term in (4.6):

$$\delta V_{\text{ext}} = \int_0^{l_0} \Big(\delta\mathbf{u}^T\mathbf{p} + \delta\bar{\boldsymbol{\varphi}}^T\mathbf{m}\Big)\,ds_0, \tag{4.36}$$

where $\mathbf{p}(s_0)$ and $\mathbf{m}(s_0)$ are the distributed force and moment, respectively, per unit original length measured by s_0. A discretized representation of the virtual translation $\delta\mathbf{u}(s_0)$ and virtual rotation $\delta\bar{\boldsymbol{\varphi}}(s_0)$ will provide the equivalent concentrated loads via the integral δV_{ext}.

A change of state, e.g. by change of the loads, may also lead to change in the external virtual work. In this process $\delta\mathbf{u}$ remains unchanged, while $\delta\bar{\boldsymbol{\varphi}}$ changes as described in detail in Section 3.3. Thus the change in δV_{ext} is

$$d(\delta V_{\text{ext}}) = \int_0^{l_0} \Big(\delta\mathbf{u}^T d\mathbf{p} + \delta\bar{\boldsymbol{\varphi}}^T d\mathbf{m}\Big) ds_0 + \int_0^{l_0} \delta\bar{\boldsymbol{\varphi}}^T (\tfrac{1}{2}\mathbf{m}\times)\, d\bar{\boldsymbol{\varphi}}\; ds_0. \quad (4.37)$$

The first integral represents the usual change in the external loads, while the second integral is a byproduct of the use of the incremental rotation $\delta\bar{\varphi}$ in the virtual work as demonstrated in Example 3.2. This second term is linear in the rotation increment $d\bar{\varphi}$ and therefore has the character of a skew-symmetric contribution to the stiffness matrix. As illustrated in Example 3.3 a conservative moment may include a dependence of the moment on $d\bar{\varphi}$, whereby this term is cancelled.

4.4 Finite element implementation

The beam theory can be implemented directly in global form as proposed by Ibrahimbegović (1995). The idea is to represent the beam by a space curve interpolating a given set of nodes, and then describe the displacement and rotation along the beam in terms of global vector components, referring to a fixed Cartesian coordinate system, common to all the elements in the analysis. This approach represents a logical extension of the use of shape functions for bar elements illustrated in Section 2.5, but is quite different from the traditional element formulation of beams, in which the properties of the beam element are obtained in a local coordinate system associated with the beam.

The present beam theory is formulated in terms of displacements $\mathbf{u}(s_0)$ and rotations $\boldsymbol{\varphi}(s_0)$, and implementation into the finite element format implies some kind of representation in terms of given interpolation or shape functions. In the following the procedure is illustrated for the element with linear interpolation of translation and rotation increments, but the procedure is fairly easily generalized to higher-order interpolation. First the stiffness matrix is derived and then the evaluation of the internal forces and moments is discussed. When the translations and rotations are represented by shape functions, it is important to realize that the total strain and the strain increments are formulated in terms of a difference between the incremental rotation vector and the derivative of the incremental translation vector. If translations and rotations are represented by polynomials of the same degree, this may lead to spurious strains in some deformation modes. This important problem, called locking, is discussed in Section 4.4.3.

4.4.1 Element stiffness matrix

The fully non-linear beam theory derived in the previous sections can be implemented in terms of two-node elements with linear interpolation. The shape functions of an element with nodes A and B and initial length l_0 are

$$h_A(s_0) = 1 - s_0/l_0, \qquad h_B(s_0) = s_0/l_0. \tag{4.38}$$

The virtual and actual increments are represented in terms of these shape functions in the form

$$\begin{aligned} d\mathbf{u}(s_0) &= h_A(s_0)\, d\mathbf{u}_A + h_B(s_0)\, d\mathbf{u}_B, \\ d\bar{\boldsymbol{\varphi}}(s_0) &= h_A(s_0)\, d\bar{\boldsymbol{\varphi}}_A + h_B(s_0)\, d\bar{\boldsymbol{\varphi}}_B, \end{aligned} \tag{4.39}$$

where subscripts A and B identify the corresponding end-point value of the translation and rotation vector increments. Similar representations are used for the virtual increments $\delta\mathbf{u}(s_0)$ and $\delta\bar{\boldsymbol{\varphi}}(s_0)$.

In order to facilitate the use of these representations it is convenient to introduce the block matrix format

$$\begin{bmatrix} d\mathbf{u}(s_0) \\ d\bar{\boldsymbol{\varphi}}(s_0) \end{bmatrix} = \begin{bmatrix} h_A(s_0)\,\mathbf{I} & \mathbf{0} & h_B(s_0)\,\mathbf{I} & \mathbf{0} \\ \mathbf{0} & h_A(s_0)\,\mathbf{I} & \mathbf{0} & h_B(s_0)\,\mathbf{I} \end{bmatrix} \begin{bmatrix} d\mathbf{u}_A \\ d\bar{\boldsymbol{\varphi}}_A \\ d\mathbf{u}_B \\ d\bar{\boldsymbol{\varphi}}_B \end{bmatrix}. \tag{4.40}$$

Following the notation introduced in Chapter 2, the components of all the displacements associated with nodes of the element are denoted $\tilde{\mathbf{u}}$. For the present element these components are found on the right-hand side of (4.40), and thus

$$d\tilde{\mathbf{u}}^T = [\, d\mathbf{u}_A^T, d\bar{\boldsymbol{\varphi}}_A^T, d\mathbf{u}_B^T, d\bar{\boldsymbol{\varphi}}_B^T \,]. \tag{4.41}$$

In this notation the representation of the displacement increments takes the form

$$\begin{bmatrix} d\mathbf{u}(s_0) \\ d\bar{\boldsymbol{\varphi}}(s_0) \end{bmatrix} = \mathbf{H}(s_0)\, d\tilde{\mathbf{u}} \tag{4.42}$$

with the shape functions contained in the matrix $\mathbf{H}(s_0)$.

It follows from the expressions (4.24) and (4.35) for the increment of the virtual work that the stiffness matrix requires the increments $d\mathbf{u}'(s_0)$,

$d\bar{\boldsymbol{\varphi}}'(s_0)$ and $d\bar{\boldsymbol{\varphi}}(s_0)$. The representation of these in terms of the nodal values follows from differentiation of (4.40) as

$$\begin{bmatrix} d\mathbf{u}'(s_0) \\ d\bar{\boldsymbol{\varphi}}'(s_0) \\ d\bar{\boldsymbol{\varphi}}(s_0) \end{bmatrix} = \begin{bmatrix} h'_A(s_0)\,\mathbf{I} & \mathbf{0} & h'_B(s_0)\,\mathbf{I} & \mathbf{0} \\ \mathbf{0} & h'_A(s_0)\,\mathbf{I} & \mathbf{0} & h'_B(s_0)\,\mathbf{I} \\ \mathbf{0} & h_A(s_0)\,\mathbf{I} & \mathbf{0} & h_B(s_0)\,\mathbf{I} \end{bmatrix} \begin{bmatrix} d\mathbf{u}_A \\ d\bar{\boldsymbol{\varphi}}_A \\ d\mathbf{u}_B \\ d\bar{\boldsymbol{\varphi}}_B \end{bmatrix}. \tag{4.43}$$

In the compact notation corresponding to (4.42), this relation is

$$\begin{bmatrix} d\mathbf{u}'(s_0) \\ d\bar{\boldsymbol{\varphi}}'(s_0) \\ d\bar{\boldsymbol{\varphi}}(s_0) \end{bmatrix} = \mathbf{G}(s_0)\, d\tilde{\mathbf{u}} \tag{4.44}$$

where the matrix $\mathbf{G}(s_0)$ is defined by (4.43).

The element tangent stiffness matrix $\mathbf{K}$ can now be expressed in a convenient matrix format. As seen from (4.18), the element stiffness consists of two contributions, one due to the constitutive changes of the internal forces in the beam and another due to the convection of the internal forces with the beam, the so-called geometric stiffness. The constitutive part of the element stiffness matrix is denoted $\mathbf{K}_c$. It is determined by substitution of the representation (4.44) of the virtual and actual increments into the constitutive part of the virtual work increment, defined by the first two terms in (4.18),

$$\delta\tilde{\mathbf{u}}^T\mathbf{K}_c\, d\tilde{\mathbf{u}} = \int_0^{l_0} \Big(\delta\varepsilon_j\, C^{\varepsilon}_{jk}\, d\varepsilon_k + \delta\kappa_j\, C^{\kappa}_{jk}\, d\kappa_k \Big) ds_0. \tag{4.45}$$

When using the expression (4.24) for the integrand, the constitutive part of the element stiffness matrix takes the form

$$\mathbf{K}_c = \int_0^{l_0} \mathbf{G}^T\mathbf{D}_c\,\mathbf{G}\; ds_0, \tag{4.46}$$

where the matrix $\mathbf{D}_c$ follows from (4.24) as

$$\mathbf{D}_c = \begin{bmatrix} \mathbf{C}^{\varepsilon}_* & \mathbf{0} & \mathbf{C}^{\varepsilon}_*\,\hat{\mathbf{x}}' \\ \mathbf{0} & \mathbf{C}^{\kappa}_* & \mathbf{0} \\ \hat{\mathbf{x}}'^T\mathbf{C}^{\varepsilon}_* & \mathbf{0} & \hat{\mathbf{x}}'^T\mathbf{C}^{\varepsilon}_*\,\hat{\mathbf{x}}' \end{bmatrix}. \tag{4.47}$$

Similarly, the geometric part of the element stiffness matrix $\mathbf{K}_g$ is determined by the last two terms of the virtual work increment (4.18),

$$\delta\tilde{\mathbf{u}}^T\mathbf{K}_g\, d\tilde{\mathbf{u}} = \int_0^{l_0} \Big(d(\delta\varepsilon_j)\, N_j + d(\delta\kappa_j)\, M_j \Big) ds_0. \tag{4.48}$$

This leads to the geometric element stiffness matrix

$$\mathbf{K}_g = \int_0^{l_0} \mathbf{G}^T \mathbf{D}_g \, \mathbf{G} \, ds_0, \tag{4.49}$$

where the matrix $\mathbf{D}_g$ follows from (4.35) as

$$\mathbf{D}_g = \begin{bmatrix} \mathbf{0} & \mathbf{0} & \hat{\mathbf{N}}^T \\ \mathbf{0} & \mathbf{0} & \frac{1}{2}\hat{\mathbf{M}}^T \\ \hat{\mathbf{N}} & \frac{1}{2}\hat{\mathbf{M}} & \begin{matrix} \frac{1}{2}(\mathbf{N}\,\mathbf{x}'^T + \mathbf{x}'\,\mathbf{N}^T) \\ -(\mathbf{N}^T\mathbf{x}')\,\mathbf{I} \end{matrix} \end{bmatrix}. \tag{4.50}$$

It is seen that the expressions for the constitutive and geometric stiffness matrices $\mathbf{K}_c$ and $\mathbf{K}_g$ are similar. This is because the incremental strain is not used explicitly in the formulation of the constitutive stiffness matrix, but is obtained by combination of terms within the matrix $\mathbf{D}_c$. Thus, in the present formulation the element stiffness matrix can be written as

$$\mathbf{K}_e = \mathbf{K}_c + \mathbf{K}_g = \int_0^{l_0} \mathbf{G}^T (\mathbf{D}_c + \mathbf{D}_g) \mathbf{G} \, ds_0. \tag{4.51}$$

Care should be exercised to control any spurious strain contributions arising from the assumed shape functions when using this formula. This problem is discussed in Section 4.4.3.

4.4.2 Loads and internal forces

The equilibrium condition, and thereby the residual, is determined from the virtual work (4.6). The contribution of distributed loads on the elements is expressed in terms of the equivalent nodal loads

$$\tilde{\mathbf{p}}^T = [\mathbf{p}_A^T, \mathbf{m}_A^T, \mathbf{p}_B^T, \mathbf{m}_B^T]. \tag{4.52}$$

These nodal loads are determined from the virtual work equation (4.6),

$$\delta\tilde{\mathbf{u}}^T \tilde{\mathbf{p}} = \int_0^{l_0} \Big(\delta\mathbf{u}^T \mathbf{p} + \delta\bar{\boldsymbol{\varphi}}^T \mathbf{m} \Big) ds_0. \tag{4.53}$$

Substitution of the representation of the virtual displacement in the form (4.42) then gives the equivalent nodal loads

$$\tilde{\mathbf{p}} = \int_0^{l_0} \mathbf{H}^T \begin{bmatrix} \mathbf{p} \\ \mathbf{m} \end{bmatrix} ds_0. \tag{4.54}$$

Concentrated loads acting at the nodes can be added directly into the global formulation and will not be associated with the elements.

The internal forces are arranged in the same format as the equivalent loads,

$$\tilde{\mathbf{q}}^T = [-\mathbf{N}_A^T, -\mathbf{M}_A^T, \mathbf{N}_B^T, \mathbf{M}_B^T], \tag{4.55}$$

where the minus signs are due to the sign convention of section forces. They are determined from the internal virtual work given by the first integral in (4.6),

$$\delta\tilde{\mathbf{u}}^T\tilde{\mathbf{q}} = \int_0^{l_0}\Big(\delta\varepsilon_j N_j + \delta\kappa_j M_j\Big)ds_0. \tag{4.56}$$

The components in this integral are local, as explained in Section 4.2. For a linear elastic beam the constitutive tangent stiffness matrices C_{jk}^{ε} and C_{jk}^{κ} introduced in (4.19) are constant. The internal virtual work (4.53) can therefore be written as

$$\delta\tilde{\mathbf{u}}^T\tilde{\mathbf{q}} = \int_0^{l_0}\Big(\delta\varepsilon_j C_{jk}^{\varepsilon}(\varepsilon_k - \varepsilon_k^0) + \delta\kappa_j C_{jk}^{\kappa}(\kappa_k - \kappa_k^0)\Big)ds_0, \tag{4.57}$$

where ε_k^0 and κ_k^0 are the strain and curvature components in the reference state, respectively. In the present formulation the global strain and curvature components $\boldsymbol{\varepsilon}$ and $\boldsymbol{\kappa}$ are used together with the global stiffness component matrices $\mathbf{C}_*^{\varepsilon}$ and $\mathbf{C}_*^{\kappa}$ introduced in (4.21). Hereby, the internal virtual work takes the form

$$\begin{aligned}\delta\tilde{\mathbf{u}}^T\tilde{\mathbf{q}} &= \int_0^{l_0}\Big(\delta\boldsymbol{\varepsilon}\mathbf{C}_*^{\varepsilon}(\boldsymbol{\varepsilon} - \boldsymbol{\varepsilon}_0) + \delta\boldsymbol{\kappa}\mathbf{C}_*^{\kappa}(\boldsymbol{\kappa} - \boldsymbol{\kappa}_0)\Big)ds_0 \\ &= \int_0^{l_0}[\delta\boldsymbol{\varepsilon}^T, \delta\boldsymbol{\kappa}^T]\begin{bmatrix}\mathbf{C}_*^{\varepsilon} & \mathbf{0}\\ \mathbf{0} & \mathbf{C}_*^{\kappa}\end{bmatrix}\begin{bmatrix}\boldsymbol{\varepsilon} - \boldsymbol{\varepsilon}_0\\ \boldsymbol{\kappa} - \boldsymbol{\kappa}_0\end{bmatrix}ds_0.\end{aligned} \tag{4.58}$$

The global components of the virtual strain and curvature follow from (4.7) and (4.11) in the form

$$\delta\boldsymbol{\varepsilon} = \delta\mathbf{u}' + \mathbf{x}' \times \delta\bar{\boldsymbol{\varphi}}, \qquad \delta\boldsymbol{\kappa} = \delta\bar{\boldsymbol{\varphi}}'. \tag{4.59}$$

The virtual strain and curvature can be represented in block matrix format as

$$\begin{bmatrix}\delta\boldsymbol{\varepsilon}\\ \delta\boldsymbol{\kappa}\end{bmatrix} = \begin{bmatrix}\mathbf{I} & \mathbf{0} & \hat{\mathbf{x}}'\\ \mathbf{0} & \mathbf{I} & \mathbf{0}\end{bmatrix}\begin{bmatrix}\delta\mathbf{u}'\\ \delta\bar{\boldsymbol{\varphi}}'\\ \delta\bar{\boldsymbol{\varphi}}\end{bmatrix}. \tag{4.60}$$

The vector on the right-hand side has already been represented in terms of shape functions in (4.44) by the matrix $\mathbf{G}(s_0)$. When this expression is used in (4.50) the internal virtual work formula (4.58) defines the internal force vector $\tilde{\mathbf{q}}$ as

$$\tilde{\mathbf{q}} = \int_0^{l_0} \mathbf{G}^T \mathbf{D}_c \begin{bmatrix} \boldsymbol{\varepsilon} - \boldsymbol{\varepsilon}_0 \\ \boldsymbol{\kappa} - \boldsymbol{\kappa}_0 \\ \mathbf{0} \end{bmatrix} ds_0, \tag{4.61}$$

where the zero vector $\mathbf{0}$ in the combined strain curvature block vector is introduced to eliminate the last block column of the constitutive matrix $\mathbf{D}_c$, which is not used here.

The global form of the total strain and curvature is given by (4.9) and (4.15) respectively,

$$\begin{bmatrix} \boldsymbol{\varepsilon} \\ \boldsymbol{\kappa} \end{bmatrix} = \begin{bmatrix} \mathbf{x}' - \mathbf{n}_1 \\ \mathbf{T}(\boldsymbol{\varphi})\, \boldsymbol{\varphi}' \end{bmatrix}. \tag{4.62}$$

In principle, these expressions are to be represented by interpolation of the current nodal values of $\mathbf{x}$ and $\boldsymbol{\varphi}$. Due to the non-linear relation between the incremental rotation $d\bar{\boldsymbol{\varphi}}$ and the increment of the rotation parameters $d\boldsymbol{\varphi}$, linear interpolation of the rotation parameters $\boldsymbol{\varphi}$ may introduce some inconsistency. In the case of the two-node element this can be avoided by use of representative mean values of $\boldsymbol{\varepsilon}$ and $\boldsymbol{\kappa}$ obtained directly from the 'mean rotation' and the rotation increment $\Delta\bar{\boldsymbol{\varphi}}$ over the element, as discussed in Section 3.5.4. The idea is to consider a point of 'mean rotation', described e.g. by the rotation pseudo-vector $\bar{\varphi}$. The mean is defined such that further rotation by $\frac{1}{2}\Delta\bar{\boldsymbol{\varphi}}$ produces the rotation $\boldsymbol{\varphi}_B$, while the reverse rotation $-\frac{1}{2}\Delta\bar{\boldsymbol{\varphi}}$ produces the rotation $\boldsymbol{\varphi}_A$. The mean strain is obtained from the vector $\mathbf{n}_1$ as determined by the mean rotation $\bar{\varphi}$, while the mean curvature is determined from the rotation increment over the element as

$$\boldsymbol{\kappa} = \frac{\Delta\bar{\boldsymbol{\varphi}}}{l_0}. \tag{4.63}$$

The direction $\mathbf{n}_1$ and the rotation increment $\Delta\bar{\boldsymbol{\varphi}}$ determined by this procedure are independent of the parametrization of the rotations and directly related to the element.

4.4.3 Shear locking

The representation of the displacement and rotation increments by linear interpolation introduces a number of approximations, of which the most important probably is *shear locking.* This phenomenon, which is known for many beam and shell elements formulated in terms of translations and rotations, will here be illustrated in a simple two-dimensional setting. The following discussion relates to the linearized theory of a straight beam of length L. The base vectors $(\mathbf{n}_1, \mathbf{n}_2, \mathbf{n}_3)$ defined by the beam cross-sections can therefore be used as a fixed global frame of reference. Extension of the beam is disregarded, and thus bending in the $\mathbf{n}_2$–$\mathbf{n}_3$ plane is described by the transverse displacement u_2 and the rotation $\varphi = \varphi_3$. The only measures of deformation are the shear strain and the curvature,

$$\varepsilon_2 = u_2' - \varphi_3, \qquad \kappa_3 = \varphi_3'. \tag{4.64}$$

With these deformation measures the elastic energy of the beam is

$$U = \frac{1}{2}\int_0^L \Big(EI\varphi_3'^{\,2} + GA(u_2' - \varphi_3)^2 \Big) ds. \tag{4.65}$$

The problem of shear locking arises when the representation of the shear strain in the last term of the energy is not able to accommodate a state of vanishing shear stress.

Example 4.1. Shear locking in homogeneous bending. The simple example in Fig. 4.5 illustrates the problem. A beam of length L is modeled by n identical elements of length $l = L/n$ and loaded in homogeneous bending by application of two opposite bending moments of magnitude M. Within each element the translation $u(s_0)$ and the rotation $\varphi(s)$ are represented by linear interpolation between the end-point values. The curvature φ' is represented correctly, giving

$$\varphi' = \frac{M}{EI}, \qquad \varphi_B = -\varphi_A = \frac{L}{2}\frac{M}{EI}.$$

The linear graph of $\varphi(s)$ is shown in Fig. 4.5(b). The shear strain is given by $u' - \varphi$. Due to the linear representation of the translation u, the derivative u' is constant within each element. In the present case each element deforms symmetrically, and thus the graph of u' intersects the graph of φ in the midpoint of each element. As a result of the different degrees of representation of the two terms, the shear strain is given by the 'sawtooth' difference between the two graphs, while in the exact solution it is identically zero.

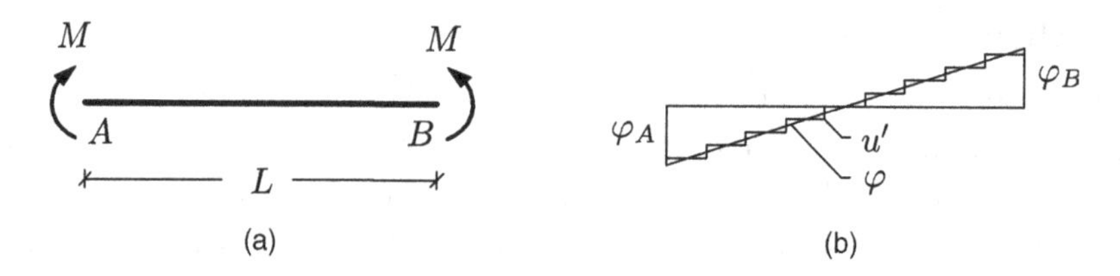

Fig. 4.5. (a) Homogeneous bending, (b) distribution of φ and u'.

The elastic strain energy of the beam as represented by the n elements with linear representation of u and φ follows by integration from (4.65),

$$U_{\text{approx}} = \tfrac{1}{2} L\, EI\, (2\varphi_A/L)^2 + \tfrac{1}{6} L\, GA\, (\varphi_A/n)^2.$$

The first term represents the exact value, while the second term is an undesirable consequence of the inconsistent representation of the shear strain. It is convenient to write the result in the form

$$U_{\text{approx}} = \Big(1 + \frac{1}{n^2 \Phi_L}\Big) U_{\text{exact}},$$

where Φ_L is the non-dimensional shear parameter of the beam

$$\Phi_L = \frac{12\, EI}{L^2\, GA}.$$

For a classical Bernoulli beam the shear stiffness is infinite, whereby $\Phi_L = 0$. It is seen that linear interpolation of both translation and rotation does not permit representation of Bernoulli beams.

The shear locking effect in homogeneous bending is seen to depend on the parameter $n^2\Phi_L$, which is the shear factor of the individual element,

$$\Phi = n^2 \Phi_L = \frac{12\, EI}{l^2\, GA},$$

where $l = L/n$ is the element length. In a displacement representation the factor on the energy corresponds directly to a factor on the stiffness. Thus, the stiffness in homogeneous bending is increased by the linear interpolation of the displacements. The error decreases with n^{-2}, but it is undesirable to use many elements to obtain a satisfactory solution to simple problems of beam theory. Solutions to the shear locking problem are described in the following.

There are three basically different ways of countering the shear locking problem of displacement-based finite elements. For simple elements, like the present beam element, the locking problem can be resolved by use of interpolation functions that lead to consistent representation of the shear stress. For the beam element this implies that the transverse components of the rotation vector $\boldsymbol{\varphi}(s_0)$ and the displacement derivative $\mathbf{u}'(s_0)$ should be interpolated to the same order, e.g. with a linear representation of $\boldsymbol{\varphi}(s_0)$ and a quadratic representation of $\mathbf{u}(s_0)$. For beam elements this can be accomplished by introducing a center-node for the translation degrees of freedom or the introduction of an additional internal mode not associated with a particular node, followed by elimination of the internal degrees of freedom at the element level, see e.g. Ibrahimbegović (1995). The second method consists in evaluating the element properties by reduced integration. The term 'reduced integration' is typically used in the finite element literature to denote a form of numerical integration that is deliberately chosen not to include certain contributions, see e.g. Hughes (1987) and Zienkiewicz and Taylor (2000). In the beam example this amounts to evaluating the shear strain contributions by their value at the element midpoint. As seen from Fig. 4.5(b), the undesired shear strain contribution vanishes here. Many shell elements make use of reduced integration. A third method consists in the use of modified stiffness parameters that are selected to compensate for the errors introduced by the interpolation. In the following, modified bending and shear stiffness parameters $\overline{EI}$ and $\overline{GA}$ are derived for the beam elements with linear interpolation. The derivation is based on small deformation bending of a simple straight element, but the resulting modified parameters can subsequently be used in the implementation of the general theory for large displacement beam theory.

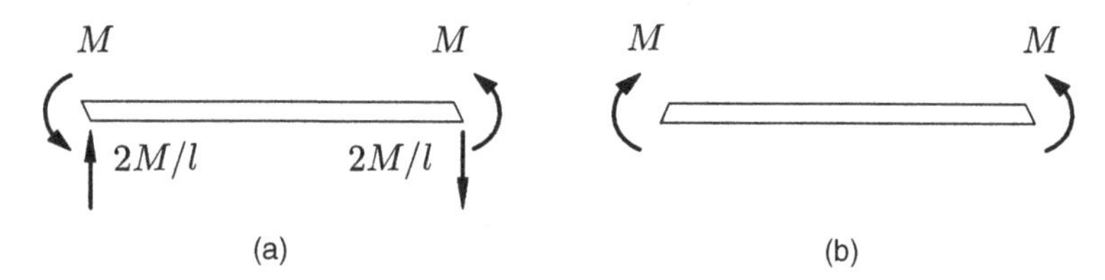

Fig. 4.6. Bending of 'linear' beam element. (a) Anti-symmetric, (b) symmetric.

For beams, modified bending and shear parameters $\overline{EI}$ and $\overline{GA}$ can be determined from the anti-symmetric and symmetric bending modes of an element. The anti-symmetric and symmetric bending modes of an element with linear interpolation of displacement and rotation are shown in Fig. 4.6. In the anti-symmetric mode of Fig. 4.6(a) the rotation φ must be equal at both ends of the element, and for linear representation the rotation must therefore be constant. In the symmetric mode of Fig. 4.6(b) the rotation φ must be of equal magnitude but opposite sign at the ends of the element. In both modes the displacement vanishes at both ends, and therefore identically. The modified stiffness parameters are now determined such that the two bending modes of the element attain the correct stiffness.

The anti-symmetric mode is treated first. Its elastic energy is evaluated by observing that the moment varies linearly between the values $-M$ and M, while there is a constant shear force of magnitude $Q = 2M/l$. For a linear elastic beam the energy follows directly from the stress state as

$$U_{\text{exact}} = \frac{1}{2}\int_0^l \Big(\frac{M(s)^2}{EI} + \frac{Q(s)^2}{GA}\Big) ds = \frac{l}{2}\Big(\frac{1}{3}\frac{M^2}{EI} + \frac{4M^2}{l^2 GA}\Big). \qquad (4.66)$$

This expression takes the form

$$U_{\text{exact}} = (1+\Phi)\,\frac{l}{6}\frac{M^2}{EI} \tag{4.67}$$

when using the shear parameter of the element

$$\Phi = \frac{12\,EI}{l^2\,GA} \tag{4.68}$$

already introduced in Example 4.1. The energy is expressed in terms of the end-point rotation φ by use of the flexibility relation $2\varphi = \partial U/\partial M$, where the factor 2 arises from the fact that both end loads contribute equally to the energy. Substitution of this expression for φ gives the energy as

$$U_{\text{exact}} = \frac{6}{1+\Phi}\,\frac{EI}{l}\,\varphi^2. \tag{4.69}$$

The energy of the beam element with linear interpolation is calculated from (4.65), using the parameter $\overline{GA}$. Due to the linear interpolation $\varphi = \text{const.}$ and the bending contribution vanishes, leaving only

$$U_{\text{approx}} = \frac{1}{2}\int_0^l \overline{GA}\,\varphi(s)^2\,ds = \frac{l}{2}\,\overline{GA}\,\varphi^2. \tag{4.70}$$

The parameter $\overline{GA}$ is now selected to make the two energy expressions (4.69) and (4.70) identical. This gives the equivalent shear stiffness

$$\overline{GA} = \frac{\Phi}{1+\Phi}\,GA. \tag{4.71}$$

This modified value of GA will produce the theoretically correct stiffness of the linearly interpolated beam element in anti-symmetric bending.

The procedure for symmetric bending, shown in Fig. 4.6(b), is similar. The symmetry of the problem and the linear interpolation imply that $u \equiv 0$ over the full element and $\varphi' = \text{const.}$ The exact elastic energy is evaluated from statics, observing that the moment is constant, whereby

$$U_{\text{exact}} = \frac{1}{2}\int_0^l \frac{M(s)^2}{EI}\,ds = \frac{l}{2}\frac{M^2}{EI}. \tag{4.72}$$

The angle at the ends of the beam is again calculated by use of the relation $2\varphi = \partial U/\partial M$, and substitution of the result gives the exact energy as

$$U_{\text{exact}} = \frac{2\,EI}{l}\,\varphi^2. \tag{4.73}$$

The energy of symmetric bending of the beam element with linear interpolation is calculated from (4.65), using the modified parameters $\overline{EI}$ and $\overline{GA}$, whereby

$$U_{\text{approx}} = \frac{1}{2}\int_0^l \Big(\overline{EI}\,\varphi'(s)^2 + \overline{GA}\,\varphi(s)^2\Big)ds = \Big(\frac{2\,\overline{EI}}{l} + \frac{l}{6}\,\overline{GA}\Big)\varphi^2. \quad (4.74)$$

The parameter $\overline{EI}$ is now selected to make the two expressions (4.73) and (4.74) identical, when using the value for $\overline{GA}$ already determined by (4.71). This gives the modified bending stiffness

$$\overline{EI} = \frac{\Phi}{1+\Phi}\,EI \qquad \text{(full integration)}. \quad (4.75)$$

Note that by this procedure the correction factors for EI and GA are identical and less than unity.

In the implementation of finite elements it is common to obtain the stiffness matrices by numerical integration. In the present case of the beam element with linear interpolation the anti-symmetric bending can be integrated exactly by use of one integration point located in the middle of the element. For the case of symmetric bending one-point integration will give the exact result for the constant curvature term, while the shear strain term vanishes at the midpoint, and therefore does not contribute to the result. This means that the second term, containing $\overline{GA}$, vanishes from U_{approx} in (4.74). Thus, the approximate energy will give the correct result for

$$\overline{EI} = EI \qquad \text{(reduced integration)}. \quad (4.76)$$

This particular form of modified parameters has been discussed, e.g. by Hughes (1987).

Linear element stiffness matrix

When the node displacements and rotations are collected in the array $\tilde{\mathbf{u}}$ in the form

$$\tilde{\mathbf{u}}^T = [\,u_A, \varphi_A, u_B, \varphi_B\,], \quad (4.77)$$

the element stiffness matrix $\mathbf{K}$ can be found from the elastic energy expression

$$U = \tfrac{1}{2}\tilde{\mathbf{u}}^T\mathbf{K}\,\tilde{\mathbf{u}}. \quad (4.78)$$

When linear interpolation is introduced, full integration of the energy expression (4.65) gives

$$\mathbf{K} = \frac{\overline{EI}}{l}\begin{bmatrix} 0 & 0 & 0 & 0 \\ 0 & 1 & 0 & -1 \\ 0 & 0 & 0 & 0 \\ 0 & -1 & 0 & 1 \end{bmatrix} + \frac{\overline{GA}}{6\,l}\begin{bmatrix} 6 & -3l & -6 & -3l \\ -3l & 2l^2 & 3l & l^2 \\ -6 & 3l & 6 & 3l \\ -3l & l^2 & 3l & 2l^2 \end{bmatrix} \quad \text{(full)}. \quad (4.79)$$

The terms containing l^2 arise from integration of φ^2. These terms are changed, if one-point integration is used, and the stiffness matrix obtained by this reduced integration is

$$\mathbf{K} = \frac{\overline{EI}}{l}\begin{bmatrix} 0 & 0 & 0 & 0 \\ 0 & 1 & 0 & -1 \\ 0 & 0 & 0 & 0 \\ 0 & -1 & 0 & 1 \end{bmatrix} + \frac{\overline{GA}}{4l}\begin{bmatrix} 4 & -2l & -4 & -2l \\ -2l & l^2 & 2l & l^2 \\ -4 & 2l & 4 & 2l \\ -2l & l^2 & 2l & l^2 \end{bmatrix} \quad \text{(reduced)}. \tag{4.80}$$

By substituting the equivalent shear stiffness $\overline{GA}$ from (4.71) and the appropriate equivalent bending stiffness $\overline{EI}$ from (4.75) or (4.76), both expressions for the stiffness matrix lead to the common form

$$\mathbf{K} = \frac{EI}{(1+\Phi)l^3}\begin{bmatrix} 12 & -6l & -12 & -6l \\ -6l & (4+\Phi)l^2 & 6l & (2-\Phi)l^2 \\ -12 & 6l & 12 & 6l \\ -6l & (2-\Phi)l^2 & 6l & (4+\Phi)l^2 \end{bmatrix}. \tag{4.81}$$

This is the well-known exact stiffness matrix for a beam with shear flexibility, see e.g. Krenk (2001) for a simple derivation from static equilibrium states.

EXAMPLE 4.2. SHEAR LOCKING IN CANTILEVER BEAM. The shear locking effects discussed above can be illustrated by considering a cantilever beam of length L with a transverse end force P. The exact solution for the translation u and rotation φ of the loaded end is

$$\frac{u}{L} = \frac{PL^2}{3EI} + \frac{P}{GA} = \frac{PL^2}{EI}\frac{4+\Phi_L}{12}, \qquad \varphi = -\frac{PL^2}{2EI}.$$

In this example $P = EI/L^2$ and the shear modulus is $G = 0.4E$. The cross-section is rectangular with $h = 2b$, effective shear area $A = \frac{5}{6}bh$ and moment of inertia $I = \frac{1}{12}bh^3$.

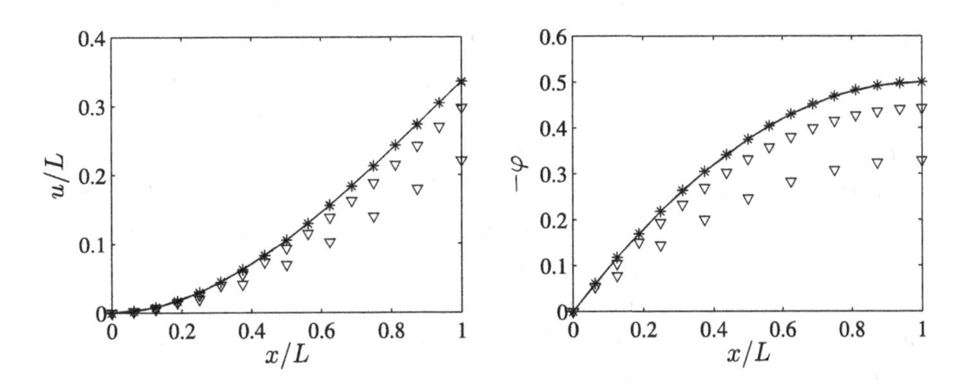

Fig. 4.7. Cantilever with transverse end force, $n = 8, 16$: – analytical, + exact element, × reduced integration, ∇ full integration with original stiffness parameters.

Figure 4.7 shows the translation and rotation for the case $L = 10h$, where the shear parameter of the beam is $\Phi_L = 0.03$. This is a fairly slender beam in which the translation contribution due to shear flexibility is less than 1%. The fully integrated beam element (4.79) without correction is approximately 40% too stiff even with 8 elements, and an increase to 16 elements still leaves a stiffness error of about $(\frac{1}{2})^2 40\% = 10\%$ in both translation and rotation. By reduced integration the mean bending within each element is exact, and only the shear flexibility effect needs further correction by (4.71). In the present example there is no contribution from shear flexibility to the rotation, which is modeled correctly by reduced integration. However, the translation still requires correction of the shear modulus by (4.71) to permit accurate analysis with few elements.

4.5 Summary of 'elastica' beam theory

An elegant way of formulating a general large deformation beam theory consists in considering the beam as a space curve, on which the cross-sections are attached. The generalized strain $\boldsymbol{\varepsilon}$ and curvature $\boldsymbol{\kappa}$ then appear as three-component vectors, conjugate to the internal force $\mathbf{N}$ and the internal moment $\mathbf{M}$, all functions of the length-coordinate s_0 along the beam. The principle of virtual work combines the deformation measures $\boldsymbol{\varepsilon}, \boldsymbol{\kappa}$ and the internal forces $\mathbf{N}, \mathbf{M}$ and provides equilibrium equations in terms of a balance between the external load and the corresponding work of the internal forces. An important consequence is that a concise statement of the tangent stiffness relation can be obtained by manipulating the internal work by the rules developed in Chapter 3. By carefully following the rules for incremental rotations it is demonstrated that the tangent relation is symmetric, and explicit expressions are obtained for the constitutive stiffness and the geometric stiffness. Thus, non-symmetric terms will only occur in the tangent stiffness relations if there are load components with directions depending on the motion of the structure, so-called 'follower-forces'.

The theory leads to an internal shear force, defined by a combination of the rotation and the lengthwise derivative of the transverse displacement. Consistent representation of the shear in the beam requires special attention to enable the beam to retain vanishing shear strain under rigid body motion and uniform bending. Three measures to ensure this are discussed: higher-degree representation of the rotation components, use of selective integration, and use of modified stiffness parameters.

4.6 Exercises

Exercise 4.1 As shown in Example 3.2 the increment of the virtual work of a moment is

$$d(\delta\bar{\varphi}^T\mathbf{M}) \;=\; \delta\bar{\varphi}^T d\mathbf{M} \;+\; \delta\bar{\varphi}^T(\tfrac{1}{2}\mathbf{M}\times d\bar{\varphi}).$$

If $\mathbf{M}$ is an external load and $d\mathbf{M}$ is given, the second term has the character of a geometric stiffness term. How will this term enter the geometric stiffness format (4.50)? Give both the matrix format and the full component format.

Exercise 4.2 Example 4.2 treated a cantilever beam with transverse end load, illustrating the effect of shear locking.

(a) Implement the three stiffness matrices corresponding to full integration, reduced integration, and the modified parameter form. Make an error analysis similar to Fig. 4.7 for the case of a concentrated moment applied to the end of the beam.

(b) How do these results relate to the general discussion of shear locking in Example 4.1?

5

Co-rotating beam elements

The beam theory of Chapter 4 and the corresponding finite element implementation was formulated in a fixed global frame of reference using the total displacements and rotations. In many cases it may be advantageous to consider the beam element with reference to a local, element-based, coordinate system. Motion of the beam then implies motion of the local frame of reference as well as deformation of the beam element within this frame. The separation of the motion of the element into two parts – a rigid body motion associated with the element-based frame of reference and a deformation of the element within this frame of reference – is called a co-rotating formulation. The co-rotating formulation has a number of advantages, provided it can be demonstrated that the tangent stiffness can be decomposed into the sum of a part associated with the rotation of the element-based frame and a part associated solely with the deformation of the element within this frame of reference. The first advantage is that displacements and rotations within the local frame of reference are small or at most moderate. Therefore, the deformation of the beam can be modeled by approximate beam theory. Secondly, the co-rotating formulation is closely associated with the idea of 'natural modes', advocated by Argyris *et al.* (1979a,b). The idea of the 'natural modes' is to consider any increment of the motion of an element as made up of a set of rigid body modes – typically translation and rotation – and a set of deformation modes – representing extension, bending and torsion of the beam element. In the co-rotating formulation the rigid body motion is associated with the motion of the local frame of reference, while the deformation is described within this frame. Considerable simplifications can be obtained by representing the deformation in terms of suitable 'natural modes'.

An important aspect of the co-rotating formulation is to establish that the contributions to the tangent stiffness involving the motion of the local frame can be expressed in a unique symmetric form, independent of the particular beam theory used to represent the deformation of the beam. For two-dimensional beam problems a simple and direct procedure can be used to obtain this part of the tangent stiffness. However, due to the non-additive nature of three-dimensional rotations, this procedure does not apply to three-dimensional co-rotational formulations. Therefore, in the following the two-dimensional problem is first presented using the classical direct method, see e.g. Crisfield (1991). The general three-dimensional problem is then formulated rigorously by introducing rigid body and deformation modes in the general theory of Chapter 4. The co-rotating formulation is demonstrated using both the classic cubic bending interpolation and a non-linear beam–column formulation originally proposed by Oran (1973a,b).

5.1 Co-rotating beams in two dimensions

Figure 5.1 shows a beam element located in a plane defined by a fixed frame of reference $\{x_1, x_2\}$, indicated in the lower left corner. The beam itself is described with respect to a local frame of reference $\{x, y\}$, following the motion of the beam. The local frame of reference is defined by the position of the end-points A and B of the beam element: the x-axis passes through the points A and B, and the origin is located with equal distance from A and B.

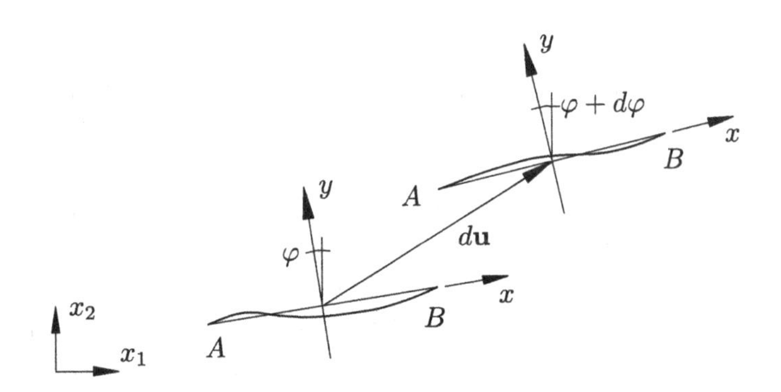

Fig. 5.1. Motion of beam element in local co-rotating frame of reference.

In the fixed global frame of reference the motion of the beam element is described by the translation and rotation of the beam end-points A and B,

$$d\mathbf{p}_A^T = [du_1^A, du_2^A, d\phi^A], \qquad d\mathbf{p}_B^T = [du_1^B, du_2^B, d\phi^B]. \tag{5.1}$$

Thus, the motion of the beam element is described by six components, and naturally there are also six conjugate generalized force components. The conjugate element forces constituting the force and moment at A and B are denoted

$$\mathbf{q}_A^T = [f_1^A, f_2^A, m^A], \qquad \mathbf{q}_B^T = [f_1^B, f_2^B, m^B]. \tag{5.2}$$

It is convenient to introduce the notation

$$d\mathbf{p}^T = [d\mathbf{p}_A^T, d\mathbf{p}_B^T], \qquad \mathbf{q}^T = [\mathbf{q}_A^T, \mathbf{q}_B^T] \tag{5.3}$$

for the complete description of the motion and the generalized forces in the fixed frame of reference. When these arrays are expressed in the local element-based frame, they are denoted $d\mathbf{p}_e$ and $\mathbf{q}_e$. In this notation the virtual work of the generalized forces at the end-points of the beam element takes the form

$$\delta V = \delta\mathbf{p}^T\mathbf{q}. \tag{5.4}$$

This relation will be used to express equilibrium of the element, using $\delta\mathbf{p}_e$ and $\mathbf{q}_e$ in the local frame, and to obtain the element tangent stiffness matrix, relating the global increments $d\mathbf{p}$ and $d\mathbf{q}$.

Fig. 5.2. Incremental rigid body modes: (a) translation, (b) rotation.

The idea of the co-rotating formulation is to separate the motion into two parts: a rigid body motion associated with the motion of the local frame of reference, and a deformation of the beam within this frame. The motion of the local frame of reference is described by the translation $\mathbf{u}^T = [u_1, u_2]$ of its origin and the rotation φ of the axes. The incremental rigid body motions corresponding to $d\mathbf{u}$ and $d\varphi$ are illustrated in Fig. 5.2.

The full description of the motion of the beam element requires six components, and when three components are used for the rigid body motion three are left for description of the deformation of the beam. These three components define three modes of deformation of the beam element. These modes can be selected in different ways, but it turns out to be convenient to use the set of 'natural deformation modes' illustrated in Fig. 5.3. The extension mode shown in Fig. 5.3(a) consists of an axial translation of magnitude $\frac{1}{2}du$ of the end-points A and B, increasing their distance by du. For a straight beam this corresponds to an extension of du. The symmetric bending mode shown in Fig. 5.3(b) is defined by a rotation $\frac{1}{2}d\varphi_s$ of the end-points, clockwise at A and counterclockwise at B. Finally, the anti-symmetric bending mode shown in Fig. 5.3(c) is defined by a counterclockwise rotation $\frac{1}{2}d\varphi_a$ of both end-points.

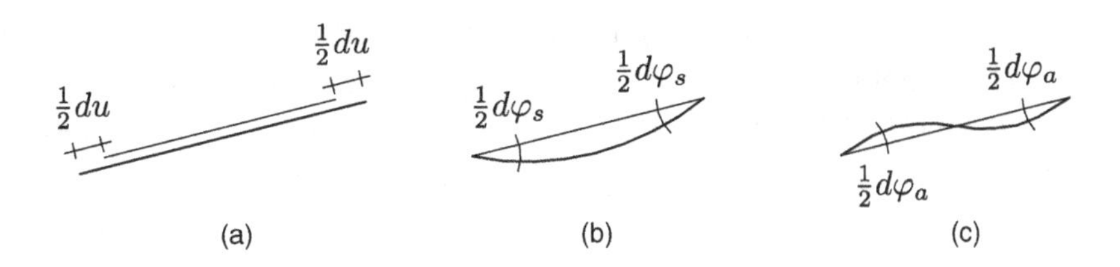

Fig. 5.3. Deformation modes: (a) extension, (b) symmetric bending, (c) anti-symmetric bending.

The six generalized force components of the beam element must satisfy three equilibrium equations. This leaves three generalized force components, conveniently selected as the generalized forces conjugate to the displacement components of the natural deformation modes of Fig. 5.3. The generalized forces corresponding to the natural deformation modes are illustrated in Fig. 5.4. The extension mode corresponds to a normal force N, shown in Fig. 5.4(a). The symmetric bending mode corresponds to a moment M_s at the end-points, clockwise at A and counterclockwise at B, Fig. 5.4(b). The anti-symmetric bending mode corresponds to a counterclockwise moment M_a at both end-points. It is easily seen that the generalized forces N and M_s are equilibrium systems, while the moment M_a in the anti-symmetric bending mode must be complemented by shear forces $Q = -2M_a/l$ as shown in Fig. 5.4(c), where l is the current distance between the end-points A and B.

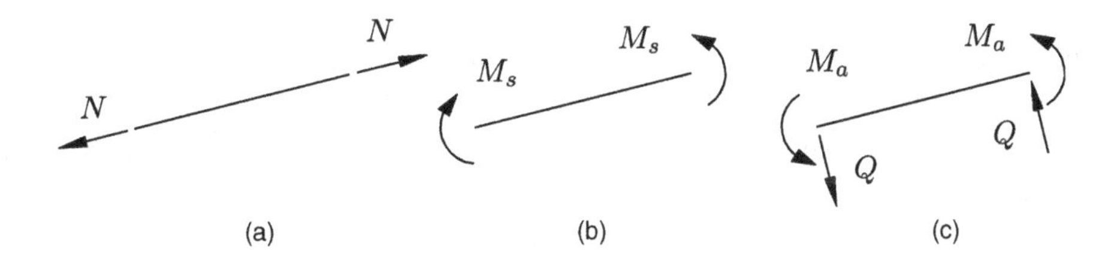

Fig. 5.4. Equilibrium force systems: (a) normal force, (b) constant moment, (c) constant shear.

It is also convenient to have a matrix notation for the components of the deformation modes and the corresponding equilibrium force systems, and therefore the following notation is introduced:

$$d\mathbf{v}^T = [\,du, d\varphi_s, d\varphi_a], \qquad \mathbf{t}^T = [\,N, M_s, M_a]. \tag{5.5}$$

In this notation the external virtual work of the generalized forces at the end-points of the beam element is

$$\delta V = \delta \mathbf{v}^T \mathbf{t}. \tag{5.6}$$

This relation is similar to (5.4) in the fixed frame of reference, but has only three components. It is seen that the factor $\frac{1}{2}$ in the definition of the deformation modes shown in Fig. 5.3 is necessary to make the generalized forces $\mathbf{t}$ conjugate to the incremental displacements $d\mathbf{v}$.

5.1.1 Co-rotation form of the tangent stiffness

The co-rotating formulation involves two transformations: one from the reduced set of internal variables $d\mathbf{v}$ and $\mathbf{t}$ to a complete set of variables $d\mathbf{p}_e$ and $\mathbf{q}_e$ in a coordinate system aligned with the element, and then a transformation of these components to the fixed frame of reference by a rotation. The rotation of components $\mathbf{q}_e$ in a local frame of reference to the fixed frame of reference is described by the rotation matrix

$$\mathbf{R} = \begin{bmatrix} \cos\varphi & -\sin\varphi & \\ \sin\varphi & \cos\varphi & \\ & & 1 \end{bmatrix}. \tag{5.7}$$

When using the combined format (5.3), the compound rotation matrix

$$\mathbf{R}_e = \begin{bmatrix} \mathbf{R} & \\ & \mathbf{R} \end{bmatrix} \tag{5.8}$$

is introduced to give the relation

$$\mathbf{q} = \mathbf{R}_e \, \mathbf{q}_e. \tag{5.9}$$

Clearly, this relation rotates the components at A and the components at B by the rotation matrix $\mathbf{R}$.

The first transformation, giving the full set of generalized forces in a local frame, is expressed as

$$\mathbf{q}_e = \mathbf{S}\,\mathbf{t}, \tag{5.10}$$

where the 6×3 transformation matrix $\mathbf{S}$ is

$$\mathbf{S} = \begin{bmatrix} \mathbf{S}_1 \\ \mathbf{S}_2 \end{bmatrix} = \begin{bmatrix} -1 & 0 & 0 \\ 0 & 0 & 2/l \\ 0 & -1 & 1 \\ 1 & 0 & 0 \\ 0 & 0 & -2/l \\ 0 & 1 & 1 \end{bmatrix}. \tag{5.11}$$

It follows from equality of the virtual work δV expressed in a set of full components by (5.4) and in internal components by (5.6) that

$$\delta V = \delta \mathbf{p}_e^T \mathbf{q}_e = \delta \mathbf{p}_e^T \mathbf{S}\,\mathbf{t} = \delta \mathbf{v}^T \mathbf{t}. \tag{5.12}$$

The last equality must hold for arbitrary internal 'stresses' $\mathbf{t}$, and the internal incremental displacements $d\mathbf{v}$ must therefore be related to the displacement components $d\mathbf{p}_e$ by

$$d\mathbf{v} = \mathbf{S}^T d\mathbf{p}_e = \mathbf{S}^T \mathbf{R}_e^T d\mathbf{p}. \tag{5.13}$$

Thus, the transformation matrix $\mathbf{S}$ serves to expand the reduced set of internal forces to full format by (5.10), while the transpose $\mathbf{S}^T$ serves to extract the modal deformation components from the full displacement representation by (5.13).

As discussed in Chapters 2 and 4, the standard procedure for obtaining the tangent stiffness consists in considering the increment of the virtual work used to express equilibrium. In the case of co-rotational elements only the external virtual work is used. In the present case this leads to consideration of the increment of the external virtual work $\delta V = \delta \mathbf{p}^T \mathbf{q}$. When calculating this increment the virtual displacement vector $\delta \mathbf{p}$ can be considered constant, and thus

$$d(\delta V) = d(\delta \mathbf{p}^T \mathbf{q}) = \delta \mathbf{p}^T d\mathbf{q}. \tag{5.14}$$

Thus, the virtual work increment $d(\delta V)$ is of the same form as the virtual work δV, when the generalized forces $\mathbf{q}$ are replaced by their increment $d\mathbf{q}$. This implies that the incremental equilibrium equations for $d\mathbf{q}$ are of the same form as those for the total forces $\mathbf{q}$. This is often taken for granted, but does not hold for general three-dimensional rotations, where the increment of the virtual rotation gives a separate contribution that must also be included in the incremental equilibrium equations as described in Section 5.2.

For the two-dimensional problems considered here the tangent stiffness can be derived from the increment of the generalized force relation

$$\mathbf{q} = \mathbf{R}_e \mathbf{S}\,\mathbf{t}. \tag{5.15}$$

The incremental relation involves a change in the internal forces $\mathbf{t}$, a change of $\mathbf{S}$ due to a change in l, and finally a change of $\mathbf{R}_e$ due to a change in φ:

$$d\mathbf{q} = \mathbf{R}_e \mathbf{S}\,d\mathbf{t} + \mathbf{R}_e\,d\mathbf{S}\,\mathbf{t} + d\mathbf{R}_e \mathbf{S}\,\mathbf{t}. \tag{5.16}$$

The change of the internal force vector is related to a change in the state of deformation, and it is therefore given by a relation of the form

$$d\mathbf{t} = \mathbf{K}_d\,d\mathbf{v}, \tag{5.17}$$

where $\mathbf{K}_d$ is the tangent stiffness matrix of the deformation modes of the element, expressed in the reduced 3×3 internal format. When this relation is substituted into the first term of (5.16), and $d\mathbf{v}$ is expressed by (5.13), the increment of the force takes the form

$$d\mathbf{q} = \mathbf{R}_e \Big[\mathbf{S}\,\mathbf{K}_d \mathbf{S}^T\, d\mathbf{p}_e + \big(d\mathbf{S} + \mathbf{R}_e^T d\mathbf{R}_e \mathbf{S} \big)\, \mathbf{t} \Big]. \qquad (5.18)$$

Multiplication with $\mathbf{R}_e^T$ gives the stiffness relation in terms of the full components in the local frame of reference,

$$d\mathbf{q}_e = \mathbf{S}\,\mathbf{K}_d \mathbf{S}^T\, d\mathbf{p}_e + \big(d\mathbf{S} + \mathbf{R}_e^T d\mathbf{R}_e \mathbf{S} \big)\, \mathbf{t}. \qquad (5.19)$$

The first term represents the stiffness due to the local deformation, while the second term containing the internal forces $\mathbf{t}$ represents the combined effect of the co-rotating frame of reference and the fact that the shear forces, determined from the moments by equilibrium, may change due to a change of the distance l between the end-points of the beam element.

The rotation and extension increments are expressed in terms of the displacement components in the local frame of reference,

$$d\varphi = (du_y^B - du_y^A)/l, \qquad dl = du_x^B - du_x^A. \qquad (5.20)$$

The local tangent stiffness relation can then be rearranged into the form

$$d\mathbf{q}_e = \mathbf{K}_e\, d\mathbf{p}_e, \qquad (5.21)$$

where the element stiffness matrix $\mathbf{K}_e$ is given by

$$\mathbf{K}_e = \mathbf{S}\,\mathbf{K}_d \mathbf{S}^T + \mathbf{K}_r. \qquad (5.22)$$

The first part is the stiffness of the deformation modes of the element, given by $\mathbf{K}_d$, while the second part $\mathbf{K}_r$ represents the combined effect of the co-rotating frame of reference and the change of the shear force due to changes in l. The co-rotation stiffness matrix is evaluated from the second term in (5.19), see Exercise 5.2. The result is conveniently given in block matrix format as

$$\mathbf{K}_r = \begin{bmatrix} \mathbf{K}_{11}^r & \mathbf{K}_{12}^r \\ \mathbf{K}_{21}^r & \mathbf{K}_{22}^r \end{bmatrix}, \qquad (5.23)$$

with

$$\mathbf{K}_{11}^r = \mathbf{K}_{22}^r = -\mathbf{K}_{12}^r = -\mathbf{K}_{21}^r = \frac{1}{l} \begin{bmatrix} 0 & -Q & 0 \\ -Q & N & 0 \\ 0 & 0 & 0 \end{bmatrix}. \qquad (5.24)$$

It is interesting to note that the stiffness matrix $\mathbf{K}_r$ includes the effect of extension dl on the shear force in addition to the effect of the rotation $d\varphi$

of the frame of reference. $\mathbf{K}_r$ would not be symmetric if the extension effect were omitted.

5.1.2 Element deformation stiffness

In the co-rotation procedure the element stiffness matrix $\mathbf{K}_e$ is composed of two parts: one from incremental deformation of the element represented in terms of the deformation modes by $\mathbf{K}_d$, and the matrix $\mathbf{K}_r$ caused by the co-rotation of the frame of reference. It is important to note that this separation is different from the previously used separation of the total stiffness into a constitutive and a geometric part. In the co-rotational formulation the geometric stiffness associated with the rotation of the local frame of reference and changes of the shear force due to extension are accounted for by $\mathbf{K}_r$. Whether or not there are additional geometric stiffness contributions in the deformation stiffness matrix $\mathbf{K}_d$ depends on the level of approximation of the beam model. At the lowest level is a linear beam theory without initial stress terms, the next level includes geometric stiffness contributions via assumed shape functions, and the last level incorporates a non-linear deformation model as discussed in Section 5.3.

Constitutive stiffness

The constitutive properties of the beam element are contained in the deformation stiffness matrix $\mathbf{K}_d$, introduced in (5.17) as the coefficient matrix of the incremental relation between the static and kinematic variables of the displacement modes,

$$\begin{bmatrix} dN \\ dM_s \\ dM_a \end{bmatrix} = \begin{bmatrix} & & \\ & \mathbf{K}_d & \\ & & \end{bmatrix} \begin{bmatrix} du \\ d\varphi_s \\ d\varphi_a \end{bmatrix}. \tag{5.25}$$

This relation involves the constitutive stiffness and may in addition include a geometric contribution. Here the constitutive contribution of a straight homogeneous elastic beam element is considered. In this case the stiffnesses of the deformation modes do not couple, whereby $\mathbf{K}_d$ becomes a diagonal matrix and each of the three deformation modes can be considered separately.

The first deformation mode is extension. For a straight homogeneous elastic beam the extension stiffness is

$$dN = \frac{EA}{l}\, du, \tag{5.26}$$

giving the first diagonal element of $\mathbf{K}_d$.

The bending modes are illustrated in Fig. 5.5. As the beam is assumed to be linear elastic, the total moments and angles are used for simplicity of notation. The stiffness of the symmetric deformation mode follows from evaluation of the complementary virtual work of a constant moment distribution

$$M_s \varphi_s = \int_0^l \frac{M(s)^2}{EI} ds = l \frac{M_s^2}{EI}, \tag{5.27}$$

corresponding to the incremental stiffness relation

$$dM_s = \frac{EI}{l} d\varphi_s. \tag{5.28}$$

This gives the second diagonal term of $\mathbf{K}_d$.

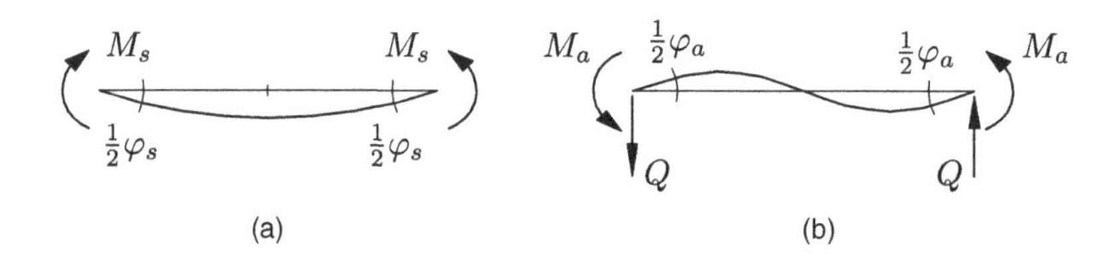

Fig. 5.5. Bending modes: (a) symmetric, (b) anti-symmetric.

The anti-symmetric bending mode is illustrated in Fig. 5.5(b). The static components are a bending moment of linear variation from $-M_a$ to M_a and a constant shear force

$$Q = -2M_a/l \tag{5.29}$$

determined from equilibrium. There is no transverse displacement at the ends, and complementary virtual work then gives

$$M_a \varphi_a = \int_0^l \Big(\frac{M(s)^2}{EI} + \frac{Q^2}{GA} \Big) ds = l \Big(\frac{1}{3} \frac{M_a^2}{EI} + \frac{Q^2}{GA} \Big). \tag{5.30}$$

When introducing the shear flexibility parameter

$$\psi_a = \frac{1}{1+\Phi}, \qquad \Phi = \frac{12\, EI}{l^2 GA}, \tag{5.31}$$

substitution of the shear force from (5.29) gives the incremental stiffness relation

$$dM_a = 3\psi_a \frac{EI}{l} d\varphi_a. \tag{5.32}$$

This is the final diagonal element in $\mathbf{K}_d$. Thus, the constitutive part of the stiffness matrix of the deformation modes is

$$\mathbf{K}_d = \frac{1}{l}\begin{bmatrix} EA & & \\ & EI & \\ & & 3\psi_a EI \end{bmatrix}. \tag{5.33}$$

This matrix enters the local element stiffness matrix $\mathbf{K}_e$ in the form $\mathbf{S}\mathbf{K}_d\mathbf{S}^T$, which may be evaluated numerically or explicitly, depending on programming taste.

In essence, the constitutive stiffness was calculated from the statics of the beam element. This procedure is easily generalized to non-homogeneous and curved elastic elements, see e.g. Krenk (1994).

Local geometric stiffness

In addition to the constitutive part of the deformation stiffness matrix $\mathbf{K}_e$ evaluated above there may be local geometric stiffness contributions. The geometric contribution to the stiffness of the local deformation modes can be evaluated under the simplifying assumption of vanishing shear strain, whereby the results become very simple and usually quite representative.

In the local xy-frame, bending of a beam with vanishing shear strain and with a normal force N is governed by the differential equation

$$(EI\, u_y'')'' - (N\, u_y')' = 0. \tag{5.34}$$

The corresponding internal virtual work is found by multiplication with δu_y, followed by integration by parts. The result is

$$\delta V_{\text{in}} = \int_0^l \Big(\delta u_y''\, EI\, u_y'' + \delta u_y'\, N\, u_y' \Big) ds. \tag{5.35}$$

In the absence of three-dimensional rotation effects, the incremental form follows directly by $d(\delta u_y) = 0$ as

$$d(\delta V_{\text{in}}) = \int_0^l \Big(\delta u_y''\, EI\, du_y'' + \delta u_y'\, N\, du_y' \Big) ds. \tag{5.36}$$

The local stiffness matrix follows from this expression by substitution of suitable shape function representations of δu_y and du_y.

It is convenient to represent the arc-length s by the non-dimensional coordinate ξ via

$$s = \tfrac{1}{2} l(1+\xi), \qquad -1 \le \xi \le 1. \tag{5.37}$$

In terms of this coordinate the symmetric bending mode is

$$du_y = -\tfrac{1}{8} l(1-\xi^2)\, d\varphi_s, \tag{5.38}$$

while the anti-symmetric mode is

$$du_y = -\tfrac{1}{8}l(1-\xi^2)\xi\, d\varphi_a. \tag{5.39}$$

The local stiffness matrix follows from substitution of the representations (5.38) and (5.39) into the incremental virtual work $d(\delta V)$ given by (5.36). The displacement representations do not couple as they are even and odd functions of ξ, respectively, and the stiffness of symmetric and anti-symmetric bending deformation modes can therefore be calculated independently. It is easily verified – see e.g. Exercise 5.3 – that the constitutive stiffness contribution from (5.36) gives the special case $\psi_a = 1$ of the result already obtained by a direct method including shear flexibility in (5.33).

The geometric stiffness of the symmetric bending mode follows from substitution of the derivative of u_y, given by (5.38), into the second term of the integrand in (5.36), whereby

$$k_{ss}^d = \tfrac{1}{2}l\int_{-1}^{1}(\tfrac{1}{2}\xi)\,N\,(\tfrac{1}{2}\xi)\,d\xi = \tfrac{1}{12}\,l\,N. \tag{5.40}$$

The geometric stiffness of the anti-symmetric mode follows similarly from substitution of the derivative of u_y, given by (5.39), as

$$k_{aa}^d = \tfrac{1}{2}l\int_{-1}^{1}\tfrac{1}{4}(1-3\xi^2)\,N\,\tfrac{1}{4}(1-3\xi^2)\,d\xi = \tfrac{1}{20}\,l\,N. \tag{5.41}$$

Thus the geometric stiffness from the deformation bending modes with cubic interpolation is

$$\mathbf{K}_d = l\begin{bmatrix} 0 & & \\ & \tfrac{1}{12}N & \\ & & \tfrac{1}{20}N \end{bmatrix}. \tag{5.42}$$

This matrix should be added to the constitutive part (5.33) to give the full stiffness associated with the deformation modes.

5.1.3 Total tangent stiffness

In the co-rotating format the element stiffness matrix in the local frame of reference is given by (5.22). In this format the stiffness of the deformation modes, with a constitutive and possibly a geometric part, is transformed into the local frame, and a contribution from the co-rotation of the local frame is added. It is simple to program this basic relation directly, using the stiffness matrix of the deformation modes $\mathbf{K}_d$, the rotation stiffness matrix $\mathbf{K}_r$, and the transformation matrix $\mathbf{S}$ relating the components of the deformation modes to a full set of variables in the local element frame. However, the transformation matrix is quite simple and involves many zero entries, and the full element stiffness matrix is therefore easily evaluated analytically by matrix multiplication. In addition to its potential use in a computer program, this form also facilitates comparison with other formulations and derivations.

The local element stiffness matrix has the block matrix format

$$\mathbf{K}_e = \begin{bmatrix} \mathbf{K}_{11}^e & \mathbf{K}_{12}^e \\ \mathbf{K}_{21}^e & \mathbf{K}_{22}^e \end{bmatrix} \tag{5.43}$$

already introduced for the rotation matrix in (5.23). When using the block matrix format (5.11) of the transformation matrix $\mathbf{S}$, the co-rotation format (5.22) can be written in sub-matrix format as

$$\mathbf{K}_{ij}^e = \mathbf{S}_i \mathbf{K}_d \mathbf{S}_j^T + \mathbf{K}_{ij}^r \tag{5.44}$$

with $i = 1, 2$ and $j = 1, 2$. The sub-matrices $\mathbf{K}_{ij}^r$ were given in (5.24), and the stiffness matrix of the modes $\mathbf{K}_d$ is the sum of the constitutive part (5.33) and the geometric part (5.42).

When carrying out the matrix multiplications in (5.44), the constitutive part of the element stiffness matrix is found to be

$$\mathbf{K}_{11}^e = \frac{1}{l^3} \begin{bmatrix} EAl^2 & 0 & 0 \\ 0 & 12\psi_a EI & 6\psi_a EIl \\ 0 & 6\psi_a EIl & (3\psi_a+1)EIl^2 \end{bmatrix}, \tag{5.45}$$

$$\mathbf{K}_{22}^e = \frac{1}{l^3} \begin{bmatrix} EAl^2 & 0 & 0 \\ 0 & 12\psi_a EI & -6\psi_a EIl \\ 0 & -6\psi_a EIl & (3\psi_a+1)EIl^2 \end{bmatrix}, \tag{5.46}$$

$$\mathbf{K}_{12}^e = \mathbf{K}_{21}^{eT} = \frac{1}{l^3} \begin{bmatrix} -EAl^2 & 0 & 0 \\ 0 & -12\psi_a EI & 6\psi_a EIl \\ 0 & -6\psi_a EIl & (3\psi_a-1)EIl^2 \end{bmatrix}. \tag{5.47}$$

Similarly the sub-matrices of the geometric stiffness matrix of the element are

$$\mathbf{K}_{11}^e = \frac{1}{l} \begin{bmatrix} 0 & -Q & 0 \\ -Q & \frac{6}{5}N & \frac{1}{10}Nl \\ 0 & \frac{1}{10}Nl & \frac{2}{15}Nl^2 \end{bmatrix}, \tag{5.48}$$

$$\mathbf{K}_{22}^e = \frac{1}{l} \begin{bmatrix} 0 & -Q & 0 \\ -Q & \frac{6}{5}N & -\frac{1}{10}Nl \\ 0 & -\frac{1}{10}Nl & \frac{2}{15}Nl^2 \end{bmatrix}, \tag{5.49}$$

$$\mathbf{K}_{12}^e = \mathbf{K}_{21}^{eT} = \frac{1}{l} \begin{bmatrix} 0 & Q & 0 \\ Q & -\frac{6}{5}N & \frac{1}{10}Nl \\ 0 & -\frac{1}{10}Nl & -\frac{1}{30}Nl^2 \end{bmatrix}, \tag{5.50}$$

where the shear force is determined from the antisymmetric bending moment as

$$Q = -2M_a/l. \tag{5.51}$$

It is seen by comparison with (5.24) that the deformation modes only contribute modestly to the geometric stiffness matrix.

5.1.4 Finite element implementation

In the co-rotation format the element properties are first obtained in a local element-based frame of reference and are subsequently transformed into the global frame. For two-dimensional co-rotating beam element problems the formation of element forces and stiffness matrix can be arranged as described in the following. In order to illustrate the procedure in the simplest possible setting a total displacement description is used, and the beam element is assumed to be homogeneous and linear elastic.

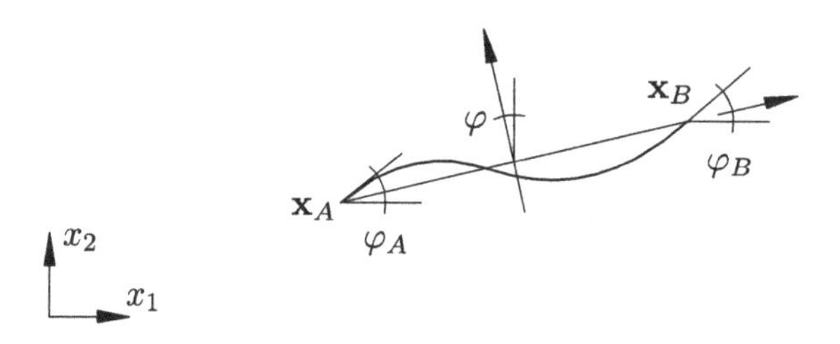

Fig. 5.6. Deformed beam element in two-dimensional problem.

The current state of the beam element is illustrated in Fig. 5.6, showing the current coordinate vectors $\mathbf{x}_A$, $\mathbf{x}_B$ and rotation angles φ_A, φ_A at the element end-points. The element angle φ with the global axes is conveniently calculated from the trigonometric relations

$$\cos\varphi = (x_1^B - x_1^A)/l, \qquad \sin\varphi = (x_2^B - x_2^A)/l \tag{5.52}$$

by using the relation for $\tan(\frac{1}{2}\varphi)$, whereby

$$\varphi = 2\arctan\left(\frac{1-\cos\varphi}{\sin\varphi}\right) = 2\arctan\left(\frac{l-\Delta x_1}{\Delta x_2}\right). \tag{5.53}$$

This expression is singular for $\Delta x_2 = 0$, where $\varphi = 0$ for $\Delta x_1 = l$ and $\varphi = \pi$ for $\Delta x_1 = -l$.

The next step is to calculate the kinematic measures of deformation: the elongation u, the angle φ_s of the symmetric deformation mode, and the angle

ALGORITHM 5.1. Two-dimensional co-rotating beam elements.

Element angle:

$$\varphi = 2\arctan\left(\frac{l-\Delta x_1}{\Delta x_2}\right) \qquad (5.53)$$

Deformation parameters $\mathbf{v}$:

$$u = l - l_0$$
$$\varphi_s = \varphi_B - \varphi_A \qquad (5.54)$$
$$\varphi_a = \varphi_B + \varphi_A - 2(\varphi - \varphi_0)$$

Modulus for absolute angle:

$$\varphi_a = \underset{2\pi}{\mathrm{mod}}(\varphi_a + \pi) - \pi \qquad (5.55)$$

Internal 'stresses':

$$\mathbf{t} = \mathbf{K}_d\,\mathbf{v} \qquad (5.33)$$

Element forces in global frame:

$$\mathbf{q} = \mathbf{R}_e\mathbf{S}\,\mathbf{t} \qquad (5.9), (5.10)$$

Element tangent stiffness in global frame:

$$\mathbf{K}_{ij} = \mathbf{R}\,\mathbf{K}^e_{ij}\mathbf{R}^T \qquad (5.58)$$

φ_a of the anti-symmetric deformation mode. These quantities are calculated from the relations

$$u = l - l_0, \quad \varphi_s = \varphi_B - \varphi_A, \quad \varphi_a = \varphi_B + \varphi_A - 2(\varphi - \varphi_0). \qquad (5.54)$$

In order to be robust for arbitrary angles, the expression for the anti-symmetric rotation should be calculated using the modulus function

$$\varphi_a := \underset{2\pi}{\mathrm{mod}}(\varphi_a + \pi) - \pi. \qquad (5.55)$$

The modulus function places the argument in the interval $[0, 2\pi[$, and the last term re-establishes symmetry with respect to zero. Omission of this step may lead to problems, when the beam element has rotated $\pm\pi, \pm 2\pi, \ldots$

The internal 'stresses' of the beam element are the normal force N, the symmetric moment M_s, and the anti-symmetric moment M_a. For the present linear theory the total relations follow from (5.25) and (5.33) as

$$N = \frac{EA}{l}\,u, \quad M_s = \frac{EI}{l}\,\varphi_s, \quad M_a = 3\psi_a\frac{EI}{l}\,\varphi_a. \qquad (5.56)$$

In the factors no distinction has been made between current length l and initial length l_0. With the internal 'stresses' $\mathbf{t}$ determined by (5.56), the full set of element forces $\mathbf{q}_e$ is determined from (5.10) and the local element stiffness matrix $\mathbf{K}_e$ from (5.43) and (5.45)–(5.51), or the underlying product format (5.22).

The final step before assembling the element contributions into the global matrices is the transformation from the local element frame to the global frame of reference. This is accomplished by use of the compound rotation matrix $\mathbf{R}_e$ given by (5.7)–(5.8). The element force vector is transformed by (5.9), while the element matrix is transformed by

$$\mathbf{K} = \mathbf{R}_e \mathbf{K}_e \mathbf{R}_e^T . \tag{5.57}$$

This transformation contains many zero entries, and the diagonal block matrix structure of the transformation matrix $\mathbf{R}_e$ implies that the sub-matrices of the stiffness matrix can be transformed directly by the rotation matrix $\mathbf{R}$,

$$\mathbf{K}_{ij} = \mathbf{R}\, \mathbf{K}_{ij}^e\, \mathbf{R}^T \tag{5.58}$$

with $i = 1,2$ and $j = 1,2$. The element sub-matrices were given in explicit form in (5.45)–(5.51).

The computational procedure for obtaining the element forces and the element tangent stiffness matrix in the global frame is summarized in Algorithm 5.1.

EXAMPLE 5.1. BENDING OF A BEAM. A simple example that illustrates the beam element's ability to handle large rotations is the roll-up of a cantilever beam by a bending moment applied to its free end. Ideally the beam will curve into a circular arc with curvature $\kappa = 1/R = M/EI$. Thus, the beam ends meet, when $M = 2\pi EI/L$. In the graphics of Fig. 5.7 the 10 elements are plotted as straight, although they are curved due to the deformation described by the shape functions. However, in the element developed here the distance between the two end-points of the element is not changed by

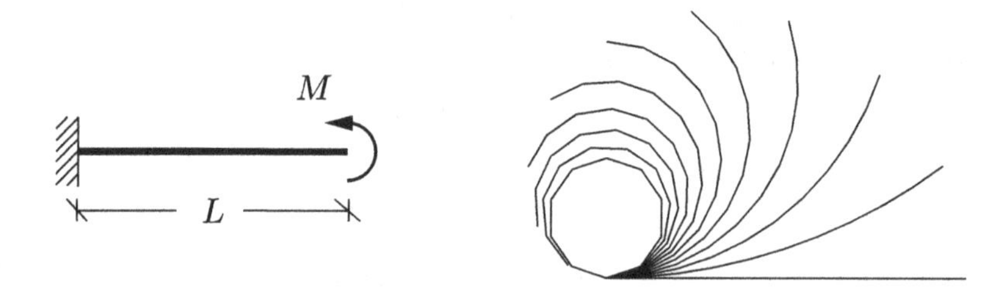

Fig. 5.7. Roll-up of cantilever beam.

curvature, and thus the nodes are located at the apexes of a polygon with sides corresponding to the initial length of the beam.

The rotation of the end node is proportional to the load, and thus Newton–Raphson iteration is fully adequate. The full circle is reached in 10 steps, and with a relative tolerance on the residual of 10^{-6} each step required six equilibrium iterations. In this example the only internal force is the moment M, and thus there is no contribution from geometric stiffness.

EXAMPLE 5.2. LARGE DEFORMATION OF CANTILEVER. An example that illustrates very large deformation as well as the influence of the geometric stiffness matrix is a cantilever beam of length L with a transverse force P applied at the free end, shown in Fig. 5.8(a). An accurate table of the displacement and rotation of the end-point has been obtained by Mattiasson (1981) using elliptic integrals. The force is assumed to retain its direction while following the position of the end-point. The load is described by the non-dimensional parameter PL^2/EI, and Fig. 5.8(b) shows the non-dimensional displacements u/l, v/L and φ at the end of the cantilever as functions of the load. The load is applied in 20 equal steps leading to a final load of $PL^2/EI = 10$. At this load level the end-point has attained a horizontal displacement of more than half the length of the beam.

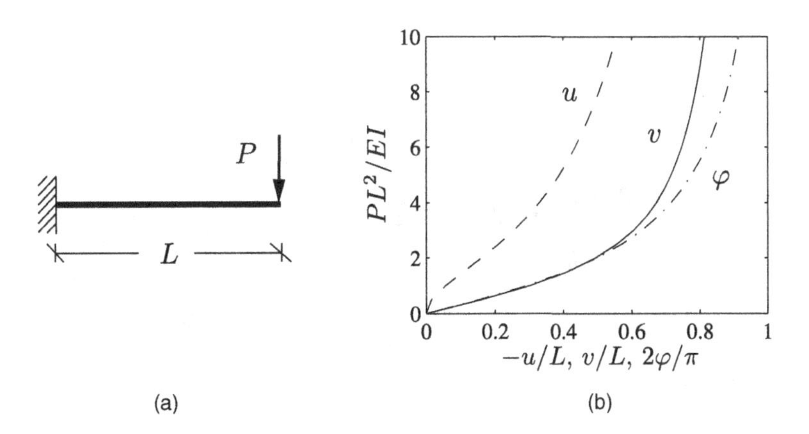

Fig. 5.8. Cantilever with conservative transverse end force.

Four models were used with 2, 4, 6, and 8 identical elements, respectively. With a relative tolerance on the residual of 10^{-6} the average number of iterations was around 7.5. The results at the final load level and half of this are shown in Table 5.1 with the analytical results in the last row. It is seen that good accuracy, even in this highly deformed state, is obtained with four elements. In this problem the geometric stiffness matrix is important, and

Table 5.1. *Cantilever beam with conservative end force.*

	$PL^2/EI = 5$			$PL^2/EI = 10$		
N_{elem}	$-u/L$	v/L	φ	$-u/L$	v/L	φ
2	0.38941	0.73858	1.21137	0.56761	0.84771	1.47759
4	0.38711	0.71915	1.22314	0.55572	0.81878	1.44037
6	0.38732	0.71607	1.21873	0.55506	0.81400	1.43460
8	0.38744	0.71506	1.21723	0.55498	0.81247	1.43268
–	0.38763	0.71379	1.21537	0.55500	0.81061	1.43029

the convergence is lost around half the final load if the geometric stiffness matrix is omitted in the iterations.

Example 5.3. Shallow angle beam. The angle beam shown in Fig. 5.9(a) includes an additional effect, the shortening of the beam due to bending. This is not included in the present beam element formulation, and it is therefore necessary to use several beam elements for each of the two straight beams. In this way the finite displacement of the nodes leads to a representation of the shortening effect.

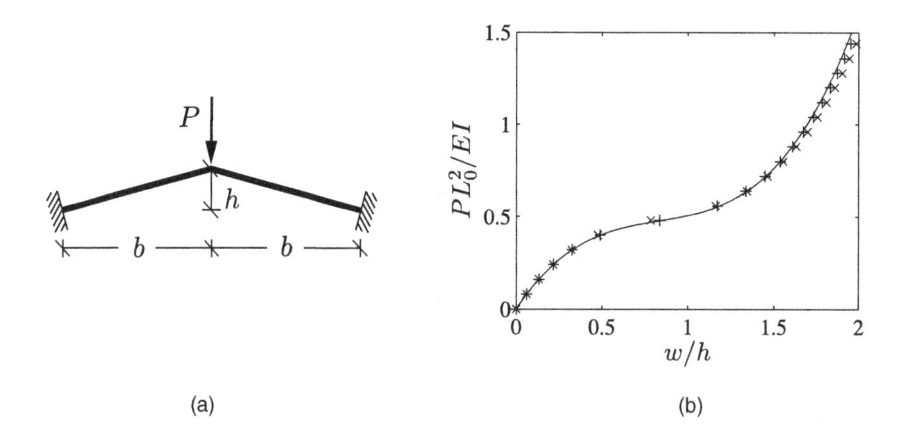

Fig. 5.9. Angle beam with clamped supports: (×) 5 elements, (+) 10 elements.

An approximate analytical solution based on second-order theory has been obtained by Williams (1964). A more direct solution can be obtained by considering each of the beams as a beam-column with finite displacement connecting the support and the apex in a state of anti-symmetric bending. The corresponding beam-column solution is well known, see e.g. Timoshenko and

Gere (1961) or Krenk (2001). The solution depends on the inclination h/b of the two beams and their slenderness, represented by the non-dimensional parameter L_0^2A/I. Figure 5.9(b) shows the non-dimensional vertical load PL_0^2/EI as a function of the vertical displacement w/h of the central node for $h/b = 0.024$ and $L_0^2A/I = 3.0 \times 10^4$. Two finite element models with straight beam elements were made, using 5 and 10 elements, respectively, for each straight beam. With equal load steps, Newton–Raphson iteration, and a relative tolerance on the residual of 10^{-6}, the average numbers of iterations were 5.6 and 4.8, respectively. It is seen that the error is negligible for $w \lesssim h$, but for larger displacements even the model with 10 elements per beam exhibits some error. This indicates the usefulness of elements in which the shortening due to bending is included in the individual element, as discussed in Section 5.3. For more slender beams or larger inclination h/b a snap-through mechanism will develop, and bifurcation into a non-symmetric deformation mode may also occur. These phenomena can be computed by the more general path-tracing methods described in Chapter 8.

5.2 Co-rotating beams in three dimensions

The derivation of the co-rotating beam element format for two-dimensional problems presented above is attractive because it only requires a constitutive relation for the deformation modes of the element, and a relation between these modes and the global set of element forces. Thus, finite rotations are accounted for without the need for a fully non-linear beam theory. This attractive feature is due to the fact that the tangent stiffness is derived from the external virtual work of the element, and not from the internal work of a fully non-linear theory as in the beam theory of Chapter 4. Furthermore, the derivation was arranged in such a way that only incremental relations between internal and global displacement components were needed. When working with finite rotations in three dimensions it is important to note the difference between a small rotation represented by the pseudo-vector $\delta\bar{\boldsymbol{\varphi}}$ and the corresponding set of component increments $\delta\boldsymbol{\varphi}$ of the total rotation $\boldsymbol{\varphi}$. This difference was discussed in detail in Section 3.3, where it was demonstrated that it is indeed possible to work directly with the small rotation pseudo-vector $\delta\bar{\boldsymbol{\varphi}}$, provided that the increment of the variation $d(\delta\bar{\boldsymbol{\varphi}})$ is properly accounted for by use of (3.30). When this term is included the need for an explicit parametrization of the finite rotation, used e.g. by Pacoste and Eriksson (1997) and Battini (2002), is avoided, and the procedure

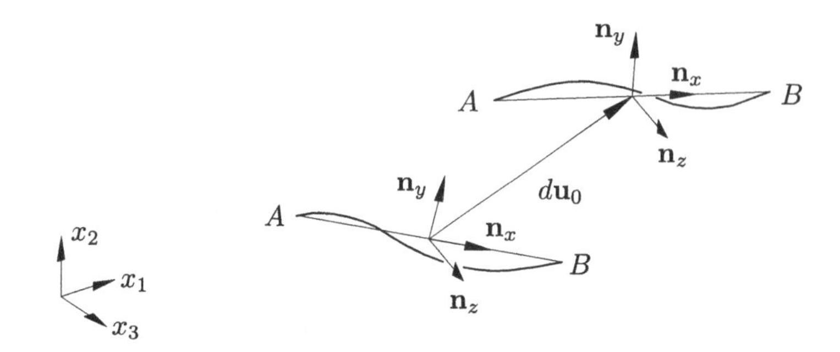

Fig. 5.10. Three-dimensional motion of beam in co-rotating frame.

of the previous section for two-dimensional beam elements can be extended directly.

Figure 5.10 shows a beam element in three-dimensional space described by the fixed frame of reference $\{x_1, x_2, x_3\}$. The beam is described in a local co-rotating frame of reference $\{x, y, z\}$ with base vectors $\{\mathbf{n}_x, \mathbf{n}_y, \mathbf{n}_z\}$. The x-axis passes through the end-points A and B of the beam element. The y- and z-axes are defined by the mean rotation of A and B as described in detail later. The incremental motion is described in the global frame of reference by the array

$$d\mathbf{p}^T = [\, d\mathbf{u}_A^T, d\bar{\boldsymbol{\varphi}}_A^T, d\mathbf{u}_B^T, d\bar{\boldsymbol{\varphi}}_B^T \,] \tag{5.59}$$

and the corresponding conjugate element forces are

$$\mathbf{q}^T = [\, \mathbf{f}_A^T, \mathbf{m}_A^T, \mathbf{f}_B^T, \mathbf{m}_B^T \,]. \tag{5.60}$$

The components of $d\mathbf{p}$ and $\mathbf{q}$ refer to the fixed frame of reference. In terms of these components the external virtual work of the beam element is

$$\delta V = \delta\mathbf{p}^T\mathbf{q} = \sum_{A,B} \Big(\delta\mathbf{u}_*^T \, \mathbf{f}_* + \delta\bar{\boldsymbol{\varphi}}_*^T \, \mathbf{m}_* \Big). \tag{5.61}$$

The increment of this relation is used to obtain the tangent stiffness.

The components of the displacement increments and the element forces in the local co-rotating frame are denoted $d\mathbf{p}_e$ and $\mathbf{q}_e$. The current orientation of the local frame of reference is defined by the rotation matrix $\mathbf{R}$. The relation between the local and global components is then given by the compound rotation matrix

$$\mathbf{R}_e = \begin{bmatrix} \mathbf{R} & & & \\ & \mathbf{R} & & \\ & & \mathbf{R} & \\ & & & \mathbf{R} \end{bmatrix} \tag{5.62}$$

as

$$d\mathbf{p} = \mathbf{R}_e d\mathbf{p}_e, \qquad \mathbf{q} = \mathbf{R}_e \mathbf{q}_e. \tag{5.63}$$

While it is necessary to have an expression for the rotation $\mathbf{R}$ of the local frame relative to the fixed frame in order to transform force components and stiffness matrix from the local to the global frame, the expressions for both local element forces and the local element stiffness matrix are independent of $\mathbf{R}$.

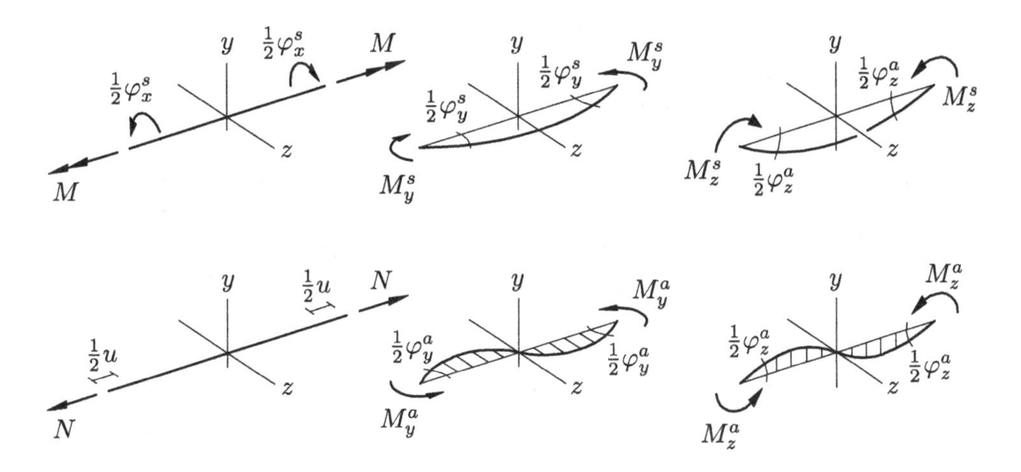

Fig. 5.11. The six natural deformation modes of a beam element.

The six natural deformation modes of a symmetric beam element are illustrated in Fig. 5.11. The top row shows three modes with constant moment, while the bottom row shows three modes with constant internal force. The constant moment modes are: a constant torsion moment M corresponding to opposing incremental angles of twist $\pm\frac{1}{2}d\varphi_x^s$, a constant moment produced by symmetric end moments M_y^s with incremental end-point rotations $\pm\frac{1}{2}d\varphi_y^s$, and similar symmetric moments M_z^s with incremental end-point rotations $\pm\frac{1}{2}d\varphi_z^s$. The constant internal force modes are: a constant normal force N corresponding to the incremental extension du, opposing bending moments $\pm M_y^a$ corresponding to incremental end-point rotations $\frac{1}{2}d\varphi_y^a$, and opposing bending moments $\pm M_z^a$ corresponding to incremental end-point rotations $\frac{1}{2}d\varphi_z^a$. This corresponds to the incremental deformation mode vector

$$d\mathbf{v}^T = [\, d\varphi_x^s, d\varphi_y^s, d\varphi_z^s, du, d\varphi_y^a, d\varphi_z^a \,] \tag{5.64}$$

and the conjugate internal 'stress' vector

$$\mathbf{t}^T = [\, M, M_y^s, M_z^s, N, M_y^a, M_z^a \,]. \tag{5.65}$$

In later computations it will be convenient to use also the shear forces

$$Q_y = -2M_z^a/l, \qquad Q_z = 2M_y^a/l \tag{5.66}$$

following from the equilibrium conditions.

The relation between the internal deformation variables and the full set of element variables is conveniently expressed in terms of a set of unit vectors $\{\mathbf{n}_x, \mathbf{n}_y, \mathbf{n}_z\}$ constituting the base vectors of the local $\{x, y, z\}$ co-ordinate system. When using these vectors it is seen directly from Fig. 5.11 that the element forces can be expressed as

$$\begin{bmatrix} \mathbf{f}_A \\ \mathbf{m}_A \\ \mathbf{f}_B \\ \mathbf{m}_B \end{bmatrix} = \underbrace{\begin{bmatrix} \mathbf{0} & \mathbf{0} & \mathbf{0} & -\mathbf{n}_x & -2\mathbf{n}_z/l & 2\mathbf{n}_y/l \\ -\mathbf{n}_x & -\mathbf{n}_y & -\mathbf{n}_z & \mathbf{0} & \mathbf{n}_y & \mathbf{n}_z \\ \mathbf{0} & \mathbf{0} & \mathbf{0} & \mathbf{n}_x & 2\mathbf{n}_z/l & -2\mathbf{n}_y/l \\ \mathbf{n}_x & \mathbf{n}_y & \mathbf{n}_z & \mathbf{0} & \mathbf{n}_y & \mathbf{n}_z \end{bmatrix}}_{\mathbf{S}} \begin{bmatrix} M \\ M_y^s \\ M_z^s \\ N \\ M_y^a \\ M_z^a \end{bmatrix}. \tag{5.67}$$

Thus the element forces $\mathbf{q}$ are given in terms of the internal 'stresses' by

$$\mathbf{q} = \mathbf{S}\,\mathbf{t}, \tag{5.68}$$

where the 12×6 transformation matrix $\mathbf{S}$ is defined by (5.67). This relation represents the global components $\mathbf{q}$ if the base vectors $\mathbf{n}_x, \mathbf{n}_y, \mathbf{n}_z$ are represented in the global frame, and the local components $\mathbf{q}_e$ if the base vectors $\mathbf{n}_x, \mathbf{n}_y, \mathbf{n}_z$ are represented in the local frame as $\mathbf{n}_x^T = [1, 0, 0]$, etc. This means that the rotation of the element is accounted for directly via the rotation of the unit vectors $\mathbf{n}_x, \mathbf{n}_y, \mathbf{n}_z$. This simplifies the derivation of the tangent stiffness, because an incremental rotation of a vector can be represented by a vector cross product as discussed in Chapter 3.

5.2.1 Co-rotation form of the tangent stiffness

In the co-rotation context the tangent stiffness is derived from the increment of the external virtual work of the element, given by (5.61). When allowance is made for the increment of the variation of the rotations $\delta\bar{\varphi}_A$ and $\delta\bar{\varphi}_B$, this variation is

$$d(\delta V) = \delta\mathbf{p}^T d\mathbf{q} + d(\delta\bar{\varphi}_A)^T \mathbf{m}_A + d(\delta\bar{\varphi}_B)^T \mathbf{m}_B. \tag{5.69}$$

Note that in this relation the transpose is just an alternative way of writing a scalar product. The tangent stiffness matrix is identified by writing this in the form

$$d(\delta V) = \delta\mathbf{p}^T \mathbf{K}\, d\mathbf{p}. \tag{5.70}$$

For ease of notation the fixed coordinate system is temporarily assumed to coincide with the element coordinate system at the start of the increment.

The element stiffness matrix $\mathbf{K}_e$ will be described in the block matrix format

$$\mathbf{K}_e = \begin{bmatrix} \mathbf{K}_{11}^e & \mathbf{K}_{12}^e & \mathbf{K}_{13}^e & \mathbf{K}_{14}^e \\ \mathbf{K}_{21}^e & \mathbf{K}_{22}^e & \mathbf{K}_{23}^e & \mathbf{K}_{24}^e \\ \mathbf{K}_{31}^e & \mathbf{K}_{32}^e & \mathbf{K}_{33}^e & \mathbf{K}_{34}^e \\ \mathbf{K}_{41}^e & \mathbf{K}_{42}^e & \mathbf{K}_{43}^e & \mathbf{K}_{44}^e \end{bmatrix}, \tag{5.71}$$

where $\mathbf{K}_{ij}^e$ are 3×3 sub-matrices. A similar block matrix format is used for the rotation part $\mathbf{K}_r$ of the stiffness matrix. It follows from the formula (3.28) for the increment of the rotation variation

$$d(\delta\bar{\varphi}) = -\tfrac{1}{2}\,\delta\bar{\varphi} \times d\bar{\varphi} \tag{5.72}$$

that the two last terms of (5.69) give contributions $\mathbf{K}_{22}^r$ and $\mathbf{K}_{44}^r$, respectively, of the form

$$\mathbf{K}_{22}^r = \frac{1}{2}\begin{bmatrix} 0 & -m_z^A & m_y^A \\ m_z^A & 0 & -m_x^A \\ -m_y^A & m_x^A & 0 \end{bmatrix}, \quad \mathbf{K}_{44}^r = \frac{1}{2}\begin{bmatrix} 0 & -m_z^B & m_y^B \\ m_z^B & 0 & -m_x^B \\ -m_y^B & m_x^B & 0 \end{bmatrix}. \tag{5.73}$$

The remaining contributions to the stiffness matrix can be calculated directly from $d\mathbf{q}$.

The increment of the element forces $\mathbf{q}$ in the fixed frame of reference is obtained by taking the increment of (5.68),

$$d\mathbf{q} = \mathbf{S}\, d\mathbf{t} + (d\mathbf{S}_{dl} + d\mathbf{S}_{d\bar{\varphi}})\,\mathbf{t}. \tag{5.74}$$

In this expression the first term corresponds to change in the internal 'stresses' due to a change in the deformation modes of the beam, while the last term accounts for a change of element length dl and element orientation $d\bar{\varphi}$. Although seemingly innocent, it is important to note that this step does not account for the effect of the deformation modes for constant internal 'stresses'. However, there is no means of representing this effect without a finite deformation model of the beam, so it is left out for the moment to be clarified at the end of the derivation of the tangent stiffness.

The change in internal 'stresses' is governed by an incremental relation of the form

$$d\mathbf{t} = \mathbf{K}_d \, d\mathbf{v}, \tag{5.75}$$

where $\mathbf{K}_d$ is the tangent stiffness matrix of the deformation modes of the element, expressed in the reduced 6×6 internal format. As in the two-dimensional case a virtual work argument shows that

$$d\mathbf{v} = \mathbf{S}^T d\mathbf{p}, \tag{5.76}$$

where $\mathbf{S}$ and $\mathbf{p}$ are either both local or global. Substitution of this into (5.74) gives the element stiffness matrix in the form

$$\mathbf{K}_e = \mathbf{S}\,\mathbf{K}_d\mathbf{S}^T + \mathbf{K}_r, \tag{5.77}$$

where the 'rotation' matrix $\mathbf{K}_r$ contains the contributions (5.73) and the contributions from change of length and orientation given by the last term in (5.74). The format (5.77) is the same as that of the two-dimensional case, but here $\mathbf{K}_r$ contains additional contributions from the increment of the variational rotations $\delta\boldsymbol{\varphi}_A$ and $\delta\boldsymbol{\varphi}_B$ given by (5.73).

The effect of a change of length dl follows by differentiation of (5.67) as

$$d\mathbf{S}_{dl}\,\mathbf{t} = -\frac{2}{l^2}\begin{bmatrix} -\mathbf{n}_z & \mathbf{n}_y \\ \mathbf{0} & \mathbf{0} \\ \mathbf{n}_z & -\mathbf{n}_y \\ \mathbf{0} & \mathbf{0} \end{bmatrix}\begin{bmatrix} M_y^a \\ M_z^a \end{bmatrix} dl. \tag{5.78}$$

The shear force components Q_y and Q_z are now introduced from (5.66) to give

$$d\mathbf{S}_{dl}\,\mathbf{t} = \frac{dl}{l}\begin{bmatrix} Q_y\mathbf{n}_y + Q_z\mathbf{n}_z \\ \mathbf{0} \\ -Q_y\mathbf{n}_y - Q_z\mathbf{n}_z \\ \mathbf{0} \end{bmatrix} = \frac{dl}{l}\begin{bmatrix} \mathbf{Q} \\ \mathbf{0} \\ -\mathbf{Q} \\ \mathbf{0} \end{bmatrix}, \tag{5.79}$$

where the shear force vector has been introduced as $\mathbf{Q} = Q_y\mathbf{n}_y + Q_z\mathbf{n}_z$. When the beam extension is introduced as $dl = du_x^B - du_x^A$, the elongation is seen to contribute to the first and seventh columns of the element stiffness matrix via the sub-matrix contributions

$$\mathbf{K}_{11}^r = \mathbf{K}_{33}^r = -\mathbf{K}_{13}^r = -\mathbf{K}_{31}^r = -\frac{1}{l}\begin{bmatrix} 0 & 0 & 0 \\ Q_y & 0 & 0 \\ Q_z & 0 & 0 \end{bmatrix}. \tag{5.80}$$

The shear force is seen to constitute the first column of these sub-matrices.

The final contribution to the element stiffness matrix is from the rotation

$d\bar{\boldsymbol{\varphi}}$ of the local frame, denoted $d\mathbf{S}_{d\bar{\varphi}}\mathbf{t}$ in (5.74). This implies that each of the unit vectors in the block format (5.67) of the transformation matrix $\mathbf{S}$ is rotated. However, rotating each of the base vectors while retaining their coefficients corresponds to rotating the resulting vectors. Thus, the contribution from rotation may be expressed directly as

$$d\mathbf{S}_{d\bar{\varphi}}\,\mathbf{t} \;=\; \begin{bmatrix} d\bar{\boldsymbol{\varphi}} \times \mathbf{f}_A \\ d\bar{\boldsymbol{\varphi}} \times \mathbf{m}_A \\ d\bar{\boldsymbol{\varphi}} \times \mathbf{f}_B \\ d\bar{\boldsymbol{\varphi}} \times \mathbf{m}_B \end{bmatrix}. \tag{5.81}$$

The computation is similar for each of the four terms. It is illustrated by the force vector $\mathbf{f}_B$. The vector $\mathbf{f}_B$ is expressed in an axial and a transverse component

$$\mathbf{f}_B \;=\; N\,\mathbf{n}_x \,+\, \mathbf{Q}, \tag{5.82}$$

where N is the normal force and $\mathbf{Q}$ is the shear force vector, orthogonal to $\mathbf{n}_x$. The incremental rotation $d\bar{\boldsymbol{\varphi}}$ is also expressed in terms of its axial and transverse components,

$$d\bar{\boldsymbol{\varphi}} \;=\; d\varphi\,\mathbf{n}_x \,+\, l^{-1}\mathbf{n}_x \times \Delta\mathbf{u}_\perp. \tag{5.83}$$

The axial component is determined by the average rotation about the x-axis, $d\varphi = \frac{1}{2}(d\varphi_x^A + d\varphi_x^B)$, while the transverse rotation component is determined from the transverse component of the displacement difference between B and A, $\Delta\mathbf{u}_\perp = \mathbf{u}_\perp^B - \mathbf{u}_\perp^A$. The representations (5.82) and (5.83) are now substituted into the cross product, and when using the triple product formula (3.25), the result takes the form

$$d\bar{\boldsymbol{\varphi}} \times \mathbf{f}_B \;=\; \mathbf{n}_x\!\times\!\mathbf{Q}\; d\varphi \,+\, l^{-1}N\,\Delta\mathbf{u}_\perp \,-\, l^{-1}\mathbf{n}_x(\mathbf{Q}^T\Delta\mathbf{u}_\perp). \tag{5.84}$$

When using the components of $d\varphi$ and $\Delta\,\mathbf{u}_\perp$ this expression directly gives the rotation contribution to the sub-matrices $\mathbf{K}_{31}^r$, $\mathbf{K}_{32}^r$, $\mathbf{K}_{33}^r$ and $\mathbf{K}_{34}^r$. When it is observed that $\mathbf{f}_A = -\mathbf{f}_B$ this also determines the contributions to $\mathbf{K}_{11}^r$, $\mathbf{K}_{12}^r$, $\mathbf{K}_{13}^r$ and $\mathbf{K}_{14}^r$.

The representation of the moment vector $\mathbf{m}_B$ in terms of an axial and a transverse component is

$$\mathbf{m}_B \;=\; M\,\mathbf{n}_x \,+\, \mathbf{m}_\perp^B. \tag{5.85}$$

The rotation cross product then follows from (5.84) as

$$d\bar{\boldsymbol{\varphi}} \times \mathbf{m}_B \;=\; \mathbf{n}_x\!\times\!\mathbf{m}_\perp^B\; d\varphi \,+\, l^{-1}M\,\Delta\mathbf{u}_\perp \,-\, l^{-1}\mathbf{n}_x(\mathbf{m}_\perp^{B\,T}\Delta\mathbf{u}_\perp). \tag{5.86}$$

This determines the rotation contribution to the sub-matrices $\mathbf{K}_{41}^r$, $\mathbf{K}_{42}^r$, $\mathbf{K}_{43}^r$ and $\mathbf{K}_{44}^r$. The similar result for $\mathbf{m}_A$ follows by changing the superscript from B to A, and replacing M with $-M$. This determines the rotation contribution to the sub-matrices $\mathbf{K}^r_{21}$, $\mathbf{K}^r_{22}$, $\mathbf{K}^r_{23}$ and $\mathbf{K}^r_{24}$.

Initial non-symmetric form of K_r

The matrix $\mathbf{K}_r$ representing the effect of co-rotation of the local frame of reference can now be assembled from the three contributions: the increment of the variation of the nodal rotations (5.73), the extension of the element (5.80), and the effect of the incremental rigid body rotation $d\bar{\boldsymbol{\varphi}}$. In this process it is convenient to collect the sub-matrices in groups.

The first sub-matrix column $\mathbf{K}^r_{i1}$ represents the effect of the translation $d\mathbf{u}_A$ of node A, while the third sub-matrix column $\mathbf{K}^r_{i3}$ represents the effect of the translation $d\mathbf{u}_B$ of node B. It follows from invariance to a rigid body translation that $\mathbf{K}_{i3} = -\mathbf{K}_{i1}$. The contributions to the force increments $d\mathbf{f}_A$ and $d\mathbf{f}_B$ from the translations are given by the sub-matrices

$$\mathbf{K}^r_{11} = \mathbf{K}^r_{33} = -\mathbf{K}^r_{13} = -\mathbf{K}^r_{31} = \frac{1}{l}\begin{bmatrix} 0 & -Q_y & -Q_z \\ -Q_y & N & 0 \\ -Q_z & 0 & N \end{bmatrix}. \tag{5.87}$$

This group of sub-matrices is seen to satisfy symmetry. The contributions to the moment increment $d\mathbf{m}_A$ from translations are given by the sub-matrices

$$\mathbf{K}^r_{21} = -\mathbf{K}^r_{23} = \frac{1}{l}\begin{bmatrix} 0 & m_y^A & m_z^A \\ 0 & M & 0 \\ 0 & 0 & M \end{bmatrix}, \tag{5.88}$$

while the contributions to $d\mathbf{m}_B$ are given by

$$\mathbf{K}^r_{41} = -\mathbf{K}^r_{43} = \frac{1}{l}\begin{bmatrix} 0 & m_y^B & m_z^B \\ 0 & -M & 0 \\ 0 & 0 & -M \end{bmatrix}. \tag{5.89}$$

The sub-matrix columns $\mathbf{K}^r_{i2}$ and $\mathbf{K}^r_{i4}$ represent the effect of the incremental rotation of node A and node B, respectively. The contributions to the force increments are given by the sub-matrices

$$\mathbf{K}^r_{12} = \mathbf{K}^r_{14} = -\mathbf{K}^r_{32} = -\mathbf{K}^r_{34} = \frac{1}{l}\begin{bmatrix} 0 & 0 & 0 \\ M_y^a & 0 & 0 \\ M_z^a & 0 & 0 \end{bmatrix}, \tag{5.90}$$

where the anti-symmetric moments M_y^a and M_z^a have been introduced in place of the shear forces by use of (5.66). The contributions to $d\mathbf{m}_A$ are given by

$$\mathbf{K}^r_{22} = \frac{1}{2}\begin{bmatrix} 0 & -m_z^A & m_y^A \\ 0 & 0 & M \\ 0 & -M & 0 \end{bmatrix}, \quad \mathbf{K}^r_{24} = \frac{1}{2}\begin{bmatrix} 0 & 0 & 0 \\ -m_z^A & 0 & 0 \\ m_y^A & 0 & 0 \end{bmatrix}, \tag{5.91}$$

while the contributions to $d\mathbf{m}_B$ are given by

$$\mathbf{K}^r_{44} = \frac{1}{2}\begin{bmatrix} 0 & -m^B_z & m^B_y \\ 0 & 0 & -M \\ 0 & M & 0 \end{bmatrix}, \quad \mathbf{K}^r_{42} = \frac{1}{2}\begin{bmatrix} 0 & 0 & 0 \\ -m^B_z & 0 & 0 \\ m^B_y & 0 & 0 \end{bmatrix}. \tag{5.92}$$

It is seen that the sub-matrix group given by (5.91)–(5.92) does not satisfy symmetry, and neither does the group (5.90) satisfy symmetry in combination with (5.88)–(5.89).

The lack of symmetry is a consequence of the use of the virtual work principle on the rigid body motion. In rigid body motion only the total force and moment on the body contribute to equilibrium, and thus the distribution of forces and moments on the individual nodes of the element is not uniquely determined. This problem is addressed in the following section.

Symmetric form of $\mathbf{K}_r$

The derivation of the tangent stiffness relation from the external virtual work may lead to a non-symmetric formulation, because the matrix $\mathbf{K}_r$ is constructed from extension and rigid body rotations alone. The need to use the internal virtual work of a full non-linear beam theory would to some extent defy the purpose of the co-rotating format, namely to derive a universal stiffness matrix $\mathbf{K}_r$ for a given class of elements, such as e.g. beam elements with two nodes, independent of the specific details of the local element model. It is therefore interesting that a systematic procedure can be devised to establish full symmetry by introducing terms corresponding to the missing effect of the deformation modes in the derivation of $\mathbf{K}_r$.

The procedure relies on the fact that in a rigid body rotation the rotation increments at both ends are equal, $d\bar{\boldsymbol{\varphi}}_A = d\bar{\boldsymbol{\varphi}}_B$. Thus, for a rigid body motion any contribution of an element force or moment of the form

$$\mathbf{K}^d_i \, (d\bar{\boldsymbol{\varphi}}_A - d\bar{\boldsymbol{\varphi}}_B) \tag{5.93}$$

will vanish. If the two rotations are not equal, their difference describes a deformation mode, and the result would represent a non-constitutive contribution from the deformation modes. Thus, non-constitutive contributions from the deformation modes can be identified by modifying the second and fourth sub-matrix columns of the non-symmetric $\mathbf{K}_r$ according to the format

$$\mathbf{K}_r := \begin{bmatrix} \mathbf{K}^r_{11} & \mathbf{K}^r_{12} + \mathbf{K}^d_1 & \mathbf{K}^r_{13} & \mathbf{K}^r_{14} - \mathbf{K}^d_1 \\ \mathbf{K}^r_{21} & \mathbf{K}^r_{22} + \mathbf{K}^d_2 & \mathbf{K}^r_{23} & \mathbf{K}^r_{24} - \mathbf{K}^d_2 \\ \mathbf{K}^r_{31} & \mathbf{K}^r_{32} + \mathbf{K}^d_3 & \mathbf{K}^r_{33} & \mathbf{K}^r_{34} - \mathbf{K}^d_3 \\ \mathbf{K}^r_{41} & \mathbf{K}^r_{42} + \mathbf{K}^d_4 & \mathbf{K}^r_{43} & \mathbf{K}^r_{44} - \mathbf{K}^d_4 \end{bmatrix}. \tag{5.94}$$

Here the assignment operator $:=$ has been used to indicate the assignment of a new value in terms of the original definition.

The four 3×3 matrices $\mathbf{K}_i^d$ are defined such that the final matrix $\mathbf{K}_r$ becomes symmetric. The matrices $\mathbf{K}_1^d$ and $\mathbf{K}_3^d$ are determined directly by symmetry. The final two matrices $\mathbf{K}_2^d$ and $\mathbf{K}_4^d$ are determined such that they make $\mathbf{K}_{24}^r$ symmetric with respect to $\mathbf{K}_{42}^r$ and make both the diagonal matrices $\mathbf{K}_{22}^r$ and $\mathbf{K}_{44}^r$ symmetric. The result is most easily obtained directly from (5.91)–(5.92) by moving the content of the matrices $\mathbf{K}_{24}^r$ and $\mathbf{K}_{42}^r$ to the matrices $\mathbf{K}_{22}^r$ and $\mathbf{K}_{44}^3$ on the diagonal, and then moving the anti-symmetric torsion moment terms the other way.

The result is the fully symmetric form of the co-rotation matrix $\mathbf{K}_r$ given by the sub-matrices

$$\mathbf{K}_{11}^r = \mathbf{K}_{33}^r = -\mathbf{K}_{13}^r = -\mathbf{K}_{31}^r = \frac{1}{l}\begin{bmatrix} 0 & -Q_y & -Q_z \\ -Q_y & N & 0 \\ -Q_z & 0 & N \end{bmatrix}, \tag{5.95}$$

$$\mathbf{K}_{12}^r = \mathbf{K}_{21}^{rT} = -\mathbf{K}_{32}^r = -\mathbf{K}_{23}^{rT} = \frac{1}{l}\begin{bmatrix} 0 & 0 & 0 \\ m_y^A & M & 0 \\ m_z^A & 0 & M \end{bmatrix}, \tag{5.96}$$

$$\mathbf{K}_{14}^r = \mathbf{K}_{41}^{rT} = -\mathbf{K}_{34}^r = -\mathbf{K}_{43}^{rT} = \frac{1}{l}\begin{bmatrix} 0 & 0 & 0 \\ m_y^B & -M & 0 \\ m_z^B & 0 & -M \end{bmatrix}, \tag{5.97}$$

$$\mathbf{K}_{22}^r = \frac{1}{2}\begin{bmatrix} 0 & -m_z^A & m_y^A \\ -m_z^A & 0 & 0 \\ m_y^A & 0 & 0 \end{bmatrix}, \quad \mathbf{K}_{44}^r = \frac{1}{2}\begin{bmatrix} 0 & -m_z^B & m_y^B \\ -m_z^B & 0 & 0 \\ m_y^B & 0 & 0 \end{bmatrix}, \tag{5.98}$$

$$\mathbf{K}_{24}^r = \mathbf{K}_{42}^{rT} = \frac{1}{2}\begin{bmatrix} 0 & 0 & 0 \\ 0 & 0 & M \\ 0 & -M & 0 \end{bmatrix}. \tag{5.99}$$

It is interesting to note that the symmetric form (5.95)–(5.99) is not uniquely determined. In fact, any symmetric matrix $\mathbf{K}_0^d$ can be added to $\mathbf{K}_{22}^r$ and $\mathbf{K}_{44}^r$ if it is also subtracted from $\mathbf{K}_{24}^r$ and $\mathbf{K}_{42}^r$. The co-rotation procedure, in which the stiffness matrix is determined from the external virtual work of the element without specific account of the element deformation properties, does not provide any means of determining such a matrix $\mathbf{K}_0^d$. However, a derivation of the geometric matrix from the general beam theory derived in Chapter 4 demonstrates the consistency of the present symmetric form.

5.2.2 Element deformation stiffness

The element deformation stiffness matrix $\mathbf{K}_d$ introduced in (5.75) contains the tangent stiffness of the deformation modes and their conjugate 'stresses'. It may be represented with different orders of accuracy, the simplest being the constitutive relation alone, the next including geometric terms. In two dimensions it was easy to obtain the linear geometric matrix from partial differentiation of the differential equation for bending. In three dimensions a complete linear geometric matrix must include coupling between bending and torsion, and therefore requires a more elaborate system of differential equations as a basis. In the following the geometric stiffness of the deformation modes is extracted from the general formulation of Chapter 4.

Constitutive stiffness

In the case of a straight homogeneous beam the stiffness of the modes is uncoupled, and in terms of linear elastic parameters it may be written as

$$\begin{bmatrix} dM \\ dM_y^s \\ dM_z^s \\ dN \\ dM_y^a \\ dM_z^a \end{bmatrix} = \underbrace{\frac{1}{l}\left[\begin{array}{ccc|ccc} GJ & & & & & \\ & EI_y & & & & \\ & & EI_z & & & \\ \hline & & & EA & & \\ & & & & 3\psi_y^a EI_y & \\ & & & & & 3\psi_z^a EI_z \end{array}\right]}_{\mathbf{K}_d} \begin{bmatrix} d\varphi \\ d\varphi_y^s \\ d\varphi_z^s \\ du \\ d\varphi_y^a \\ d\varphi_z^a \end{bmatrix}. \tag{5.100}$$

GJ is the St. Venant torsion stiffness, and EA the axial stiffness. EI_y and EI_z represent the stiffness of symmetric bending about the y- and the z-axis, respectively. The last two coefficients $3\psi_y^a EI_y$ and $3\psi_z^a EI_z$ are the stiffness of the anti-symmetric bending modes about the y- and the z-axis, respectively. The theory of shear flexibility was given in Section 5.1.2. In the present notation for the three-dimensional problem the shear coefficient of anti-symmetric bending about the y-axis is given by

$$\psi_y^a = \frac{1}{1+\Phi_y}, \qquad \Phi_y = \frac{12\,EI_y}{l^2 GA_z}, \tag{5.101}$$

while the shear coefficients for anti-symmetric bending about the z-axis are

$$\psi_z^a = \frac{1}{1+\Phi_z}, \qquad \Phi_z = \frac{12\,EI_z}{l^2 GA_y}. \tag{5.102}$$

Here A_y is the effective area of the shear force Q_y, and A_z is the effective area of the shear force Q_z.

For curved or non-homogeneous beams, coupling terms will appear in $\mathbf{K}_d$. These terms are conveniently calculated by use of the complementary energy, using equilibrium moment distributions (Krenk, 1994). Coupling between torsion and extension for twisted beams can be accounted for by a coupling term (Krenk, 1983a,b).

Local geometric stiffness

In the computation of geometric stiffness it is often useful to consider the local motion as composed of rotation and strain. The rotation changes the orientation of the internal stresses in space while the strain mainly plays a role by permitting the geometric stiffness to be expressed in the form of internal virtual work. In the case of beams with finite rotations and small strains it is convenient to calculate the geometric stiffness based on the assumption of negligible shear strains. According to the general 'elastica' formulation in (4.7), the relation between the displacement derivative $d\mathbf{u}' = d(d\mathbf{u})/ds_0$ and the rotation of the beam cross-sections $d\bar{\boldsymbol{\varphi}}$ is given by

$$d\mathbf{u}' = d\bar{\boldsymbol{\varphi}}\times\mathbf{x}' + d\boldsymbol{\varepsilon}. \tag{5.103}$$

When the transverse components of the incremental strain $d\boldsymbol{\varepsilon}$ are neglected, the displacement derivative can be written in the form of an axial component du' and a transverse component $d\mathbf{u}'_\perp$,

$$d\mathbf{u}' = \mathbf{n}_x\, du' + d\mathbf{u}'_\perp, \tag{5.104}$$

where the transverse displacement component $d\mathbf{u}'_\perp$ is expressed in terms of the transverse components of the rotation,

$$d\mathbf{u}'_\perp = d\bar{\boldsymbol{\varphi}}\times\mathbf{x}'. \tag{5.105}$$

The derivatives are with respect to the initial arc length s_0, and $\mathbf{x}(s_0)$ describes the current configuration. Thus, the unit vector $\mathbf{n}_x$ is defined by $l\mathbf{n}_x = l_0\mathbf{x}'$.

The geometric stiffness follows from that part of the incremental virtual work that is proportional to the current value of the internal forces. This part of the incremental internal work was given in (4.35) for the general elastica beam theory. When the assumption of vanishing shear strains is introduced, this expression can be reduced by use of the relations for triple cross products. The following notation is introduced for a combination of internal forces,

$$[\mathbf{N}] = N\,\mathbf{I} - \tfrac{1}{2}(\mathbf{N}\,\mathbf{n}_x^T + \mathbf{n}_x\,\mathbf{N}^T), \tag{5.106}$$

where $N = \mathbf{N}^T\mathbf{n}_x$ is the axial force of the beam. The relation (4.35) can then be written as

$$
\begin{aligned}
d(\delta\varepsilon_j)\,N_j + d(\delta\kappa_j)\,M_j \;=\; & -\delta\mathbf{u}_\perp'^T\mathbf{N}\,(l_0/l)\,du' - \delta u'\,(l_0/l)\,\mathbf{N}^T d\mathbf{u}_\perp' \\
& + [\,\delta\bar{\boldsymbol{\varphi}}'\;\delta\bar{\boldsymbol{\varphi}}\,]\begin{bmatrix} \mathbf{0} & \tfrac{1}{2}\hat{\mathbf{M}}^T \\ \tfrac{1}{2}\hat{\mathbf{M}} & (l/l_0)[\mathbf{N}] \end{bmatrix}\begin{bmatrix} d\bar{\boldsymbol{\varphi}}' \\ d\bar{\boldsymbol{\varphi}} \end{bmatrix},
\end{aligned}
\qquad (5.107)
$$

where the terms with explicit dependence on the axial displacements have been extracted, and the notation $\hat{\mathbf{M}}$ has been used for the equivalent skew-symmetric matrix. In the present connection the integral over the length of the beam is needed for displacements and rotations representing the deformation modes of the beam. The internal forces $\mathbf{N}$ as well as the derivatives $\delta u'$ and du' of the axial displacement are constant. Furthermore, the transverse displacements $\delta\mathbf{u}_\perp$ and $d\mathbf{u}_\perp$ of the deformation modes of the beam vanish by definition at the beam ends, and therefore the two first terms containing the displacements do not contribute to the integral.

It is convenient to express the shape functions in terms of the non-dimensional coordinate ξ defined by

$$
s_0 = \tfrac{1}{2}\,l_0(1+\xi), \qquad -1 \le \xi \le 1. \qquad (5.108)
$$

The incremental rotation $d\bar{\boldsymbol{\varphi}}(s_0)$ and its derivative with respect to s_0 are now expressed in terms of $d\bar{\boldsymbol{\varphi}}_s$ and $d\bar{\boldsymbol{\varphi}}_a$, representing symmetric and anti-symmetric deformation, respectively. There are three symmetric modes $d\bar{\boldsymbol{\varphi}}_s^T = [d\bar{\varphi}_x^s, d\bar{\varphi}_y^s, d\bar{\varphi}_z^s]$ corresponding to torsion and bending, respectively, while there are only two anti-symmetric bending modes, $d\bar{\boldsymbol{\varphi}}_a^T = [0, d\bar{\varphi}_y^a, d\bar{\varphi}_z^a]$. The polynomial shape function representation is

$$
\begin{bmatrix} d\bar{\boldsymbol{\varphi}}'(s_0) \\ d\bar{\boldsymbol{\varphi}}(s_0) \end{bmatrix} = \begin{bmatrix} \mathbf{I}/l_0 & 3\xi\,\mathbf{I}/l_0 \\ \tfrac{1}{2}\xi\,\mathbf{I} & \tfrac{1}{4}(3\xi^2-1)\,\mathbf{I} \end{bmatrix}\begin{bmatrix} d\bar{\boldsymbol{\varphi}}_s \\ d\bar{\boldsymbol{\varphi}}_a \end{bmatrix}. \qquad (5.109)
$$

The bending modes can be represented as derivatives of the transverse displacement by (5.105), and it therefore follows that their integral over the element must vanish. For the symmetric mode this follows immediately from the representation as an odd function of ξ, while for the anti-symmetric mode it determines the relative magnitude of the quadratic and the constant term.

The integration of the rotation terms follows after substitution of the representations (5.109). When it is observed that the internal forces $\mathbf{N}$ are constant within the beam element, while the moments $\mathbf{M}$ may contain linear and constant terms, the integration gives

$$\int_0^{l_0} \Big(d(\delta\varepsilon_j)\, N_j \;+\; d(\delta\kappa_j)\, M_j \Big) ds_0$$
$$= [\, \delta\bar{\varphi}_s \; \delta\bar{\varphi}_a \,] \begin{bmatrix} \frac{1}{12} l\, [\mathbf{N}] & \frac{1}{4}\hat{\bar{\mathbf{M}}} \\ \frac{1}{4}\hat{\bar{\mathbf{M}}}^T & \frac{1}{20} l\, [\mathbf{N}] \end{bmatrix} \begin{bmatrix} d\bar{\varphi}_s \\ d\bar{\varphi}_a \end{bmatrix}, \tag{5.110}$$

where $\bar{\mathbf{M}}$ is the mean value of the internal moment $\mathbf{M}(\xi)$ over the beam element, i.e. the value at its center.

It is noted that the contributions of the internal forces $\mathbf{N}$ through the deformation modes appear with the coefficients $\frac{1}{12}$ and $\frac{1}{20}$ already found for the two-dimensional element in (5.42). In addition, there are non-trivial coupling terms between the symmetric and anti-symmetric deformation modes. These are important, e.g. for phenomena like torsional buckling.

5.2.3 Total tangent stiffness

The element stiffness matrix is conveniently described in the block matrix format (5.71) in terms of sub-matrices $\mathbf{K}^e_{ij}$, $i, j = 1, \dots, 4$. When the individual four-block rows in the definition (5.67) of the transformation matrix $\mathbf{S}$ are denoted $\mathbf{S}_i$, $i = 1, \dots, 4$, the transformation from deformation modes to local element forces can be written in the form

$$\mathbf{K}^e_{ij} \;=\; \mathbf{S}_i \mathbf{K}_d \mathbf{S}_j^T \;+\; \mathbf{K}^r_{ij} \tag{5.111}$$

with $i, j = 1, \dots, 4$. The sub-matrices $\mathbf{K}^r_{ij}$ were given in (5.95)–(5.99), and the deformation stiffness matrix $\mathbf{K}_d$ contains the constitutive terms given in (5.100) and the geometric contributions given by (5.110).

The relation (5.111) is easily programmed directly, but the transformation contains a large number of zero entries and for the simple straight beam considered here the result takes a quite simple form when computed analytically. For the diagonal deformation stiffness matrix (5.100), the block matrices containing the constitutive stiffness in full element force format are

$$\mathbf{K}^e_{11} \;=\; \mathbf{K}^e_{33} \;=\; -\mathbf{K}^e_{13} \;=\; -\mathbf{K}^e_{31} \;=\; \frac{1}{l^3} \begin{bmatrix} EAl^2 & 0 & 0 \\ 0 & 12\psi^a_z EI_z & 0 \\ 0 & 0 & 12\psi^a_y EI_y \end{bmatrix}, \tag{5.112}$$

$$\mathbf{K}^e_{22} \;=\; \mathbf{K}^e_{44} \;=\; \frac{1}{l} \begin{bmatrix} GJ & 0 & 0 \\ 0 & (3\psi^a_y\!+\!1)EI_y & 0 \\ 0 & 0 & (3\psi^a_z\!+\!1)EI_z \end{bmatrix}, \tag{5.113}$$

$$\mathbf{K}^e_{24} = \mathbf{K}^e_{42} = \frac{1}{l}\begin{bmatrix} -GJ & 0 & 0 \\ 0 & (3\psi^a_y-1)EI_y & 0 \\ 0 & 0 & (3\psi^a_z-1)EI_z \end{bmatrix} \tag{5.114}$$

and

$$\mathbf{K}^e_{12} = \mathbf{K}^e_{14} = \mathbf{K}^e_{23} = \mathbf{K}^e_{43} =$$

$$-\mathbf{K}^e_{21} = -\mathbf{K}^e_{41} = -\mathbf{K}^e_{32} = -\mathbf{K}^e_{34} = \frac{6}{l^2}\begin{bmatrix} 0 & 0 & 0 \\ 0 & 0 & \psi^a_z EI_z \\ 0 & -\psi^a_y EI_y & 0 \end{bmatrix}. \tag{5.115}$$

These relations are well known from the linear theory of beam bending.

The total geometric contribution to the element stiffness matrix is found as the sum of the contributions from $\mathbf{K}_r$ given in (5.95)–(5.99) and the contribution (5.110) from the deformation modes. The result is expressed in terms of the normal force N, the shear forces $[Q_y, Q_z]$, the torsion moment M, and the *external* bending moments at the nodes $[m^A_y, m^A_z]$ and $[m^B_y, m^B_z]$:

$$\mathbf{K}^e_{11} = \mathbf{K}^e_{33} = -\mathbf{K}^e_{13} = -\mathbf{K}^e_{31} = \frac{1}{l}\begin{bmatrix} 0 & -Q_y & -Q_z \\ -Q_y & \frac{6}{5}N & 0 \\ -Q_z & 0 & \frac{6}{5}N \end{bmatrix}, \tag{5.116}$$

$$\mathbf{K}^e_{12} = \mathbf{K}^{eT}_{21} = -\mathbf{K}^e_{32} = -\mathbf{K}^{eT}_{23} = \frac{1}{l}\begin{bmatrix} 0 & 0 & 0 \\ m^A_y & M & \frac{1}{10}lN \\ m^A_z & -\frac{1}{10}lN & M \end{bmatrix}, \tag{5.117}$$

$$\mathbf{K}^e_{14} = \mathbf{K}^{eT}_{41} = -\mathbf{K}^e_{34} = -\mathbf{K}^{eT}_{43} = \frac{1}{l}\begin{bmatrix} 0 & 0 & 0 \\ m^B_y & -M & \frac{1}{10}lN \\ m^B_z & -\frac{1}{10}lN & -M \end{bmatrix}, \tag{5.118}$$

$$\mathbf{K}^e_{24} = \mathbf{K}^{eT}_{42} = \frac{1}{6}\begin{bmatrix} 0 & lQ_y & lQ_z \\ lQ_y & -\frac{1}{5}lN & 3M \\ lQ_z & -3M & -\frac{1}{5}lN \end{bmatrix}, \tag{5.119}$$

$$\mathbf{K}^e_{22} = \frac{1}{6}\begin{bmatrix} 0 & -2m^A_z + m^B_z & 2m^A_y - m^B_y \\ -2m^A_z + m^B_z & \frac{4}{5}lN & 0 \\ 2m^A_y - m^B_y & 0 & \frac{4}{5}lN \end{bmatrix}, \tag{5.120}$$

$$\mathbf{K}^e_{44} = \frac{1}{6}\begin{bmatrix} 0 & -2m^B_z + m^A_z & 2m^B_y - m^A_y \\ -2m^B_z + m^A_z & \frac{4}{5}lN & 0 \\ 2m^B_y - m^A_y & 0 & \frac{4}{5}lN \end{bmatrix}. \tag{5.121}$$

In these expressions the shear force components Q_y and Q_z appear explicitly, although they can be expressed in terms of the external moments at the nodes as

$$Q_y = -(m_z^A + m_z^B)/l, \qquad Q_z = (m_y^A + m_y^B)/l. \tag{5.122}$$

In particular, these relations have been used to obtain the present form of $\mathbf{K}_{22}^g$ and $\mathbf{K}_{44}^g$.

EXAMPLE 5.4. LATERAL BUCKLING. Figure 5.12 shows a slender beam of length L with simple supports at both ends. The supports prevent translation and rotation about the axis of the beam. The load consists of identical bending moments of magnitude M, applied at the ends. For a slender beam the bending stiffness about a horizontal axis is large, and the deformation in bending is therefore small. However, as the bending stiffness about a vertical axis and the torsional stiffness are much smaller, the beam may fail by lateral buckling. For sufficiently slender beams only little vertical deformation will develop, and the stability problem may be solved to a good approximation in linearized form as an eigenvalue problem. This example illustrates the ability of the cubic bending element to capture this phenomenon in a model with two elements. The shear flexibility for bending in the horizontal plane is assumed negligible, i.e. $\psi_z^a = 1$.

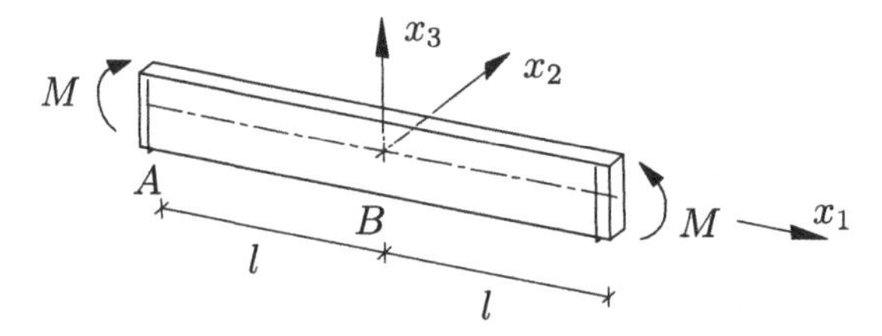

Fig. 5.12. Lateral buckling of slender beam with simple supports.

Let the beam be represented by two identical elements of length $l = \frac{1}{2}L$. Due to symmetry of the problem only the element AB representing the left half of the beam need be considered. The buckling mode involves a combination of torsion with bending out of the plane of the beam. This deformation is described by the nodal displacement components $[d\varphi_3^A, du_2^B, d\varphi_1^B]$. These components correspond to the full element displacement components with index 6, 8 and 10, respectively. The reduced constitutive matrix follows from (5.112)–(5.115) as

$$\mathbf{K}_e = \begin{bmatrix} 4EI_3/l & -6EI_3/l^2 & 0 \\ -6EI_3/l^2 & 12EI_3/l^3 & 0 \\ 0 & 0 & GJ/l \end{bmatrix}.$$

There is no axial or shear force in the beam, and the reduced geometric stiffness matrix then follows from (5.116)–(5.121) as

$$\mathbf{K}_g = \begin{bmatrix} 0 & 0 & 0 \\ 0 & 0 & M/l \\ 0 & M/l & 0 \end{bmatrix}.$$

When combining these matrices into an eigenvalue problem it is seen that the first row constitutes a relation between $d\varphi_3^A$ and du_2^B that does not involve the load M,

$$du_2^B = \tfrac{2}{3}l\, d\varphi_3^A.$$

This relation defines the shape of the lateral bending. When this relation is used to eliminate the component $d\varphi_3^A$, the remaining eigenvalue problem is

$$\begin{bmatrix} 3EI_3/l & M/l \\ M/l & GJ/l \end{bmatrix} \begin{bmatrix} du_2^B \\ d\varphi_1^B \end{bmatrix} = \begin{bmatrix} 0 \\ 0 \end{bmatrix}.$$

The eigenvalue is

$$M = \pm\frac{\sqrt{3}}{l}\sqrt{EI_3\,GJ} = \pm\frac{1.732}{l}\sqrt{EI_3\,GJ}.$$

The analytical solution of the linearized lateral buckling problem gives the factor as $\pi/L = 1.571/l$, see e.g. Timoshenko and Gere (1961). Thus, in this case a single element gives a fair result. However, the accuracy is limited by the low-order representation of the torsion angle, and for accurate solution of torsional and lateral buckling problems several elements should be used.

5.2.4 Finite element implementation

In three dimensions rotations are no longer algebraically additive. The current rotation of the nodes is calculated by tracing the equilibrium path of the structure as in the case of truss structures or two-dimensional beams. The rotation of a node A can be represented in several ways, either by the pseudo-vector $\boldsymbol{\varphi}_A$, by a set of quaternion parameters $(r_A, \mathbf{r}_A)$, or by the rotation matrix $\mathbf{R}_A$. The finite element formulation presented here and in Chapter 4 is based on rotation increments $d\bar{\boldsymbol{\varphi}}_A$ in the form of an infinitesimal vector. In the computation this is replaced by a small but finite increment $d\bar{\boldsymbol{\varphi}}_A$. The individual components of this representation are non-additive as explained in Chapter 3, and it is therefore convenient to represent the accumulated current rotation by its quaternion parameters $(r_A, \mathbf{r}_A)$. The rotation increment is represented by quaternion parameters $(dr_A, d\mathbf{r}_A)$, and it is consistent with the linearization implied by the tangent stiffness formulation to introduce the first-order approximation

$$d\mathbf{r}_A = \tfrac{1}{2}\, d\bar{\boldsymbol{\varphi}}_A, \qquad dr_A = \left(1 - d\mathbf{r}_A^T d\mathbf{r}_A\right)^{1/2} \tag{5.123}$$

following from the quaternion definition (3.48). The quaternion representation of the accumulated rotation is then updated by the non-linear quaternion addition formula (3.63) as

$$r_A := dr_A\, r_A - d\mathbf{r}_A^T \mathbf{r}_A, \quad \mathbf{r}_A := dr_A\, \mathbf{r}_A + r_A\, d\mathbf{r}_A + d\mathbf{r}_A \times \mathbf{r}_A. \tag{5.124}$$

The quaternion representation $(r_A, \mathbf{r}_A)$ of the node A can then be used to extract the total rotation components $\boldsymbol{\varphi}_A$ or to form the corresponding rotation tensor $\mathbf{R}_A$ by use of (3.50) or (3.52).

In the two-dimensional co-rotating beam problem the local element frame of reference is fully determined by the end-points of the element. In the three-dimensional problem the end-points only determine the location of the local x-axis, while the orientation of the y- and the z-axis must be determined by use of the rotation of the nodes. The orientation of the beam coordinate system in space can be formulated in two ways: by updating its position incrementally, or by placing it with reference to the total accumulated rotations of the beam element nodes. In the incremental procedure a linearized form of the rotation increments can be used. However, this in turn implies the use of steps with limited magnitude of the rotation increments. The formulation in terms of total accumulated rotation is independent of the size of the individual load steps, but must then be formulated in terms of finite rotations. For most cases of practical analysis the incremental method will be sufficient.

Incremental formulation

The co-rotating beam is described in a local frame of reference, defined by the base vectors $[\mathbf{n}_x, \mathbf{n}_y, \mathbf{n}_z]$. In the incremental procedure this set of base vectors is updated from its value in the last step by use of the displacement increments $d\mathbf{u}_A, d\mathbf{u}_B$ and the rotation increments $d\bar{\boldsymbol{\varphi}}_A, d\bar{\boldsymbol{\varphi}}_B$ of the beam nodes A and B. The symmetric deformation modes, shown in the top row of Fig. 5.11, are described in terms of the local components of the difference between the rotations of node B and node A. When the current element base vectors are used to define the rotation matrix $\mathbf{R} = [\mathbf{n}_x, \mathbf{n}_y, \mathbf{n}_z]$, the linearized form of the symmetric rotation increment is

$$d\boldsymbol{\varphi}_s = \mathbf{R}^T (d\bar{\boldsymbol{\varphi}}_B - d\bar{\boldsymbol{\varphi}}_A). \tag{5.125}$$

The pre-multiplication by $\mathbf{R}^T$ performs projection of the global components on the local base vectors, thereby yielding the local components $d\boldsymbol{\varphi}_s^T = [d\varphi_x^s, d\varphi_y^s, d\varphi_z^s]$. The corresponding anti-symmetric rotation components are found from the sum of the nodal rotation increments by subtracting twice the incremental rotation of the beam axis,

$$d\boldsymbol{\varphi}_a = \mathbf{R}^T[(d\bar{\boldsymbol{\varphi}}_B + d\bar{\boldsymbol{\varphi}}_A) - 2\mathbf{n}_x \times (d\mathbf{u}_B - d\mathbf{u}_A)/l]. \tag{5.126}$$

The last term defines the incremental rotation of the beam axis, $d\bar{\boldsymbol{\varphi}} = \mathbf{n}_x \times (d\mathbf{u}_B - d\mathbf{u}_A)/l$. Pre-multiplication by $\mathbf{R}^T$ gives the local components $d\boldsymbol{\varphi}_a^T = [d\varphi_x^a, d\varphi_y^a, d\varphi_z^a]$. The component $d\varphi_x^a$ does not have a counterpart in the two-dimensional theory, as it does not contribute to the deformation of the beam, but describes the rigid body rotation around the beam axis.

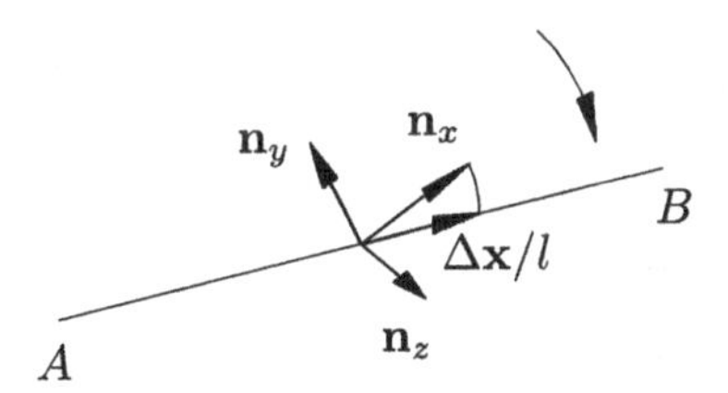

Fig. 5.13. Rotation of element base vectors $[\mathbf{n}_x, \mathbf{n}_y, \mathbf{n}_z]$ to element axis $\Delta\mathbf{x}$.

The element base vectors $[\mathbf{n}_x, \mathbf{n}_y, \mathbf{n}_z]$ are updated by a rotation of magnitude $d\varphi_x^a$ around the beam axis, followed by a rotation of the full basis to align the base vector $\mathbf{n}_x$ with the updated beam axis as shown in Fig. 5.13. The rotation of a set of base vectors through the minimum angle bringing one of the vectors into a given new direction was described in Section 3.2. In the present case the vector $\mathbf{n}_x$ is to be rotated into the direction of the beam axis, described by the unit vector $\Delta\mathbf{x}/l$. First mean direction is defined by the unit vector

$$\mathbf{n} = (\mathbf{n}_x + \Delta\mathbf{x}/l)/|\mathbf{n}_x + \Delta\mathbf{x}/l|. \tag{5.127}$$

The full basis can then be updated by a reflection in the plane orthogonal to $\mathbf{n}$,

$$[\,\mathbf{n}_x, \mathbf{n}_y, \mathbf{n}_z\,] = (\mathbf{I} - 2\,\mathbf{n}\,\mathbf{n}^T)[\,-\mathbf{n}_x, \mathbf{n}_y, \mathbf{n}_z\,]. \tag{5.128}$$

For the vector $\mathbf{n}_x$ being realigned the transformation also involves a change of sign.

When the deformation parameters $\mathbf{v}^T = [\,\varphi_x^s, \varphi_y^s, \varphi_z^s, u, \varphi_y^a, \varphi_z^a]$ have been determined in the element frame of reference, the internal forces $\mathbf{t}$ follow from the constitutive equation (5.100). The nodal forces $\mathbf{q}$ are found from

ALGORITHM 5.2. Incremental update of 3D co-rotating beam elements.

Element extension:

$$u = l - l_0$$

Symmetric and anti-symmetric rotation increments:

$$\mathbf{R} = [\,\mathbf{n}_x, \mathbf{n}_y, \mathbf{n}_z\,]$$

$$d\boldsymbol{\varphi}_s = \mathbf{R}^T(d\bar{\boldsymbol{\varphi}}_B - d\bar{\boldsymbol{\varphi}}_A) \qquad (5.125)$$

$$d\boldsymbol{\varphi}_a = \mathbf{R}^T[(d\bar{\boldsymbol{\varphi}}_B + d\bar{\boldsymbol{\varphi}}_A) - 2\mathbf{n}_x \times (d\mathbf{u}_B - d\mathbf{u}_A)/l] \qquad (5.126)$$

$$\boldsymbol{\varphi}_s = \boldsymbol{\varphi}_s + d\boldsymbol{\varphi}_s, \quad \boldsymbol{\varphi}_a = \boldsymbol{\varphi}_a + d\boldsymbol{\varphi}_a$$

Rotate element basis around axis:

$$[\,\mathbf{n}_y, \mathbf{n}_z\,] = [\,\mathbf{n}_y, \mathbf{n}_z\,]\begin{bmatrix} \cos(d\varphi_x^a) & -\sin(d\varphi_x^a) \\ \sin(d\varphi_x^a) & \cos(d\varphi_x^a) \end{bmatrix}$$

Rotate basis to new element axis and update $\mathbf{R}$:

$$\mathbf{n} = \mathbf{n}_x + \Delta\mathbf{x}/l, \qquad \mathbf{n} = \mathbf{n}/|\mathbf{n}|$$

$$[\,\mathbf{n}_x, \mathbf{n}_y, \mathbf{n}_z\,] = (\mathbf{I} - 2\,\mathbf{n}\,\mathbf{n}^T)[\,-\mathbf{n}_x, \mathbf{n}_y, \mathbf{n}_z\,] \qquad (5.128)$$

$$\mathbf{R} = [\,\mathbf{n}_x, \mathbf{n}_y, \mathbf{n}_z\,]$$

Internal element forces:

$$\mathbf{t} = \mathbf{K}_d\,\mathbf{v} \qquad (5.75)$$

Nodal element forces in global frame:

$$\mathbf{q} = \mathbf{R}_e \mathbf{S}\,\mathbf{t} \qquad (5.63),\ (5.68)$$

Element tangent stiffness in global frame:

$$\mathbf{K}_{ij} = \mathbf{R}\,\mathbf{K}_{ij}^e \mathbf{R}^T \qquad (5.129)$$

(5.68) and transformed to global components by use of the updated element basis $\mathbf{R} = [\mathbf{n}_x, \mathbf{n}_y, \mathbf{n}_z]$. The local components of the element stiffness matrix are given in terms of 3×3 block matrices by (5.112)–(5.122) and transformed to global components by use of the updated element basis $\mathbf{R}$,

$$\mathbf{K}_{ij} = \mathbf{R}\,\mathbf{K}_{ij}^e\,\mathbf{R}^T, \quad i, j = 1, \ldots, 4. \qquad (5.129)$$

This completes the incremental update of element deformations, local frame of reference, nodal element forces and element stiffness matrix. The procedure is summarized in pseudo-code format as Algorithm 5.2. The update is based on linearized rotation increments, and can therefore be carried out independently from the calculation of the total rotation of the nodes.

Total formulation

At the end of each load increment the new orientation in space of each beam element must be determined. In the incremental procedure described above this was done in terms of linearized rotation increments. This leads to a fairly simple procedure that is independent of any description of the accumulated nodal rotations. The price of this independence is the need for sufficiently small steps to justify the linearization of the rotation increments. Alternatively, the orientation of the beam element in space as well as its deformation parameters can be determined from the current position and total rotation of the nodes. In such a procedure the element axes are defined by coincidence of the base vector $\mathbf{n}_x$ with the element axis, and orientation of the base vectors $\mathbf{n}_y, \mathbf{n}_z$ via a suitable mean value of the nodal rotations.

A procedure based on total displacements and rotations can be implemented in several ways. The present procedure generalizes the incremental formulation by using a quaternion representation of the rotation of the nodes. Thus, the first step is to update the quaternions $(r_A, \mathbf{r}_A)$ and $(r_B, \mathbf{r}_B)$ for the rotations of the nodes by use of (5.123)–(5.124). The updated node quaternions are used to find a mean rotation quaternion $(r, \mathbf{r})$ and the quaternion $(s, \mathbf{s})$ defining the rotation from node A to the mean, and from the mean to B. The mean and difference quaternions were discussed in Section 3.5, and explicit formulae were given in (3.70), (3.71) and (3.75). They are included in the pseudo-code Algorithm 5.3 for the total formulation.

The updated set of base vectors $[\mathbf{n}_x, \mathbf{n}_y, \mathbf{n}_z]$ is found from the initial element basis $[\mathbf{n}_x^0, \mathbf{n}_y^0, \mathbf{n}_z^0]$ in two steps. First the initial basis is rotated by the mean rotation of the nodes

$$[\mathbf{n}_x, \mathbf{n}_y, \mathbf{n}_z] = \mathbf{R}(r, \mathbf{r})[\mathbf{n}_x^0, \mathbf{n}_y^0, \mathbf{n}_z^0], \tag{5.130}$$

where the rotation tensor $\mathbf{R}(r, \mathbf{r})$ is defined in terms of the mean quaternion $(r, \mathbf{r})$ in (3.50). This step produces the intermediate base vectors $[\mathbf{n}_x, \mathbf{n}_y, \mathbf{n}_z]$ shown in Fig. 5.13. This intermediate basis is then rotated through the smallest angle that aligns the vector $\mathbf{n}_x$ with the beam axis $\Delta\mathbf{x}$. This step is identical to that of the incremental formulation. First a unit vector $\mathbf{n}$ is defined by (5.127) as the mean between the initial and final position of $\mathbf{n}_x$, and the rotation of the basis is then set up as a reflection in the plane orthogonal to $\mathbf{n}$ by (5.128). This gives the updated element base vectors, and a corresponding updated rotation tensor $\mathbf{R} = [\mathbf{n}_x, \mathbf{n}_y, \mathbf{n}_z]$.

The deformation of the element consists of the extension $u = l - l_0$, three symmetric and two anti-symmetric rotation components. The extension follows directly from the displacements of the element nodes, while the five

ALGORITHM 5.3. Total update of 3D co-rotating beam elements.

Scalar part of difference quaternion:

$$s = \tfrac{1}{2}\big[\,(r_A + r_B)^2 + |\,\mathbf{r}_A + \mathbf{r}_B|^2\,\big]^{1/2} \qquad (3.71)$$

Mean rotation quaternion:

$$r = \tfrac{1}{2}(r_A + r_B)/s, \quad \mathbf{r} = \tfrac{1}{2}(\mathbf{r}_A + \mathbf{r}_B)/s \qquad (3.70)$$

Vector part of difference quaternion:

$$\mathbf{s} = \tfrac{1}{2}(r_A\mathbf{r}_B - r_B\mathbf{r}_A + \mathbf{r}_A \times \mathbf{r}_B)/s \qquad (3.75)$$

Rotate initial element basis:

$$[\,\mathbf{n}_x, \mathbf{n}_y, \mathbf{n}_z\,] = \mathbf{R}(r, \mathbf{r})[\,\mathbf{n}_x^0, \mathbf{n}_y^0, \mathbf{n}_z^0\,] \qquad (3.50),\ (5.130)$$

Rotate basis to element axis and redefine $\mathbf{R}$:

$$\mathbf{n} = \mathbf{n}_x + \Delta\mathbf{x}/l, \qquad \mathbf{n} = \mathbf{n}/|\mathbf{n}|$$

$$[\,\mathbf{n}_x, \mathbf{n}_y, \mathbf{n}_z\,] := (\mathbf{I} - 2\,\mathbf{n}\,\mathbf{n}^T)[\,-\mathbf{n}_x, \mathbf{n}_y, \mathbf{n}_z\,] \qquad (5.128)$$

$$\mathbf{R} = [\,\mathbf{n}_x, \mathbf{n}_y, \mathbf{n}_z\,]$$

Symmetric and anti-symmetric rotation components:

$$\boldsymbol{\varphi}_s = 4\,\mathbf{R}^T\,\mathbf{s} \qquad (5.131)$$

$$\boldsymbol{\varphi}_a = 4\,\mathbf{R}^T\,(\mathbf{n}_x \times \mathbf{n}) \qquad (5.132)$$

Element extension:

$$u = l - l_0$$

Internal element forces:

$$\mathbf{t} = \mathbf{K}_d\,\mathbf{v} \qquad (5.100)$$

Nodal element forces in global frame:

$$\mathbf{q} = \mathbf{R}_e\mathbf{S}\,\mathbf{t} \qquad (5.63),\ (5.68)$$

Element tangent stiffness in global frame:

$$\mathbf{K}_{ij} = \mathbf{R}\,\mathbf{K}_{ij}^e\mathbf{R}^T \qquad (5.129)$$

---

remaining components depend on the rotations. The symmetric deformation modes shown in the top row of Fig. 5.11 are described by the difference between the rotation of nodes A and B. The vector part of the difference quaternion $(s, \mathbf{s})$ describes half the angle $\frac{1}{2}\boldsymbol{\varphi}_s$ in the global frame of reference. This angle is a relative angle, describing the element deformation and the sine function involved in the definition of the quaternion can therefore be linearized, whereby

$$\boldsymbol{\varphi}_s \simeq 4\,\mathbf{s}. \qquad (5.131)$$

The local components are obtained by pre-multiplication with $\mathbf{R}^T$. The anti-symmetric part of the rotations arises because the intermediate base vectors, defined by the mean of the node rotations and shown in Fig. 5.13, are generally not aligned with the beam axis. In the algorithm it is inconvenient to store the intermediate basis, and the rotation required for alignment is therefore calculated from the mean vector $\mathbf{n}$, defined by (5.127). When linearizing the involved sine function, the anti-symmetric rotation components are defined by the vector product

$$\varphi_a = 4\,\mathbf{n}_x \times \mathbf{n}. \tag{5.132}$$

The local components of the anti-symmetric rotations are found by pre-multiplication with $\mathbf{R}^T$. The vector product (5.132) produces a vector that is orthogonal to the beam axis, and thus the local component φ_x^a vanishes, leaving only the two components describing anti-symmetric bending. The beam deformations are included in Algorithm 5.3, together with the calculation of nodal forces and the tangent stiffness matrix.

5.3 Summary and extensions

In the co-rotating formulation the total motion is separated into two parts: the motion of the element-based local co-rotating frame of reference, and the deformation of the element in this local frame. The advantage of the method is that the contributions from the motion of the local frame are independent of the particular element formulation, and the element formulation is limited to the deformation modes in the local frame of reference. These local deformation modes can be modeled at different levels of sophistication, ranging from simple linear theory, over small finite deformation theories with simple geometric stiffness, to fairly advanced theories with non-linear column effects and shortening due to bending. In Sections 5.1.2 and 5.2.2 the local deformation modes were modeled at the intermediate level corresponding to including geometric stiffness effects based on cubic shape functions. This puts certain restrictions on the effects that can be represented accurately within an individual element.

Several extensions of the co-rotating beam theory have found application for offshore structures where column effects, initial imperfections, etc. are important (Skallerud and Amdahl, 2002). A brief indication of the basic idea of a beam-column with special effects like shortening due to bending and the option of initial curvature is given here. The basic idea is to consider the axial force N, or the associated axial strain ε, as the source of geometric effects (Oran, 1973a,b). This suggests an equilibrium formulation for a beam-column element of the form

$$N = EA\,\varepsilon, \qquad \mathbf{M} = \mathbf{A}(\varepsilon)\,\boldsymbol{\varphi}, \tag{5.133}$$

where N and ε are the normal force and the corresponding axial strain, while $\mathbf{M}$ and $\boldsymbol{\varphi}$ are arrays containing the moments and the corresponding rotation angles describing the natural deformation modes shown in Fig. 5.11. The axial stiffness EA is assumed constant, while the moment stiffness matrix $\mathbf{A}(\varepsilon)$ is a function of the normal force or strain. It is an important consequence of the format (5.133) that it implies a relation between the axial strain φ and the elongation u of the beam element of the form

$$u = l\,\varepsilon - \tfrac{1}{2}\boldsymbol{\varphi}^T\mathbf{B}(\varepsilon)\,\boldsymbol{\varphi}, \tag{5.134}$$

where the last term represents the apparent shortening due to bending. The change of separation of the end-points of the beam with strain then follows as

$$\frac{\partial u}{\partial \varepsilon} = l - \tfrac{1}{2}\boldsymbol{\varphi}^T\frac{d\mathbf{B}}{d\varepsilon}\,\boldsymbol{\varphi} = L, \tag{5.135}$$

where L appears as the 'apparent length' of the beam-column element with respect to the extension–strain relation.

It is an important aspect of the present beam-column theory that the 'shortening matrix' $\mathbf{B}(\varepsilon)$ is determined explicitly by the moment stiffness matrix $\mathbf{A}(\varepsilon)$ (Krenk, 1995b). The energy of the beam element follows from first increasing the normal force to its final value while maintaining $\boldsymbol{\varphi} = \mathbf{0}$, and then bending the beam with ε constant. This gives the internal energy

$$\Phi(\varepsilon, \boldsymbol{\varphi}) = \tfrac{1}{2} l\,\varepsilon\,N + \tfrac{1}{2}\boldsymbol{\varphi}^T\mathbf{M} - \tfrac{1}{2}\boldsymbol{\varphi}^T\mathbf{B}\,\boldsymbol{\varphi}\,N. \tag{5.136}$$

The normal force N is conjugate to the extension given by u, and thus

$$N = \frac{\partial \Phi}{\partial \varepsilon}\frac{\partial \varepsilon}{\partial u} = \frac{1}{L}\frac{\partial \Phi}{\partial \varepsilon}. \tag{5.137}$$

Differentiation of the expression (5.136) gives

$$\frac{\partial \Phi}{\partial \varepsilon} = L\,N + \tfrac{1}{2}\boldsymbol{\varphi}^T\Big(\frac{d\mathbf{A}}{d\varepsilon} - EA\,\mathbf{B}\Big)\boldsymbol{\varphi}. \tag{5.138}$$

It follows from a comparison of the two last equations that consistency requires the matrix $\mathbf{B}$ to be defined by

$$\mathbf{B} = \frac{1}{EA}\frac{d\mathbf{A}}{d\varepsilon} = \frac{d\mathbf{A}}{dN}. \tag{5.139}$$

Thus, the beam-column theory is completely defined by the constitutive relations (5.133) and the kinematic relation (5.134) with the matrix $\mathbf{B}$ defined as the derivative of the moment stiffness matrix $\mathbf{A}(N)$.

The constitutive relations (5.133) are in total form. In the computation also the tangent stiffness relation is needed. It follows by differentiation of the total relations, using the strain–elongation relation (5.135),

$$\begin{bmatrix} dN \\ d\mathbf{M} \end{bmatrix} = \left(\begin{bmatrix} 0 & \mathbf{0}^T \\ \mathbf{0} & \mathbf{A} \end{bmatrix} + \frac{EA}{L} \begin{bmatrix} 1 \\ \mathbf{B}\boldsymbol{\varphi} \end{bmatrix} [1, \boldsymbol{\varphi}^T\mathbf{B}] \right) \begin{bmatrix} du \\ d\boldsymbol{\varphi} \end{bmatrix}. \tag{5.140}$$

It is seen that the apparent length L defined by (5.135) acts as the length of the beam-column in the incremental stiffness relation. The matrix in the parentheses corresponds to the element deformation matrix $\mathbf{K}_d$ for which the linearized theory was given in Section 5.2.2. The components have been reordered to accommodate the special status of the normal force in the present beam-column theory. The details of the formulation for the classic beam-column shape functions with added shear flexibility have been given by Krenk (1995b) and Krenk *et al.* (1999).

A number of special effects can be incorporated into the beam-column theory. Initial imperfections in the form of 'out of straightness' can be included explicitly in the formulation by adding a hypothetical step in which the beam-column element is straightened at zero normal force by application of suitable end moments. This is then accounted for by explicit terms in the bending relation (5.133b) and the strain–elongation relation (5.134). This feature enables representation of the column effect for design purposes without additional degrees of freedom. Many frame structures are designed to accommodate a certain amount of yielding – typically at joints. This effect can be incorporated into the individual beam-column element by including the extra local deformation at the nodes in the form of plastic hinges (Ueda and Yao, 1982; Powell and Chen, 1986). Furthermore, plastic hardening can be accounted for with the extent of the plastic yield zone estimated from the variation of the moments along the element (Fujikubo *et al.*, 1991). A particularly simple formulation is obtained when using a flexibility formulation for the local element properties as in Section 5.1.2. In that case the local plastic straining appears as a purely additive contribution to the flexibility matrix (Krenk *et al.*, 1999). By including these additional features the co-rotating beam formulation can be developed into a highly efficient analysis tool for frame structures.

5.4 Exercises

Exercise 5.1 Consider a two-dimensional beam element in the x_1x_2-plane as shown in Fig. 5.1. The bending deformation modes are described in terms of the parameters φ_s and φ_a as illustrated in Fig. 5.3. Denote the end-points of the beam by A and B, and let their incremental transverse displacement and rotation be given by $[du_y^A, d\varphi^A]$ and $[du_y^B, d\varphi^B]$, respectively. These four displacement components include a rigid body translation du_2^0 and a rigid body rotation $d\varphi_3^0$. Show that the incremental rigid body motion is found as

$$\begin{bmatrix} du_y^0 \\ d\varphi_3^0 \end{bmatrix} = \begin{bmatrix} 1 & 1 \\ -1/l & 1/l \end{bmatrix} \begin{bmatrix} du_y^A \\ du_y^B \end{bmatrix}.$$

Show that the deformation modes are found by subtracting the rigid body rotation of the beam ends to give

$$\begin{bmatrix} \frac{1}{2}d\varphi_s \\ \frac{1}{2}d\varphi_a \end{bmatrix} = \begin{bmatrix} 0 & -\frac{1}{2} & 0 & \frac{1}{2} \\ 1/l & \frac{1}{2} & -1/l & \frac{1}{2} \end{bmatrix} \begin{bmatrix} du_y^A \\ d\varphi^A \\ du_y^B \\ d\varphi^B \end{bmatrix}.$$

These relations are special cases of the transformation relations (5.13a) between deformation modes and local nodal displacements.

Exercise 5.2 Use the definition of the co-rotation stiffness matrix

$$\mathbf{K}_r\, d\mathbf{p}_e \;=\; \left(d\mathbf{S} + \mathbf{R}_e^T d\mathbf{R}_e\, \mathbf{S}\right)\mathbf{t}$$

to evaluate its components in the two-dimensional case and obtain (5.23)–(5.24). It is convenient first to express the first term in terms of (N, M_s, M_a) and dl and the second term in terms of (N, M_s, M_a) and $d\varphi$. The increments dl and $d\varphi$ are then introduced from (5.20) and the expression rearranged in matrix format.

Exercise 5.3 Consider a plane straight beam element with end-points A and B. Let $\mathbf{n}_1$ denote a unit vector in the direction from A to B and $\mathbf{n}_2$ a transverse unit vector. In the local frame of reference these vectors have the components $\mathbf{n}_1^T = [1, 0, 0]$ and $\mathbf{n}_2^T = [0, 1, 0]$.

Show that the basic block matrix $\mathbf{K}_{11}^r$ of the co-rotation stiffness matrix $\mathbf{K}_r$, given in (5.24), can be expressed in the form

$$\mathbf{K}_{11}^r \;=\; \frac{N}{l}\,\mathbf{n}_2\mathbf{n}_2^T \;-\; \frac{Q}{l}\left(\mathbf{n}_1\mathbf{n}_2^T + \mathbf{n}_2\mathbf{n}_1^T\right)$$

when the unit vectors $\mathbf{n}_1$ and $\mathbf{n}_2$ are expressed in the local coordinate system.

Explain why the formula remains valid and gives the co-rotation part of the stiffness matrix in a general coordinate system, when the unit vectors are expressed in this coordinate system.

Exercise 5.4 Show that the bending part of the constitutive stiffness matrix for the deformation modes given by (5.33) with $\psi_a = 1$ can be derived from the expression (5.36) for the incremental virtual work with the shape functions for the symmetric and anti-symmetric bending modes given by (5.38) and (5.39). The advantage of the flexibility method based on energy is that it allows inclusion of the shear flexibility effect as a simple additional term.

Exercise 5.5* The geometric stiffness matrix is formulated for a state of equilibrium. During iterations the state of the individual elements does not correspond to equilibrium of the structure. In some cases it is found that retaining the last established equilibrium stress state in the geometric stiffness matrix during the equilibrium iterations reduces the number of iterations. Perform the calculations of Example 5.2 with full update of the internal stresses and with update only at equilibrium. Compare and comment on the resulting number of iterations.

Exercise 5.6* Figure 5.14 shows an initially square frame with side length L and member bending stiffness EI. Use the two-dimensional co-rotating element formulation of Section 5.1 to analyze the tension and compression problem in the interval $0 \leq PL^2/EI \leq 10$. Make plots of u/L, v/L and φ as a function of the load.

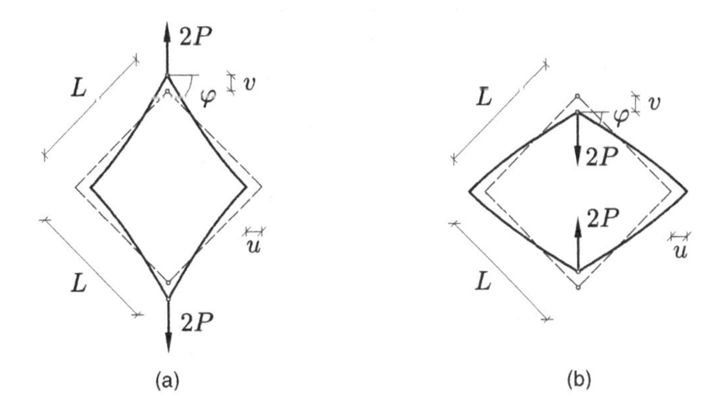

Fig. 5.14. Diamond shaped frame: (a) tension, (b) compression.

Table 5.2. *Displacements of diamond-shaped frame.*

	Tension			Compression		
PL^2/EI	u/L	v/L	φ	u/L	v/L	φ
1	0.13960	0.11252	1.05144	0.17046	0.24754	0.32789
2	0.23184	0.16429	1.20263	0.24224	0.58236	−0.19539
5	0.37322	0.21931	1.40209	0.07735	1.08927	−0.98149
10	0.46601	0.24380	1.50351	−0.12724	1.30578	−1.34277

This problem was initially analyzed by Jenkins *et al.* (1966) using elliptic integrals. Accurate results, tabulated by Mattiasson (1981), are given in Table 5.2. Discuss the accuracy of the finite element solution with two and four elements per side member of the frame.

Exercise 5.7* The results for the angle beam in Fig. 5.9 were obtained for the parameters $A = 1$ and $I = \frac{1}{12}$ corresponding to a square cross-section of unit side length and a frame half-width of $b = 50$. If the height at the center of the angle beam is increased above the value $h = 1.2$ used in Example 5.3 the load–displacement curve becomes non-monotonic, indicating a snap-through. Make a finite element model for the case $h = 1.6$, and obtain the solution by applying a vertical spring at the apex as explained in Exercise 1.5. More efficient solution techniques are developed in Chapter 8.

Exercise 5.8 The procedure described in Algorithm 5.2 for the calculation of the element rotation matrix $\mathbf{R}$ and the deformation parameters $\varphi, \varphi_y^s, \varphi_z^s, u, \varphi_y^a, \varphi_z^a$ can be formulated in several alternative ways. The important thing is that the total rotation of the element is described in terms of accurate formulae for finite rotations, while differences within the element may be handled approximately, assuming small angles.

Describe and illustrate an alternative procedure, in which the rotation matrices $\mathbf{R}'_A$ and $\mathbf{R}'_B$ define nodal base vectors $[\mathbf{n}_x^A, \mathbf{n}_y^A, \mathbf{n}_z^A]$ and $[\mathbf{n}_x^B, \mathbf{n}_y^B, \mathbf{n}_z^B]$ that are rotated such that $\mathbf{n}_x^A$ and $\mathbf{n}_x^B$ coincide with the beam axis, represented by $\mathbf{n}_x$, see Fig. 5.12. In this procedure the element base vectors and the deformation parameters are then defined by mean values and differences between these base vectors at the nodes A and B.

Exercise 5.9 Define the nodal displacement vector $d\mathbf{u} = [\,d\mathbf{u}_A, d\boldsymbol{\varphi}_A, d\mathbf{u}_B, d\boldsymbol{\varphi}_B\,]$ corresponding to a rigid body rotation $d\boldsymbol{\varphi}_0$ and find the corresponding nodal forces $\mathbf{K}_r d\mathbf{u}$ generated by the co-rotation stiffness matrix given by (5.95)–(5.99).

Show that the effect corresponds to rotating the force at each end by $d\boldsymbol{\varphi}_0$ and the moment at each end by $\frac{1}{2}d\boldsymbol{\varphi}_0$, corresponding to the semi-tangential property of the moment. Recall the explanation given in Section 3.3 of the fact that moments appear to rotate by only half when treated within an incremental form of a virtual work equation.

Exercise 5.10 In the co-rotating formulation the total geometric stiffness matrix is considered as consisting of two parts $\mathbf{K}_g = \mathbf{K}_r + \mathbf{K}_l$, where $\mathbf{K}_r$ is the co-rotating part determined by the procedure of Section 5.2. Determine the local part $\mathbf{K}_l$, and show that the contribution from the local part vanishes for a rigid body displacement.

6

Deformation and equilibrium of solids

Kinematic non-linearity is a recognized part of continuum mechanics often termed 'large' or 'finite' displacement theory. The non-linearity arises because equilibrium is considered in the current, and initially unknown, state of the body. In order to describe finite deformation of a continuous body it is necessary to have a non-linear measure of strain and a stress definition that can be used in the deformed state. It turns out that the Green strain, introduced for axial strain in Chapter 2, can be generalized to multi-component form describing the deformation of a continuous body. This is the subject of Section 6.1.

For any continuum mechanics theory it is very desirable to use stresses and strains that satisfy some form of virtual work principle. It was demonstrated in Chapter 2 that the use of the Green strain, which was a convenient quadratic strain measure with exact invariance with respect to arbitrary rigid body motion, led to a slightly modified interpretation of the normal force N appearing in the principle of virtual work. In a similar way the use of the Green strain for a continuous body leads to a special stress definition, the second Piola–Kirchhoff stress. For small strains this stress definition has a simple physical interpretation, precisely as N in the case of a bar element. This stress is introduced in Section 6.2, and it is demonstrated how it serves as a convenient reference for other stress measures of practical importance. In particular the change of stress, the so-called stress rate, is discussed with a view to its use in plasticity theory in Chapter 7.

The change of virtual work as an instrument to study a change of state in a solid body is presented in Section 6.3. This establishes the equilibrium equations in a form suitable for finite element representation with the external forces defined via the external work and the internal forces via the internal work. For elastic materials an integral to the virtual work can be found and identified as the potential energy of the body. The principle of virtual work is the instrument for checking equilibrium and defining the 'out of balance' forces if equilibrium has not been established. In order to approach equilibrium in the case of unbalance a representative stiffness is needed. Again the tangent stiffness is particularly useful because it represents the exact solution for very small changes, and because the tangent stiffness matrix for elastic bodies is symmetric. This is discussed in Section 6.4.

Up until this stage all relations have been kept in general form with arbitrary displacement fields. In practice, the discretization of the problem in terms of finite elements is a very important step. In Section 6.4.3 the introduction of shape functions and their role in the formulation of geometrically non-linear isoparametric solid elements is discussed. The results appear as a systematic generalization of the linear isoparametric element.

There are numerous books on continuum mechanics, many of which cover stresses and strains in finite deformation theory. Malvern (1969), Gurtin (1981) and Truesdell (1991) are standard references presenting a broad overview, while Holzapfel (2000) provides an introduction to more recent developments. The presentation of deformation and equilibrium of solids given here is influenced by Washizu (1974).

6.1 Deformation and strain

Figure 6.1 shows a continuous body in the initial undeformed configuration, serving as reference configuration. In this configuration each point of the body has a position vector $\mathbf{x}_0$. A set of orthogonal unit base vectors $\mathbf{i}_1, \mathbf{i}_2, \mathbf{i}_3$ is introduced, forming a Cartesian coordinate system. The initial position vector can then be written in component form as

$$\mathbf{x}_0 = x_1^0\mathbf{i}_1 + x_2^0\mathbf{i}_2 + x_3^0\mathbf{i}_3. \tag{6.1}$$

In the following the components will either be given in vector notation as $\mathbf{x}_0^T = [x_1^0, x_2^0, x_3^0]$, or in index form as x_α^0, where the range of the subscript α is $1, 2, 3$ for three-dimensional problems and $1, 2$ for two-dimensional problems. Greek subscripts will be reserved for spatial components, and thus always have a range corresponding to the spatial dimension. In terms of the index notation the coordinate decomposition (6.1) is written as

$$\mathbf{x}_0 = x_\alpha^0\mathbf{i}_\alpha. \tag{6.2}$$

The fact that the subscript α occurs twice in the same term is used to indicate a sum over the range of α. This so-called summation convention is a part of tensor analysis, the theory of vectors and their combination in space, see e.g. Malvern (1969), but its use here will be quite straightforward.

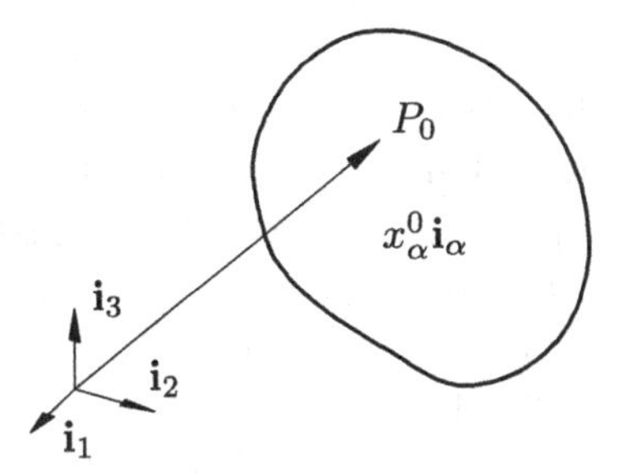

Fig. 6.1. Initial configuration with orthogonal unit base vectors $\mathbf{i}_\alpha$.

Each coordinate set x^0_α defines a material particle by its position in the initial configuration. Now, each material particle is displaced from its initial position $\mathbf{x}_0$ to its current position $\mathbf{x}$ by the displacement vector $\mathbf{u}$, see Fig. 6.2,

$$\mathbf{x} = \mathbf{x}(\mathbf{x}_0) = \mathbf{x}_0 + \mathbf{u}. \tag{6.3}$$

In the following each point is identified by its material coordinates x^0_γ. Thus, the coordinates x_α of the current position are considered as functions of the material coordinates x^0_γ, and similarly for the displacement components. The coordinate relation corresponding to (6.3) then is

$$x_\alpha(x^0_\gamma) = x^0_\alpha + u_\alpha(x^0_\gamma). \tag{6.4}$$

This type of formulation in which the displacement, and thereby the current position, is expressed in terms of the coordinates of the initial position is called *material* or *Lagrangian.* It is the commonly used formulation in solid mechanics, whereas fluid mechanics problems are often formulated in terms of the current position giving a *spatial* or *Eulerian* description.

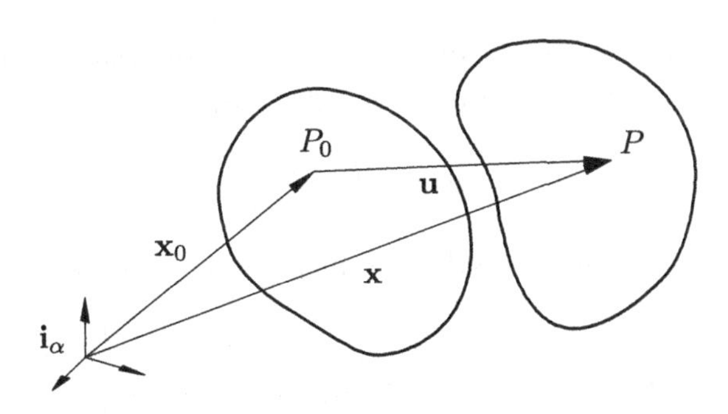

Fig. 6.2. Displacement $\mathbf{u}$ from initial configuration $\mathbf{x}_0$ to current configuration $\mathbf{x}$.

6.1.1 Non-linear strain

A strain measure in a solid body must characterize the state of deformation at each point of the body. This includes the ability to describe different elongations in different directions and also that vanishing strain should correspond to undeformed rigid body motion. It turns out that such a strain measure can be obtained by generalizing the simple length formula (2.13) for the axial Green strain. In the case of a continuum strain is a point property, and therefore only infinitesimal lengths can be used. Also the direction in which the length is measured must be incorporated. Consider a vector $d\mathbf{x}_0$ of infinitesimal length ds_0 in the initial configuration, shown in Fig. 6.3. The length ds_0 is determined by

$$ds_0^2 = d\mathbf{x}_0^T d\mathbf{x}_0. \tag{6.5}$$

The definition of ds_0 as the length of the vector $d\mathbf{x}_0$ implies that the normalized vector $d\mathbf{x}_0/ds_0$ is a unit vector, indicating the direction being considered.

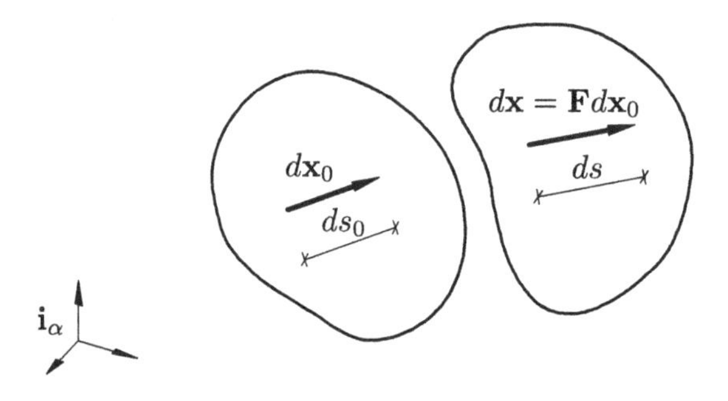

Fig. 6.3. Motion of infinitesimal vector $d\mathbf{x}$.

The current coordinates $d\mathbf{x}$ of the initial vector $d\mathbf{x}_0$ are defined by the partial derivatives of the function $\mathbf{x}(\mathbf{x}_0)$ in (6.3) as

$$d\mathbf{x} = \mathbf{F}\, d\mathbf{x}_0, \tag{6.6}$$

where $\mathbf{F}$ is the deformation gradient tensor, with components

$$\mathbf{F} = \begin{bmatrix} \partial x_1/\partial x_1^0 & \partial x_1/\partial x_2^0 & \partial x_1/\partial x_3^0 \\ \partial x_2/\partial x_1^0 & \partial x_2/\partial x_2^0 & \partial x_2/\partial x_3^0 \\ \partial x_3/\partial x_1^0 & \partial x_3/\partial x_2^0 & \partial x_3/\partial x_3^0 \end{bmatrix}. \tag{6.7}$$

The partial derivatives of the displacement vector $\mathbf{u}$ are arranged similarly in the displacement gradient tensor $\mathbf{D}$ with components

$$\mathbf{D} = \begin{bmatrix} \partial u_1/\partial x_1^0 & \partial u_1/\partial x_2^0 & \partial u_1/\partial x_3^0 \\ \partial u_2/\partial x_1^0 & \partial u_2/\partial x_2^0 & \partial u_2/\partial x_3^0 \\ \partial u_3/\partial x_1^0 & \partial u_3/\partial x_2^0 & \partial u_3/\partial x_3^0 \end{bmatrix}. \tag{6.8}$$

It follows from the linear relation (6.3) that the deformation gradient tensor can be expressed in terms of the displacement gradient tensor as

$$\mathbf{F} = \mathbf{I} + \mathbf{D}, \tag{6.9}$$

where $\mathbf{I}$ is the unit tensor with components given by the unit matrix.

The deformation gradient components (6.7) have a very direct interpretation. Consider the unit vector $\mathbf{i}_1 = d\mathbf{x}_0/dx_1^0 = [1,0,0]^T$ along the x_1-axis. It follows from (6.6) that the current position of this vector is given by the first column in the deformation gradient matrix (6.7). Thus, the three columns $\partial\mathbf{x}/\partial x_\alpha^0$ of the deformation gradient matrix $\mathbf{F}$ give the current components of the three original unit vectors $\mathbf{i}_1$, $\mathbf{i}_2$ and $\mathbf{i}_3$. This relation is illustrated in Fig. 6.4, showing displacement and deformation of the three edges of a unit cube, originally aligned with the coordinate axes. It is seen that the displacement vector describes a translation, while the nine components of the deformation gradient combine a rotation and deformation of the cube. While the deformation depends on the loads and the constitutive behavior of the material, the rotation does not change the state of the material, and it is therefore important to separate these two parts of the motion.

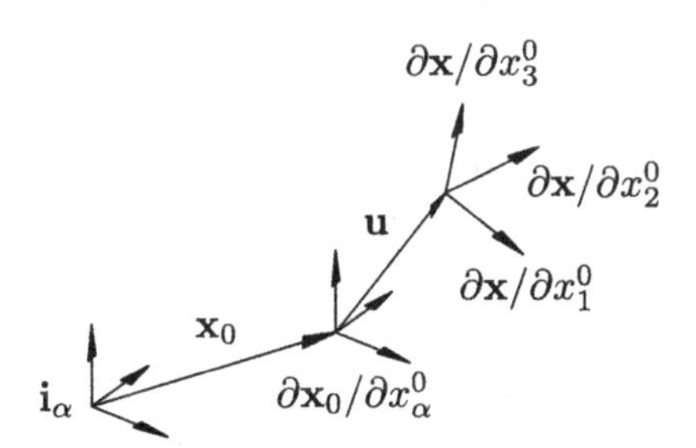

Fig. 6.4. Displacement vector $\mathbf{u}$ and deformation gradient $\mathbf{F}$.

Clearly, a change of the length between two points of a solid body is independent of any rigid body motion. It therefore appears natural to investigate the axial Green strain ε_G using a generic line element with the initial vector representation $d\mathbf{x}_0$ of infinitesimal length ds_0. The current vector representation is $d\mathbf{x}$ with length ds. The axial Green strain introduced in Chapter 2 is

$$\varepsilon_G = \frac{ds^2 - ds_0^2}{2\,ds_0^2} = \frac{d\mathbf{x}^T d\mathbf{x} \;-\; d\mathbf{x}_0^T d\mathbf{x}_0}{2\,ds_0^2}. \tag{6.10}$$

Substitution of the current increment $d\mathbf{x}$ from (6.6) into this expression yields the axial Green strain as

$$\varepsilon_G = \frac{d\mathbf{x}_0^T}{ds_0}\,\tfrac{1}{2}\big(\mathbf{F}^T\mathbf{F} - \mathbf{I}\big)\,\frac{d\mathbf{x}_0}{ds_0}. \tag{6.11}$$

The first and last factor define the initial direction, and the factor in the middle is now defined as the Green strain tensor,

$$\mathbf{E} = \tfrac{1}{2}(\mathbf{F}^T\mathbf{F} - \mathbf{I}) = \tfrac{1}{2}(\mathbf{D} + \mathbf{D}^T) + \tfrac{1}{2}\mathbf{D}^T\mathbf{D}. \tag{6.12}$$

The second expression in terms of the displacement gradient follows from use of (6.9). It is observed that the Green strain tensor $\mathbf{E}$ can be written as the symmetric part of $\frac{1}{2}(\mathbf{F}+\mathbf{I})^T\mathbf{D} = \mathbf{F}_{1/2}^T\mathbf{D}$, similar to the notion of a mean state introduced in Section 2.2.

Substitution of the components of the displacement gradient tensor (6.8) gives the components of the Green strain tensor,

$$E_{\alpha\beta} = \frac{1}{2}\left(\frac{\partial u_\alpha}{\partial x_\beta^0} + \frac{\partial u_\beta}{\partial x_\alpha^0}\right) + \frac{1}{2}\frac{\partial u_\gamma}{\partial x_\alpha^0}\frac{\partial u_\gamma}{\partial x_\beta^0}. \tag{6.13}$$

The components of the Green strain tensor constitute a symmetric matrix where each term is a sum of a linear and a quadratic term. The linear term is identical to that of 'small' displacement theory. The components have a simple interpretation as the change of the scalar products of the base vectors before and after the motion. As illustrated in Fig. 6.4, the initial base vectors are $\partial\mathbf{x}_0/\partial x_\alpha^0$ while their current coordinates are $\partial\mathbf{x}/\partial x_\alpha^0$. The components of the Green strain tensor express the change in the scalar products of the corresponding base vectors,

$$2E_{\alpha\beta} = \frac{\partial\mathbf{x}^T}{\partial x_\alpha^0}\frac{\partial\mathbf{x}}{\partial x_\beta^0} - \frac{\partial\mathbf{x}_0^T}{\partial x_\alpha^0}\frac{\partial\mathbf{x}_0}{\partial x_\beta^0}. \tag{6.14}$$

Hereby the diagonal elements of $2E_{\alpha\beta}$ correspond to the increase of the square of the length of the unit vectors, while the off-diagonal elements describe the cosines to the angles between the base vectors after the motion. However, this relation is somewhat indirect due to the change of the length of the base vectors. For small deformation the Green strain tensor components correspond to those of linear strain.

In order to establish the principle of virtual work, the variation $\delta\mathbf{E}$ of the

Green strain tensor $\mathbf{E}$ is also needed. The variation follows immediately from (6.12),

$$\delta\mathbf{E} = \tfrac{1}{2}(\mathbf{F}^T\delta\mathbf{D} + \delta\mathbf{D}^T\mathbf{F}). \tag{6.15}$$

This defines the variation of the strain components $\delta\mathbf{E}$ as the symmetric part of $\mathbf{F}^T\delta\mathbf{D}$. The corresponding component form is

$$\delta E_{\alpha\beta} = \frac{1}{2}\left(\frac{\partial x_\gamma}{\partial x_\alpha^0}\frac{\partial(\delta u_\gamma)}{\partial x_\beta^0} + \frac{\partial(\delta u_\gamma)}{\partial x_\alpha^0}\frac{\partial x_\gamma}{\partial x_\beta^0}\right). \tag{6.16}$$

It is seen that both the current Green strain tensor $\mathbf{E}$ and its variation $\delta\mathbf{E}$ are symmetric and therefore have only six independent components, while the deformation gradient and its variation are in general non-symmetric and therefore have nine independent components. The extra three components describe the rotation of the material at the point, as described in the next section.

6.1.2 Decomposition into deformation and rigid body motion

The deformation gradient tensor $\mathbf{F}$ relates the initial and current local geometry around a point as illustrated in Fig. 6.4. It follows from the interpretation of the columns of the component matrix $\mathbf{F}$ as the coordinates of the base vectors after the motion that $\det(\mathbf{F}) > 0$, and thus the component matrix is invertible. The deformation gradient matrix has nine components, while the local deformation is described by only six components. The explanation is given by the polar decomposition theorem, which states that a positive definite three-dimensional tensor can be factored in two ways as

$$\mathbf{F} = \mathbf{R}\mathbf{U} = \mathbf{V}\mathbf{R}, \tag{6.17}$$

where $\mathbf{U}$ and $\mathbf{V}$ are positive definite symmetric tensors, and $\mathbf{R}$ is a proper orthogonal tensor, describing a rotation as discussed in Section 3.1. Each of the symmetric tensors has six independent components, while the rotation tensor is defined in terms of three independent components, for example in the form of a rotation pseudo-vector.

The factors in (6.17) can be interpreted by considering a line element with initial components $d\mathbf{x}_0$. After the motion the current line element is

$$d\mathbf{x} = \mathbf{F}\,d\mathbf{x}_0 = \mathbf{R}\,(\mathbf{U}\,d\mathbf{x}_0) = \mathbf{V}\,(\mathbf{R}\,d\mathbf{x}_0). \tag{6.18}$$

In the right-hand decomposition $\mathbf{F} = \mathbf{R}\mathbf{U}$ the line element first participates in a deformation described by the symmetric matrix $\mathbf{U}$, after which it is rotated by the rotation tensor $\mathbf{R}$. Conversely, in the left-hand decomposition

$\mathbf{F} = \mathbf{VR}$ the line element is first rotated by the rotation tensor $\mathbf{R}$, and then participates in a deformation described by the symmetric tensor $\mathbf{V}$. The rotation is the same in the two cases, while the deformation tensors $\mathbf{U}$ and $\mathbf{V}$ will in general not be identical.

Proof of the decompositions (6.17) and the relation between the tensors $\mathbf{U}$ and $\mathbf{V}$ are established as follows (Gurtin, 1981). The symmetric tensor $\mathbf{U}$ is determined from the deformation gradient tensor via the product

$$\mathbf{F}^T\mathbf{F} = (\mathbf{R}\,\mathbf{U})^T\,\mathbf{R}\,\mathbf{U} = \mathbf{U}\,\mathbf{U} = \mathbf{U}^2. \tag{6.19}$$

This product defining $\mathbf{U}^2$, and thereby $\mathbf{U}$, is determined by the Green strain tensor as shown by (6.12). As $\mathbf{F}$ is invertible, the same applies to $\mathbf{U}$, and thus the rotation tensor is determined as

$$\mathbf{R} = \mathbf{F}\,\mathbf{U}^{-1}. \tag{6.20}$$

It is easily verified that $\mathbf{R}^T\mathbf{R} = \mathbf{I}$, which completes the proof of the right-hand decomposition.

The left-hand decomposition follows from introducing the symmetric tensor $\mathbf{V}$ in the form

$$\mathbf{V} = \mathbf{R}\,\mathbf{U}\,\mathbf{R}^T. \tag{6.21}$$

The product of the left-hand decomposition (6.17) then takes the form

$$\mathbf{V}\mathbf{R} = (\mathbf{R}\,\mathbf{U}\,\mathbf{R}^T)\,\mathbf{R} = \mathbf{R}\,\mathbf{U} = \mathbf{F}. \tag{6.22}$$

This completes the proof of the decomposition relations and provides explicit definitions for the tensors $\mathbf{U}$, $\mathbf{R}$ and $\mathbf{V}$.

The decomposition of a motion into a deformation and a rotation part is illustrated by the two-dimensional examples shown in Fig. 6.5. In both cases there is no displacement in the x_3-direction, i.e. $u_3 = 0$, and only a 2×2 sub-matrix of the components of $\mathbf{F}$ needs to be considered.

EXAMPLE 6.1. BIAXIAL EXTENSION. Figure 6.5(a) shows a state of biaxial extension, in which a unit vector along the x_1-axis is extended by ε_1, and a unit vector along the x_2-axis is extended by ε_2. There is no change of angle of the axes, and the deformation gradient therefore is

$$\mathbf{F} = \begin{bmatrix} 1+\varepsilon_1 & 0 \\ 0 & 1+\varepsilon_2 \end{bmatrix}.$$

The right deformation matrix $\mathbf{U}^2$ follows from (2.19) as

$$\mathbf{U}^2 = \mathbf{F}^T\mathbf{F} = \begin{bmatrix} (1+\varepsilon_1)^2 & 0 \\ 0 & (1+\varepsilon_2)^2 \end{bmatrix}.$$

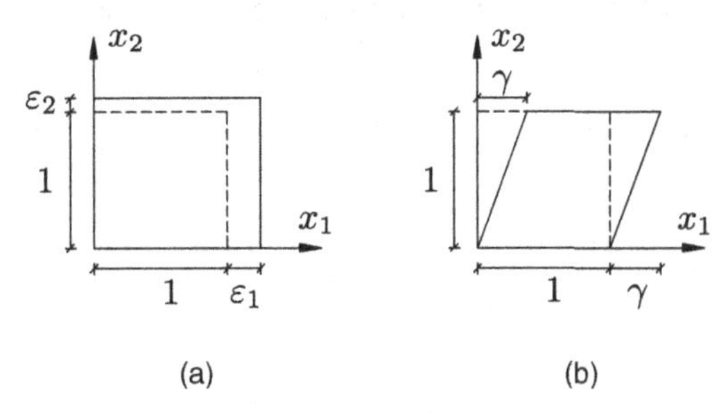

Fig. 6.5. (a) Biaxial extension, (b) shear at constant volume.

The diagonal form of $\mathbf{U}^2$ immediately gives

$$\mathbf{U} = \begin{bmatrix} 1+\varepsilon_1 & 0 \\ 0 & 1+\varepsilon_2 \end{bmatrix}.$$

It is seen that the diagonal of $\mathbf{U}$ contains the so-called stretches of the axes. As a consequence of the diagonal form of $\mathbf{U}$ there is no rotation, and $\mathbf{V} = \mathbf{U}$.

EXAMPLE 6.2. SHEAR AT CONSTANT VOLUME. Figure 6.5(b) shows a state of shear at constant volume, in which the top side of a unit square is shifted γ to the right. Within linear deformation theory γ is the angular shear strain, but in a rigorous large deformation formulation matters are slightly more complicated. The deformation gradient for this case is

$$\mathbf{F} = \begin{bmatrix} 1 & \gamma \\ 0 & 1 \end{bmatrix}$$

and $\mathbf{U}^2$ is then defined by the product

$$\mathbf{U}^2 = \mathbf{F}^T\mathbf{F} = \begin{bmatrix} 1 & \gamma \\ \gamma & 1+\gamma^2 \end{bmatrix}.$$

It is easy to verify – but somewhat more complicated to derive – that the matrix $\mathbf{U}$ then is

$$\mathbf{U} = \frac{1}{\sqrt{4+\gamma^2}} \begin{bmatrix} 2 & \gamma \\ \gamma & 2+\gamma^2 \end{bmatrix},$$

where it should be noted that the factor applies to all components in the matrix. The inverse of this matrix is

$$\mathbf{U}^{-1} = \frac{1}{\sqrt{4+\gamma^2}} \begin{bmatrix} 2+\gamma^2 & -\gamma \\ -\gamma & 2 \end{bmatrix}$$

and the rotation matrix $\mathbf{R}$ then follows from (6.20) as

$$\mathbf{R} = \mathbf{F}\mathbf{U}^{-1} = \frac{1}{\sqrt{4+\gamma^2}} \begin{bmatrix} 1 & \gamma \\ 0 & 1 \end{bmatrix} \begin{bmatrix} 2+\gamma^2 & -\gamma \\ -\gamma & 2 \end{bmatrix} = \frac{1}{\sqrt{4+\gamma^2}} \begin{bmatrix} 2 & \gamma \\ -\gamma & 2 \end{bmatrix}.$$

The rotation matrix is orthogonal with unit determinant, and can be written in terms of the rotation angle φ as

$$\mathbf{R} = \begin{bmatrix} \cos\varphi & -\sin\varphi \\ \sin\varphi & \cos\varphi \end{bmatrix}, \qquad \tan\varphi = -\tfrac{1}{2}\gamma.$$

For small deformation $\varphi \simeq -\frac{1}{2}\gamma$ but for increasing values of γ the angle approaches $\frac{1}{2}\pi$, and the deformation loses its resemblance to shear. Finally, the left deformation tensor $\mathbf{V}$ follows from (6.21). Note that the determinant of $\mathbf{U}$ and $\mathbf{V}$ is equal to unity, because the volume is unchanged by the motion.

6.2 Virtual work and stresses

The traditional way of introducing stresses is via an infinitesimal volume element. First a rectangular element is considered to establish the differential equations of equilibrium, and then a tetrahedral element is used to establish the transformation rule needed for the surface stress vector. As already mentioned, the choice of a strain measure and the requirement that a principle of virtual work must exist leads to the stress measure that is conjugate to the chosen strain, i.e. the stress that combines with the strain variation into a virtual work equation. Here, the procedure already introduced in Chapter 2 is followed, starting out from the principle of virtual work, deriving the equilibrium equations and the formula for the surface stress vector, and then giving the physical interpretation of the stress obtained in this way.

The Green strain leads to the Piola–Kirchhoff stress, described in Section 6.2.1. The definition of the Piola–Kirchhoff stress makes explicit reference to the initial state, and there are theoretical as well as practical reasons to investigate stresses defined solely with reference to the current state. This leads to the Cauchy and the Kirchhoff stress measures introduced in Section 6.2.2. These stresses refer to current directions and areas. However, the material properties – and thereby the current state of stress – are convected with the material, and it is therefore necessary to introduce a special description of the stress changes that accounts for the motion of the material. These so-called objective stress rates are discussed in Section 6.2.3.

6.2.1 Piola–Kirchhoff stress

The stress components associated with the Green strain components $\delta E_{\alpha\beta}$ are denoted $S_{\alpha\beta}$. The virtual work identity is formulated using the initial coordinates x_γ^0 and the initial volume and surface elements dV_0 and dS_0, respectively. When $p_\gamma^0 dV_0$ are the volume force components and $t_\gamma^0 dS_0$ the surface traction components corresponding to the Cartesian base vectors $\mathbf{i}_\gamma$, the virtual work equation is

$$\int_{V_0} \delta E_{\alpha\beta}\, S_{\alpha\beta}\, dV_0 = \int_{S_0} \delta u_\gamma\, t_\gamma^0\, dS_0 + \int_{V_0} \delta u_\gamma\, p_\gamma^0\, dV_0. \tag{6.23}$$

The virtual strains $\delta E_{\alpha\beta}$ are symmetric. The product with the stress components will therefore only depend on the symmetric part, and the stress components appearing in the virtual work can therefore be introduced in symmetric form, i.e. $S_{\beta\alpha} = S_{\alpha\beta}$. Due to the symmetry of the stress components the stress integral can be expressed by substituting the simple non-symmetric form of the virtual strain expression (6.16), whereby

$$\int_{V_0} \delta E_{\alpha\beta}\, S_{\alpha\beta}\, dV = \int_{V_0} \frac{\partial(\delta u_\gamma)}{\partial x_\beta^0} \frac{\partial x_\gamma}{\partial x_\alpha^0}\, S_{\alpha\beta}\, dV_0. \tag{6.24}$$

By use of the divergence theorem the derivative $\partial/\partial x_\beta^0$ is moved to the product of the second and third factors,

$$\int_{V_0} \frac{\partial(\delta u_\gamma)}{\partial x_\beta^0} \frac{\partial x_\gamma}{\partial x_\alpha^0}\, S_{\alpha\beta}\, dV_0 = \int_{S_0} \delta u_\gamma \left(\frac{\partial x_\gamma}{\partial x_\alpha^0} S_{\alpha\beta} \right) n_\beta^0\, dS_0 - \int_{V_0} \delta u_\gamma \frac{\partial}{\partial x_\beta^0} \left(\frac{\partial x_\gamma}{\partial x_\alpha^0} S_{\alpha\beta} \right) dV_0. \tag{6.25}$$

When this form is introduced into the virtual work identity (6.23), and it is observed that the displacement variation δu_γ is arbitrary, it is found that the equilibrium equation

$$\frac{\partial}{\partial x_\beta^0} \left(\frac{\partial x_\gamma}{\partial x_\alpha^0} S_{\alpha\beta} \right) + p_\gamma^0 = 0, \qquad \gamma = 1, 2, 3 \tag{6.26}$$

must be satisfied in V_0, and the surface traction must be defined by

$$t_\gamma^0 = \left(\frac{\partial x_\gamma}{\partial x_\alpha^0} S_{\alpha\beta} \right) n_\beta^0, \qquad \gamma = 1, 2, 3 \tag{6.27}$$

on the surface S_0.

The stresses $S_{\alpha\beta}$ introduced in this way are the components of the second Piola–Kirchhoff stress tensor. These stress components have the advantage of being symmetric. However, as seen from the equilibrium equation (6.26) and the definition of the surface traction (6.27) they are used in the form $P_{\gamma\beta} = (\partial x_\gamma / \partial x_\alpha^0) S_{\alpha\beta}$. This combination is the first Piola–Kirchhoff stress tensor. The first Piola–Kirchhoff stress tensor $P_{\gamma\beta}$ appears directly in the vector relations (6.26) and (6.27), while the second stress tensor $S_{\alpha\beta}$ plays the direct role in the principle of virtual work. The symmetry and the direct relation to the principle of virtual work have led to general use of the second Piola–Kirchhoff stress tensor.

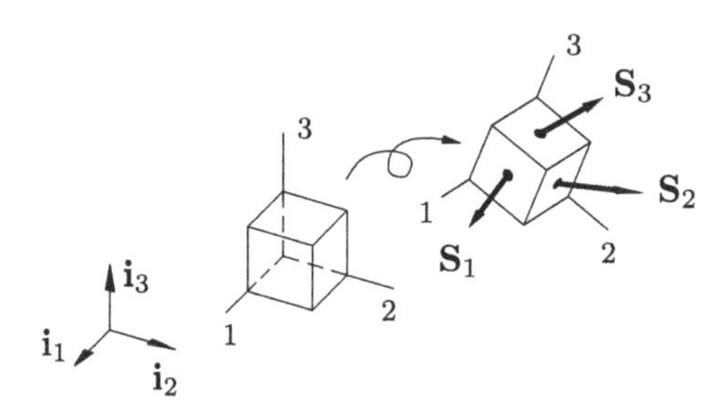

Fig. 6.6. Initial unit cube and deformed shape with face tractions $\mathbf{S}_1, \mathbf{S}_2, \mathbf{S}_3$.

The physical meaning of the two Piola–Kirchhoff stress tensors can be explained with reference to Fig. 6.6. The figure shows a small unit cube with sides $\mathbf{i}_\gamma$ located at the point x_γ^0 in the initial configuration. After the displacement the faces have the traction vectors $\mathbf{S}_1, \mathbf{S}_2, \mathbf{S}_3$ as shown. In the initial configuration the cube had unit side length, and therefore the change of the traction vectors from one side of the volume to the other can be expressed by differentiation with respect to the original Cartesian coordinates x_α^0. With the body force $\mathbf{p} = p_\gamma^0 \mathbf{i}_\gamma$ acting on the original unit volume the equilibrium equation is

$$\frac{\partial \mathbf{S}_1}{\partial x_1^0} + \frac{\partial \mathbf{S}_2}{\partial x_2^0} + \frac{\partial \mathbf{S}_3}{\partial x_3^0} + \mathbf{p}_0 = \mathbf{0}, \tag{6.28}$$

or in index notation

$$\frac{\partial \mathbf{S}_\beta}{\partial x_\beta^0} + \mathbf{p}_0 = \mathbf{0}, \tag{6.29}$$

where the subscript β identifies the face of the volume in the original configuration. The similar component form was given in (6.26). When the base vectors $\mathbf{i}_\gamma$ are introduced into (6.26), the following identity is obtained:

$$\mathbf{S}_\beta = \mathbf{i}_\gamma \left(\frac{\partial x_\gamma}{\partial x_\alpha^0} S_{\alpha\beta} \right) = \frac{\partial (x_\gamma \mathbf{i}_\gamma)}{\partial x_\alpha^0} S_{\alpha\beta}, \qquad \beta = 1, 2, 3. \tag{6.30}$$

These equalities show that the first Piola–Kirchhoff stress tensor corresponds to resolving the traction vectors $\mathbf{S}_\beta$ in components along the original unit base vectors $\mathbf{i}_\gamma$, while the second Piola–Kirchhoff stress tensor corresponds to resolving the traction vectors $\mathbf{S}_\beta$ in components along the vectors $\partial\mathbf{x}/\partial x_\alpha^0$. These vectors form the sides of the deformed cube and thus correspond to a deformed non-orthogonal set of base vectors. The curious fact is that the use of this non-orthogonal set of base vectors leads to symmetric stress components, see e.g. Washizu (1974, p. 57).

EXAMPLE 6.3. PIOLA–KIRCHHOFF STRESS IN BEAM THEORY. For small strain but arbitrarily large displacements the second Piola–Kirchhoff stress components act as convected stress components, as illustrated in Fig. 6.7 for a beam.

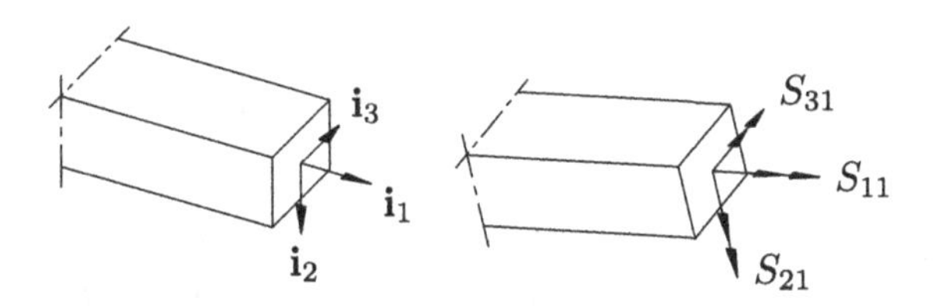

Fig. 6.7. Stresses S_{11}, S_{21}, S_{31} on beam cross-section.

In the initial configuration the axis of the beam is in the direction of the x_1-axis, and the cross-section of the beam therefore described in the x_2–x_3 plane. The main stress components of a beam theory are those acting on the cross-section. In the present case the initial position of the cross-section is identified by the normal $\mathbf{i}_1$ and thus the relevant stress components correspond to the traction vector $\mathbf{S}_1$. The stress components $S_{\alpha 1}$ correspond to a resolution in the deformed basis. In the present case there are two vectors fixed in the cross-section giving the shear stress components S_{21}, S_{31}, and the tangent vector to the center-line of the deformed beam giving the axial stress component S_{11}. The lengths of the deformed base vectors are determined by the strains E_{11}, E_{22}, E_{33} and the deviation from orthogonality by the strains E_{13}, E_{23}, E_{12}. In the case of small strain the deformed base vectors still constitute a nearly orthogonal set of vectors of nearly unit length, and the second Piola–Kirchhoff stress components are then nearly the ordinary stresses referred to this convected coordinate system. Thus, it turns out to be quite convenient that the stress components $S_{\alpha\beta}$ refer to a plane with normal direction β in the initial configuration with vector components α along the deformed base vectors.

6.2.2 Cauchy and Kirchhoff stresses

The Piola–Kirchhoff stress refers to areas and directions in the reference state, and this gives an indirect relation to the current state in some situations involving large strains. It is therefore desirable to investigate the possibility of formulations based on the current state. Here the problem is posed in terms of small displacements $d\mathbf{u}$ imposed on the current state, described by the coordinates $\mathbf{x}$. The strains introduced by this motion will be the linear part of the Green strains, but with the current state as reference state. Thus these strains are

$$d\varepsilon_{\alpha\beta} = \frac{1}{2}\left(\frac{\partial(du_\alpha)}{\partial x_\beta} + \frac{\partial(du_\beta)}{\partial x_\alpha}\right). \tag{6.31}$$

These strains may be related to the Green strain increments with components given by (6.16). When using the chain rule of differentiation on the derivatives of the displacements, the Green strain increments may be written as

$$dE_{\alpha\beta} = \frac{1}{2}\left(\frac{\partial x_\gamma}{\partial x_\alpha^0}\frac{\partial(du_\gamma)}{\partial x_\lambda}\frac{\partial x_\lambda}{\partial x_\beta^0} + \frac{\partial x_\lambda}{\partial x_\alpha^0}\frac{\partial(du_\gamma)}{\partial x_\lambda}\frac{\partial x_\gamma}{\partial x_\beta^0}\right). \tag{6.32}$$

It is seen that the deformation gradient appears as a factor twice in each term, and thereby establishes the relation between the Green strain increment $d\mathbf{E}$ and the incremental linear strain $d\boldsymbol{\varepsilon}$ as

$$d\mathbf{E} = \mathbf{F}^T d\boldsymbol{\varepsilon}\,\mathbf{F}. \tag{6.33}$$

A similar relation applies to the variations $\delta\mathbf{E}$ and $\delta\boldsymbol{\varepsilon}$.

The Cauchy stress tensor $\boldsymbol{\sigma}$ with components $\sigma_{\alpha\beta}$ can now be introduced as conjugate to the linear current strains (6.31) when using the current volume and surface. This definition implies that the virtual work of the Cauchy stress through the virtual linear strain equals the virtual work of the second Piola–Kirchhoff stress through the virtual Green strain,

$$\int_{V_0} \delta E_{\alpha\beta}\, S_{\alpha\beta}\, dV_0 = \int_V \delta\varepsilon_{\alpha\beta}\,\sigma_{\alpha\beta}\, dV. \tag{6.34}$$

The relation between the two sets of stress components is obtained by introducing the relation (6.33) between the virtual strain components and by obtaining a relation between the reference volume element dV_0 and the current volume element dV.

The current volume element dV is related to the initial volume element dV_0 by considering the sides $\partial\mathbf{x}/\partial x_\alpha^0$ of the deformed unit cube in Fig. 6.4.

The volume of the deformed cube is given by the triple product of the side vectors,

$$dV = \Big(\frac{\partial \mathbf{x}}{\partial x_1^0} \cdot \frac{\partial \mathbf{x}}{\partial x_2^0} \times \frac{\partial \mathbf{x}}{\partial x_3^0}\Big) dV_0 = \det\Big(\frac{\partial x_\alpha}{\partial x_\beta^0}\Big) dV_0. \tag{6.35}$$

Introducing the Jacobian determinant $J = \det(\mathbf{F})$, the volume relation is

$$dV = J\, dV_0 = \det(\mathbf{F})\, dV_0. \tag{6.36}$$

Introduction of this relation and the strain increment relation (6.33) into the left side of the internal virtual work equation (6.34) gives the following formula for the Cauchy stress:

$$\boldsymbol{\sigma} = \frac{1}{J} \mathbf{F}\, \mathbf{S}\, \mathbf{F}^T. \tag{6.37}$$

The corresponding component relation is

$$\sigma_{\alpha\beta} = \frac{1}{J} \frac{\partial x_\alpha}{\partial x_\gamma^0}\, S_{\gamma\delta}\, \frac{\partial x_\beta}{\partial x_\delta^0}. \tag{6.38}$$

It is seen that the role of the deformation gradient is to transform the reference of $S_{\gamma\delta}$ components from the reference coordinates to the current coordinates. It follows from the polar decomposition (6.17) that in general this is not an orthogonal transformation, but involves deformation as well.

The Cauchy stress $\boldsymbol{\sigma}$ conforms closely to the small deformation theory notion of stress with its use of current geometry. However, for processes involving large deformation the volume may change, and thereby the volume increment dV used in the principle of virtual work is not constant during the motion. This fact is accounted for by the so-called Kirchhoff stress $\boldsymbol{\tau}$, that is introduced in a similar way as the Cauchy stress but using the original volume element dV_0. The reason behind the Kirchhoff stress is that for actual processes the constitutive behavior of the material is associated with a certain mass of material. Thus, the volume integrals would actually be over mass elements $dm = \rho dV$, where ρ is the mass density. The change of density with change of volume element would neutralize the apparent effect of the changing volume increment dV. This effect is often included by representing the volume element in the form $J^{-1}dV$. Thus, the Kirchhoff stress components $\tau_{\alpha\beta}$ are introduced via the internal virtual work identity

$$\int_{V_0} \delta E_{\alpha\beta}\, S_{\alpha\beta}\, dV_0 = \int_V \delta\varepsilon_{\alpha\beta}\, \tau_{\alpha\beta}\, J^{-1}\, dV. \tag{6.39}$$

It follows directly by comparison with the Cauchy stress relation (6.34) that the Kirchhoff stress is

$$\boldsymbol{\tau} = J\,\boldsymbol{\sigma} = \mathbf{F}\,\mathbf{S}\,\mathbf{F}^T. \tag{6.40}$$

The similarity with the inverse relation (6.33) for the corresponding strain increments is evident.

6.2.3 Stress rates

The Piola–Kirchhoff stress components $S_{\alpha\beta}$ describe tractions acting on material surfaces as described in Section 6.2.1. The change in these components can therefore be attributed to the constitutive behavior of the material. The Cauchy and Kirchhoff stress components are different in this respect, as they refer to tractions on surfaces fixed in space. Thus, a solid body with a constant state of stress undergoing rigid body motion will not experience any change in the Piola–Kirchhoff stress components. However, the changing direction of the body relative to the fixed directions in space will in general imply that there is a change in the Cauchy and Kirchhoff stress components. In this section the changes in the components of Cauchy and Kirchhoff stresses that can be ascribed to changes in the state of the material are described. It is customary to imagine the associated motion as taking place in time, and the stress component changes are therefore called stress rates. The stress rates that can be attributed to changes of the state of the material are called objective stress rates. By implication the objective stress rate associated with a rigid body motion must vanish. This criterion is necessary, but not sufficient, to define an objective stress rate. Several objective stress rates have been defined in the literature, see e.g. Holzapfel (2000), but only the two most common will be used here.

A simple and direct approach to the derivation of objective stress rates starts out by observing that the Piola–Kirchhoff stress $\mathbf{S}$ is defined with reference to material surfaces and directions, and thus the rate of its components defined as the time derivative $\dot{\mathbf{S}}$ is objective. This enables derivation of objective rates of other stress definitions directly from the relation of their original components to the Piola–Kirchhoff stress. This approach leads to the so-called Truesdell stress rate.

The simplest case is the Kirchhoff stress $\boldsymbol{\tau}$. Inversion of the relation (6.40) gives

$$\mathbf{S} = \mathbf{F}^{-1}\,\boldsymbol{\tau}\,\mathbf{F}^{-T}. \tag{6.41}$$

The time derivative of this relation is

$$\dot{\mathbf{S}} = \mathbf{F}^{-1}\,\dot{\boldsymbol{\tau}}\,\mathbf{F}^{-T} + \dot{\mathbf{F}}^{-1}\,\boldsymbol{\tau}\,\mathbf{F}^{-T} + \mathbf{F}^{-1}\,\boldsymbol{\tau}\,\dot{\mathbf{F}}^{-T}. \tag{6.42}$$

Now, the basic argument is that the objective rate of the Kirchhoff stress, denoted $\overset{\circ}{\boldsymbol{\tau}}$, should follow from the rate of the Piola–Kirchhoff stress $\dot{\mathbf{S}}$ in the same way as the current components $\boldsymbol{\tau}$ follow from $\mathbf{S}$, namely by the relation (6.40). This defines the objective rate of the Kirchhoff stress as

$$\overset{\circ}{\boldsymbol{\tau}} = \mathbf{F}\,\dot{\mathbf{S}}\,\mathbf{F}^T. \tag{6.43}$$

Substitution of $\dot{\mathbf{S}}$ from (6.42) then gives

$$\overset{\circ}{\boldsymbol{\tau}} = \dot{\boldsymbol{\tau}} + \mathbf{F}\,\dot{\overline{\mathbf{F}^{-1}}}\,\boldsymbol{\tau} + \boldsymbol{\tau}\,(\mathbf{F}\,\dot{\overline{\mathbf{F}^{-1}}})^T. \tag{6.44}$$

It is seen that the effect of convection of the material is included via a factor $\mathbf{F}\dot{\overline{\mathbf{F}^{-1}}}$ and its transpose.

The factors that represent the material convection in this formula can be given a precise interpretation. By definition of the time derivative, the deformation gradient has the components

$$\dot{F}_{\alpha\beta} = \frac{\partial \dot{x}_\alpha}{\partial x^0_\beta}. \tag{6.45}$$

The time derivatives $\dot{x}_\alpha$ are the components of the velocity of the particle. Thus (6.45) is the gradient of the particle velocity field with respect to the reference coordinates x^0_β. The velocity gradient $\mathbf{L}$ with respect to the current coordinates is obtained via post-multiplication by $\mathbf{F}^{-1}$,

$$\mathbf{L} = \dot{\mathbf{F}}\,\mathbf{F}^{-1} = \frac{\partial \dot{x}_\alpha}{\partial x^0_\gamma}\frac{\partial x^0_\gamma}{\partial x_\beta} = \frac{\partial \dot{x}_\alpha}{\partial x_\beta}. \tag{6.46}$$

The velocity gradient tensor is defined in terms of derivatives with respect to the current state, in contrast to the deformation and displacement gradients, where gradient refers to derivatives with respect to the coordinates of the reference state.

The factor $\mathbf{F}\dot{\overline{\mathbf{F}^{-1}}}$ describing the convection in the objective stress rate (6.44) is obtained by taking the time derivative of the identity $\mathbf{F}\mathbf{F}^{-1} = \mathbf{I}$, whereby

$$\mathbf{F}\dot{\overline{\mathbf{F}^{-1}}} = -\dot{\mathbf{F}}\,\mathbf{F}^{-1} = -\mathbf{L}. \tag{6.47}$$

The final form of the Truesdell rate of the Kirchhoff stress in terms of the velocity gradient is

$$\overset{\circ}{\boldsymbol{\tau}} = \dot{\boldsymbol{\tau}} - \mathbf{L}\,\boldsymbol{\tau} - \boldsymbol{\tau}\,\mathbf{L}^T. \tag{6.48}$$

Note that while $\dot{\boldsymbol{\tau}}$ denotes the time derivative of the components of the Kirchhoff stress tensor, constitutive relations must be formulated in terms of objective stress rates, such as the Truesdell rate $\overset{\circ}{\boldsymbol{\tau}}$.

The Truesdell rate of the Cauchy stress can be found via the time derivatives of the second Piola–Kirchhoff stress tensor by a similar procedure. Inversion of (6.37) gives

$$\mathbf{S} = J\,\mathbf{F}^{-1}\boldsymbol{\sigma}\,\mathbf{F}^{-T} \tag{6.49}$$

and time differentiation leads to

$$\dot{\mathbf{S}} = \mathbf{F}^{-1}(J\dot{\boldsymbol{\sigma}} + \dot{J}\boldsymbol{\sigma})\mathbf{F}^{-T} + J\dot{\mathbf{F}}^{-1}\boldsymbol{\sigma}\,\mathbf{F}^{-T} + J\,\mathbf{F}^{-1}\boldsymbol{\sigma}\,\dot{\mathbf{F}}^{-T}. \tag{6.50}$$

This formula contains the time derivative $\dot{J}$ of the Jacobian determinant. The Jacobian determinant V is the current volume of the material originally occupying unit volume. Thus, $\dot{J}/J$ is the current rate of volume change, which can be expressed in terms of the time derivatives of the strains,

$$\dot{J}/J = \dot{\varepsilon}_{11} + \dot{\varepsilon}_{22} + \dot{\varepsilon}_{33} = \mathrm{tr}(\dot{\boldsymbol{\varepsilon}}). \tag{6.51}$$

Definition of the objective Cauchy stress rate $\overset{\circ}{\boldsymbol{\sigma}}$ from the original transformation formula (6.37) then leads to

$$\overset{\circ}{\boldsymbol{\sigma}} = \dot{\boldsymbol{\sigma}} + \mathrm{tr}(\dot{\boldsymbol{\varepsilon}})\,\boldsymbol{\sigma} - \mathbf{L}\boldsymbol{\sigma} - \boldsymbol{\sigma}\mathbf{L}^T. \tag{6.52}$$

The term containing $\mathrm{tr}(\dot{\boldsymbol{\varepsilon}})$ accounts for the change in the current volume element dV, an effect that has been neutralized in the definition of the Kirchhoff stress.

Note that the circle symbol for the Truesdell stress rate has here been used to denote a particular procedure for defining the objective rate, in which the stress – here Kirchhoff or Cauchy stress – is transformed to a material frame of reference by a relation of the form (6.41) or (6.49). In the material frame a direct time derivative is objective and can be used to describe material behavior. The corresponding Truesdell rate of Kirchhoff or Cauchy stress is therefore defined by a similar transformation of the increments in the material frame. This procedure corresponds to the concept of a Lie derivative discussed e.g. by Marsden and Hughes (1983) and Holzapfel (2000). It is similar to the discussion in Section 4.2 of the virtual and actual changes of the local components of strain and curvature in the development of a large deformation beam theory.

EXAMPLE 6.4. TRUESDELL STRESS RATE IN CONSTANT VOLUME SHEAR. The constant volume shear problem introduced in Example 6.2 and illustrated in Fig. 6.5(b) is described by the deformation tensor, its inverse and its time derivative:

$$\mathbf{F} = \begin{bmatrix} 1 & \gamma \\ 0 & 1 \end{bmatrix}, \quad \mathbf{F}^{-1} = \begin{bmatrix} 1 & -\gamma \\ 0 & 1 \end{bmatrix}, \quad \dot{\mathbf{F}} = \begin{bmatrix} 0 & \dot{\gamma} \\ 0 & 0 \end{bmatrix}.$$

The velocity gradient tensor is determined by (6.46) as

$$\mathbf{L} = \dot{\mathbf{F}}\,\mathbf{F}^{-1} = \begin{bmatrix} 0 & \dot{\gamma} \\ 0 & 0 \end{bmatrix}.$$

The velocity gradient tensor determines the extra terms in the objective stress rate (6.48),

$$\overset{\circ}{\boldsymbol{\tau}} = \dot{\boldsymbol{\tau}} - \mathbf{L}\boldsymbol{\tau} - \boldsymbol{\tau}\mathbf{L}^T = \dot{\boldsymbol{\tau}} - \dot{\gamma}\begin{bmatrix} \tau_{12}+\tau_{21} & \tau_{22} \\ \tau_{22} & 0 \end{bmatrix}.$$

When this formula is used to express the time derivative of the Kirchhoff stress,

$$\dot{\boldsymbol{\tau}} = \overset{\circ}{\boldsymbol{\tau}} + \dot{\gamma}\begin{bmatrix} \tau_{12}+\tau_{21} & \tau_{22} \\ \tau_{22} & 0 \end{bmatrix},$$

the last term represents the effect of the convection of the material in the spatial frame of reference. It is seen that the convection contribution is independent of the stress component τ_{11}. Intuitively this stress may be associated with tension in an imagined fiber in the x_1-direction, which does not change during the motion. On the other hand, the current tension τ_{22} in the direction of the x_2-axis contributes $\dot{\gamma}\tau_{22}$ to the time derivative of the shear strain, and a current shear stress $\tau_{12} = \tau_{21}$ contributes $2\dot{\gamma}\tau_{12}$ to the time derivative of the normal stress in the x_1-direction. Intuitively, effects of this type were to be expected.

In the present problem there is no change of volume, and therefore the stress rate formulae also apply to the Cauchy stress.

Other objective stress rates have also found use in modeling large deformation of solids. Probably the most used is the Jaumann stress rate. It is based on the idea that the change of stress components in a fixed spatial coordinate system consists of two parts: one part corresponding to a change of stress in the material, and another due to convection of the material relative to the spatial frame of reference. In the Jaumann stress rate the effect of convection is described by reference to an instantaneous angular velocity $\boldsymbol{\omega}$. This angular velocity, called the spin, is identified by resolving the velocity gradient tensor into a symmetric and an anti-symmetric part. The symmetric part is the strain rate $\dot{\boldsymbol{\varepsilon}}$, while the anti-symmetric part is the spin tensor $\mathbf{W}$:

$$\mathbf{L} = \dot{\boldsymbol{\varepsilon}} + \mathbf{W} \tag{6.53}$$

with strain rate and spin tensor defined by

$$\dot{\boldsymbol{\varepsilon}} = \tfrac{1}{2}(\mathbf{L} + \mathbf{L}^T), \qquad \mathbf{W} = \tfrac{1}{2}(\mathbf{L} - \mathbf{L}^T). \tag{6.54}$$

It follows from the discussion in Section 3.1 that any skew-symmetric tensor can be associated with a vector representing an infinitesimal rotation or angular velocity. Thus, according to (3.7) the spin tensor $\mathbf{W}$ can be expressed in terms of the angular velocity $\boldsymbol{\omega}$ as

$$\mathbf{W} = \hat{\boldsymbol{\omega}} = \begin{bmatrix} 0 & -\omega_3 & \omega_2 \\ \omega_3 & 0 & -\omega_1 \\ -\omega_2 & \omega_1 & 0 \end{bmatrix}. \tag{6.55}$$

The Jaumann stress rate is defined by a relation similar to the Truesdell rate (6.48), but now using only the skew-symmetric part of the velocity gradient. This gives the Jaumann rate

$$\overset{\triangledown}{\boldsymbol{\tau}} = \dot{\boldsymbol{\tau}} - \mathbf{W}\boldsymbol{\tau} - \boldsymbol{\tau}\mathbf{W}^T \tag{6.56}$$

of the Kirchhoff stress. The Jaumann rate only accounts for the spin, but not for the strain rate. The volume rate does not enter into the definition of the Jaumann stress rate, and thus the formula for the Jaumann rate of the Cauchy stress is of the same form as (6.56).

The Truesdell stress rates for the Kirchhoff stress and the Cauchy stress are easily expressed in terms of the Jaumann rate by substitution of the decomposition (6.53) of the velocity gradient tensor. The Kirchhoff stress rate formula follows from (6.48),

$$\overset{\circ}{\boldsymbol{\tau}} = \overset{\triangledown}{\boldsymbol{\tau}} - \dot{\boldsymbol{\varepsilon}}\boldsymbol{\tau} - \boldsymbol{\tau}\dot{\boldsymbol{\varepsilon}} \tag{6.57}$$

and the Cauchy stress rate formula from (6.52),

$$\overset{\circ}{\boldsymbol{\sigma}} = \overset{\triangledown}{\boldsymbol{\sigma}} + \mathrm{tr}(\dot{\boldsymbol{\varepsilon}})\boldsymbol{\sigma} - \dot{\boldsymbol{\varepsilon}}\boldsymbol{\sigma} - \boldsymbol{\sigma}\dot{\boldsymbol{\varepsilon}}. \tag{6.58}$$

It is seen that the difference between the Truesdell and the Jaumann stress rates is described by the current state of stress and the current rate of strain, but is independent of any rate of rotation. With a suitable constitutive relation for the material the strain rate can be expressed in terms of the current state of the material and the rate of the loads, and thus the choice between the Truesdell and Jaumann stress rates is largely a matter of taste and convenience.

EXAMPLE 6.5. JAUMANN STRESS RATE IN CONSTANT VOLUME SHEAR. The Truesdell stress rate of the constant volume shear problem was considered in Example 6.4. In the Jaumann stress rate the convection terms are represented by the spin tensor $\mathbf{W}$, defined as the symmetric part of the velocity gradient tensor $\mathbf{L}$. The strain rate $\dot{\boldsymbol{\varepsilon}}$ and the spin tensor $\mathbf{W}$ are found as the symmetric and anti-symmetric parts of the velocity gradient tensor $\mathbf{L}$, determined in Example 6.4.

$$\mathbf{L} == \begin{bmatrix} 0 & \dot{\gamma} \\ 0 & 0 \end{bmatrix}, \quad \dot{\boldsymbol{\varepsilon}} = \begin{bmatrix} 0 & \frac{1}{2}\dot{\gamma} \\ \frac{1}{2}\dot{\gamma} & 0 \end{bmatrix}, \quad \mathbf{W} = \begin{bmatrix} 0 & \frac{1}{2}\dot{\gamma} \\ -\frac{1}{2}\dot{\gamma} & 0 \end{bmatrix}.$$

The spin tensor $\mathbf{W}$ determines the extra terms in the objective stress rate (6.56),

$$\overset{\nabla}{\boldsymbol{\tau}} = \dot{\boldsymbol{\tau}} - \mathbf{W}\boldsymbol{\tau} - \boldsymbol{\tau}\mathbf{W}^T = \dot{\boldsymbol{\tau}} - \tfrac{1}{2}\dot{\gamma}\begin{bmatrix} \tau_{12}+\tau_{21} & \tau_{22}-\tau_{11} \\ \tau_{22}-\tau_{11} & -(\tau_{12}+\tau_{21}) \end{bmatrix}.$$

The time derivative of the Kirchhoff stress in terms of the Jaumann stress rate is

$$\dot{\boldsymbol{\tau}} = \overset{\nabla}{\boldsymbol{\tau}} + \tfrac{1}{2}\dot{\gamma}\begin{bmatrix} \tau_{12}+\tau_{21} & \tau_{22}-\tau_{11} \\ \tau_{22}-\tau_{11} & -(\tau_{12}+\tau_{21}) \end{bmatrix}.$$

It is seen that the Jaumann stress rate formula involves all the current stress components, while the Truesdell rate is independent of τ_{11} in this motion.

6.3 Total Lagrangian formulation

As discussed in Chapter 2, non-linear problems of solids and structures may be formulated either in the total or the updated Lagrangian format. In the total Lagrangian format the motion of the material points is described by the total displacement field $\mathbf{u}(\mathbf{x}_0)$ with reference to the initial position $\mathbf{x}_0$ of the material point. This formulation leads naturally to deformations described in terms of the Green strain tensor $\mathbf{E}$ and stresses described in terms of the second Piola–Kirchhoff stress tensor $\mathbf{S}$. An advantage of this method is that the total stress appears directly in the equations, while a disadvantage is that strain increments become non-linear, as they refer to the initial state. Alternatively, the non-linear problem may be described in the updated Lagrangian format in terms of the displacement increments $d\mathbf{u}(\mathbf{x})$ with reference to the current state $\mathbf{x}$. This formulation leads to linearized incremental strains $d\boldsymbol{\varepsilon}$ and Cauchy or Kirchhoff stresses. This gives the advantage that the strain increments may be introduced in linear form. However, the stress increments must be represented by their non-linear objective rates, as discussed in the previous section. Thus, the choice between the total and the updated Lagrangian formulations includes a choice of the specific form of the non-linearity in the equations to be solved.

This section describes three main aspects of the total Lagrangian formulation: the formulation of equilibrium equations, the formation of the tangent stiffness matrix, and the implementation in the finite element format by use of shape functions. The formulation of these steps in the updated Lagrangian format is presented in Section 6.4.

6.3.1 Equilibrium and residual forces

The total Lagrangian format makes use of the total displacement components $u_\alpha(x^0_\xi)$, considered as functions of the initial position x^0_ξ. This format makes use of the Green strain $E_{\alpha\beta}(x^0_\xi)$ corresponding to the conjugate Piola–Kirchhoff stress components $S_{\alpha\beta}(x^0_\xi)$. In order to illustrate the general procedure the virtual work relation is obtained from the equilibrium equation, reversing the argument in Section 6.2.1.

The equilibrium equation corresponding to the Piola–Kirchhoff stress components was given in (6.26) as

$$\frac{\partial}{\partial x^0_\beta}\left(\frac{\partial x_\gamma}{\partial x^0_\alpha} S_{\alpha\beta}\right) + p^0_\gamma = 0 \qquad \text{in } V_0. \tag{6.59}$$

In this equation p^0_γ stresses and displacements are considered as functions of the initial position x^0_ξ. The corresponding virtual work equation is obtained by scalar multiplication of this equation with a virtual displacement field $\delta u_\gamma(x^0_\xi)$, followed by integration over the initial volume V_0:

$$\int_{V_0} \delta u_\gamma \frac{\partial}{\partial x^0_\beta}\left(\frac{\partial x_\gamma}{\partial x^0_\alpha} S_{\alpha\beta}\right) dV_0 + \int_{V_0} \delta u_\alpha \, p^0_\alpha \, dV_0 = 0. \tag{6.60}$$

By use of the divergence theorem, expressed by (6.25), this relation may be written in the form of an equality of internal and external virtual work

$$\int_{V_0} \delta E_{\alpha\beta} \, S_{\alpha\beta} \, dV_0 = \int_{S_0} \delta u_\gamma \, t^0_\gamma \, dS_0 + \int_{V_0} \delta u_\gamma \, p^0_\gamma \, dV_0, \tag{6.61}$$

where the virtual Green strain components $\delta E_{\alpha\beta}$ have been introduced from (6.16),

$$\delta E_{\alpha\beta} = \frac{1}{2}\left(\frac{\partial x_\gamma}{\partial x^0_\alpha}\frac{\partial(\delta u_\gamma)}{\partial x^0_\beta} + \frac{\partial(\delta u_\gamma)}{\partial x^0_\alpha}\frac{\partial x_\gamma}{\partial x^0_\beta}\right) \tag{6.62}$$

and t^0_γ is the surface traction vector, related to the stress tensor components by (6.27).

The right side of the virtual work equation (6.61) defines the external virtual work, i.e. the virtual work of the external loads,

$$\delta V_{\text{ext}} = \int_{S_0} \delta u_\gamma \, t_\gamma^0 \, dS_0 + \int_{V_0} \delta u_\gamma \, p_\gamma^0 \, dV_0 \tag{6.63}$$

while the left side defines the internal virtual work of the stresses,

$$\delta V_{\text{int}} = \int_{V_0} \delta E_{\alpha\beta} \, S_{\alpha\beta} \, dV_0. \tag{6.64}$$

Equilibrium is expressed by the statement $\delta V_{\text{int}} = \delta V_{\text{ext}}$, valid for any displacement variation field $\delta x_\gamma(x_\xi^0)$. Thus, any lack of equilibrium may be considered equivalent to a distribution of residual forces $r_\gamma(x_\xi^0)$, which accounts for the difference in the external and internal virtual work. This is expressed by the relation

$$\int_{V_0} \delta u_\gamma \, r_\gamma \, dV_0 = \delta V_{\text{ext}} - \delta V_{\text{int}} \tag{6.65}$$

which, for any state of displacement $u_\gamma(x_\xi)$ and stress $S_{\alpha\beta}(x_\xi)$, defines the residual force distribution $r_\gamma(x_\xi^0)$. For a suitably discretized displacement field this relation defines a discrete set of residual forces, which are eliminated in an iteration process that gradually establishes equilibrium. The finite element discretization is described in Section 6.3.3.

6.3.2 Tangent stiffness

In the same way the virtual work equation for the total internal and external forces is used to check equilibrium and to define residual forces, an incremental form of the virtual work equation is established in order to obtain the tangent stiffness. Instead of considering the total loads t_γ^0, p_γ^0 and stresses $S_{\alpha\beta}$, two neighboring configurations separated by the increments dt_γ^0, dp_γ^0 and $dS_{\alpha\beta}$ are considered. In practice it is easier to obtain the incremental relation by differentiation of the virtual work (6.61) than by considering the two neighboring states individually. The key point is that while loads, stresses and displacements are incremented, the variation δu_γ remains unaffected. This follows from the fact that δu_γ is to be considered as an arbitrary field, independent of the actual displacement field. In the case of continua, where the displacement representation depends on rotations, this argument has to be modified as explained in Chapter 3.

The tangent stiffness follows from the increment of the internal virtual work $\delta V_{\rm int}$,

$$d(\delta V_{\rm int}) = d\int_{V_0} \delta E_{\alpha\beta}\, S_{\alpha\beta}\, dV_0. \tag{6.66}$$

The stress components are symmetric, and the virtual strain can then be substituted from the first term in (6.62). It then follows from differentiation that

$$d(\delta V_{\rm int}) = \int_{V_0} \frac{\partial(\delta u_\gamma)}{\partial x^0_\alpha} S_{\alpha\beta} \frac{\partial(du_\gamma)}{\partial x^0_\beta}\, dV_0 + \int_{V_0} \delta E_{\alpha\beta}\, dS_{\alpha\beta}\, dV_0. \tag{6.67}$$

Now, assume that there exists a linear constitutive relation between the stress increments $dS_{\alpha\beta}$ and the strain increments $dE_{\gamma\delta}$ in the form

$$dS_{\alpha\beta} = C^0_{\alpha\beta\gamma\delta}\, dE_{\gamma\delta}. \tag{6.68}$$

Specific forms of this relation corresponding to elastic and elasto-plastic materials are discussed in Chapter 7.

The increment of the internal virtual work then takes the form

$$d(\delta V_{\rm int}) = \int_{V_0} \frac{\partial(\delta u_\gamma)}{\partial x^0_\alpha} S_{\alpha\beta} \frac{\partial(du_\gamma)}{\partial x^0_\beta}\, dV_0 + \int_{V_0} \delta E_{\alpha\beta}\, C^0_{\alpha\beta\gamma\delta}\, dE_{\gamma\delta}\, dV_0. \tag{6.69}$$

Due to the symmetry of the stress and strain components the elastic tensor $C_{\alpha\beta\gamma\delta}$ must be symmetric in the first and last pair of subscripts, and the strain variation and strain increment can therefore be introduced in the non-symmetric form corresponding to the first term in (6.62). With this, the increment of the internal virtual work is

$$\begin{aligned} d(\delta V_{\rm int}) &= \int_{V_0} \frac{\partial(\delta u_\gamma)}{\partial x^0_\alpha} S_{\alpha\beta} \frac{\partial(du_\gamma)}{\partial x^0_\beta}\, dV_0 \\ &+ \int_{V_0} \frac{\partial(\delta u_\kappa)}{\partial x^0_\alpha} \frac{\partial x_\kappa}{\partial x^0_\beta}\, C^0_{\alpha\beta\gamma\delta}\, \frac{\partial x_\lambda}{\partial x^0_\delta} \frac{\partial(du_\lambda)}{\partial x^0_\gamma}\, dV_0. \end{aligned} \tag{6.70}$$

This form of the incremental virtual work is similar to the tangent stiffness relation (2.68) obtained for bar elements. It consists of a constitutive stiffness contribution and a geometric stiffness contribution, where the deformation gradients appear as a consequence of the finite change of geometry.

In order to emphasize the simple structure of the relations they are also given in matrix form using the symbol : for contraction on two subscripts,

$$d(\delta V_{\rm int}) = \int_{V_0} (\delta\mathbf{D}^T d\mathbf{D}) : \mathbf{S}\, dV_0 + \int_{V_0} \delta\mathbf{E}^T : \mathbb{C}_0 : d\mathbf{E}\, dV_0. \tag{6.71}$$

The parentheses indicate that the product $\delta\mathbf{D}^T d\mathbf{D}$ should be formed, leaving two free subscripts that are then contracted with $\mathbf{S}$. The previous formulae may help in the detailed interpretation of the matrix products.

Several special cases of the incremental virtual work (6.67) are of interest. If the initial stress contribution and the dependence on the current displacement are omitted, the traditional linear form is obtained. Note that material non-linearities can still be included via the coefficients $C^0_{\gamma\delta\alpha\beta}$. In updated formulations and in linearized stability theory the initial stress terms are included, while the dependence on the current displacement is left out, see e.g. Washizu (1974).

There is a close relation between the symmetry of the incremental virtual work and the existence of a strain energy function for the material. If the material has a strain energy density function $\varphi(E_{\alpha\beta})$, the stresses are given by the derivatives

$$S_{\alpha\beta} = \frac{\partial\varphi}{\partial E_{\alpha\beta}}. \tag{6.72}$$

Then the stress increments follow by further differentiation,

$$dS_{\alpha\beta} = \frac{\partial^2\varphi}{\partial E_{\alpha\beta}\partial E_{\gamma\delta}}\, dE_{\gamma\delta} = C^0_{\alpha\beta\gamma\delta}\, dE_{\gamma\delta}. \tag{6.73}$$

This establishes the coefficients $C^0_{\alpha\beta\gamma\delta}$ as double derivatives of the strain energy density $\varphi(E_{\alpha\beta})$, and the symmetry relation $C^0_{\gamma\delta\alpha\beta} = C^0_{\alpha\beta\gamma\delta}$ follows from interchanging the order of differentiation. With this symmetry the incremental internal virtual work is symmetric with respect to interchange of $d\mathbf{u}$ and $\delta\mathbf{u}$. This symmetry in turn leads to a symmetric tangent stiffness relation. It should be noted that several material models do not lead to a strain energy density, see e.g. Ottosen and Ristinmaa (2005). Those materials must therefore be treated by incremental relations. An important example is the incremental theory of plasticity treated in the next chapter.

The principle of virtual work may also be written as an equation expressing that the change in virtual work δV is zero for arbitrary variations of the displacement field:

$$\delta V = \int_{V_0} S_{\alpha\beta}\,\delta E_{\alpha\beta}\, dV_0 - \int_{S_0} t^0_\gamma\,\delta u_\gamma\, dS_0 - \int_{V_0} p^0_\gamma\,\delta u_\gamma\, dV_0. \tag{6.74}$$

If the material has a strain energy density function, and if the loads are conservative, this may be considered as the variation of a potential energy functional $\Phi(\mathbf{u})$ found by integration. In the case of constant loads the potential energy has the form

$$\Phi(\mathbf{u}) = \int_{V_0} \varphi(E_{\gamma\delta})\, dV_0 - \int_{S_0} t_\gamma^0 u_\gamma\, dS_0 - \int_{V_0} p_\gamma^0 u_\gamma\, dV_0. \tag{6.75}$$

In this formulation it is seen that the geometrical non-linearity is contained completely in the non-linear strain definition, while material non-linearity determines the form of the strain energy function.

6.3.3 Finite element implementation

Within the finite element method the most common way of discretizing problems of solid mechanics consists in representing the displacement field $\mathbf{u}(\mathbf{x}_0)$ in terms of the displacement $\mathbf{u}_n$ at selected nodes $\mathbf{x}_n^0$, $n = 1, 2, \ldots$ This corresponds to a representation in the form

$$\mathbf{u}(\mathbf{x}_0) = \sum_n h_n(\mathbf{x}_0)\, \mathbf{u}_n, \tag{6.76}$$

where $h_n(\mathbf{x}_0)$ is the shape function corresponding to a unit displacement at node n. Thus, the shape functions must satisfy the relation

$$h_n(\mathbf{x}_m^0) = \delta_{nm}, \tag{6.77}$$

where δ_{nm} is the Kronecker delta. This relation implies that $h_n(\mathbf{x}_0)$ is unity at the node $\mathbf{x}_n^0$ and zero at all other nodes. In practice the shape functions may not be explicitly given functions of the spatial coordinates $\mathbf{x}_0$, but be given as a mapping from a normalized set of coordinates.

In the formulation of the virtual work, the external virtual work can be evaluated directly from (6.63) by substituting a representation of the virtual displacement field of the same form as (6.76),

$$\delta\mathbf{u}(\mathbf{x}_0) = \sum_n h_n(\mathbf{x}_0)\, \delta\mathbf{u}_n. \tag{6.78}$$

The external virtual work can now be used to define equivalent external nodal forces $\mathbf{f}_n^{\text{ext}}$ via the relation

$$\delta V_{\text{ext}} = \delta\mathbf{u}_n^T \mathbf{f}_n^{\text{ext}} = \delta\mathbf{u}_n^T \Big\{ \int_{S_0} h_n(\mathbf{x}_0)\, \mathbf{t}_0\, dS_0 + \int_{V_0} h_n(\mathbf{x}_0)\, \mathbf{p}_0\, dV_0 \Big\}, \tag{6.79}$$

where summation over the nodes $n = 1, 2, \ldots$ is implied. The components of the virtual nodal displacements $\delta\mathbf{u}_n$ can be selected independently, and thus the scalar relation (6.79) defines the external nodal forces as

$$\mathbf{f}_n^{\text{ext}} = \int_{S_0} h_n(\mathbf{x}_0)\, \mathbf{t}_0\, dS_0 + \int_{V_0} h_n(\mathbf{x}_0)\, \mathbf{p}_0\, dV_0. \tag{6.80}$$

Thus, the external nodal forces are obtained by integrating the surface traction and the volume force, weighted by the corresponding shape function $h_n(\mathbf{x}_0)$.

In spite of the fact that the stresses $S_{\alpha\beta}$ and virtual strains $\delta E_{\alpha\beta}$ are components of two-dimensional tensors, it is often computationally convenient to represent these components as a one-dimensional array. In the following a column format will be used. The symmetry of the stress and strain components leads to the six-component form of the virtual work

$$\begin{aligned}\delta E_{\alpha\beta}\, S_{\alpha\beta} &= \delta E_{11}\, S_{11} + \delta E_{22}\, S_{22} + \delta E_{33}\, S_{33} \\ &+ 2\delta E_{23}\, S_{23} + 2\delta E_{31}\, S_{31} + 2\delta E_{12}\, S_{12}. \end{aligned} \tag{6.81}$$

This expression of virtual work may alternatively be formed by use of the one-dimensional arrays for the strain increments and stresses

$$\delta\mathbf{E} = [\,\delta E_{11}, \delta E_{22}, \delta E_{33}, 2\delta E_{23}, 2\delta E_{31}, 2\delta E_{12}\,]^T \tag{6.82}$$

and

$$\mathbf{S} = [\; S_{11}\,,\, S_{22}\,,\, S_{33}\,,\, S_{23}\,,\, S_{31}\,,\, S_{12}\;]^T\,, \tag{6.83}$$

where the factor two has been included in the virtual strains.

The internal virtual work (6.64) involves the virtual strain $\delta E_{\alpha\beta}$, given by (6.62) in terms of products of the deformation gradient $\mathbf{F} = [\partial x_\gamma / \partial x^0_\alpha]$ and the displacement gradient $\delta\mathbf{D} = [\partial(\delta u_\gamma)/\partial x^0_\beta]$. It is computationally desirable to arrange these tensors in a format that gives the incremental strain $\delta\mathbf{E}$ in a product format. This is accomplished by considering the columns of the component matrix of the deformation gradient

$$\mathbf{F} = [\partial x_\alpha / \partial x^0_\beta] = [\,\mathbf{f}_1, \mathbf{f}_2, \mathbf{f}_3\,] \tag{6.84}$$

and the virtual displacement gradient

$$\delta\mathbf{D} = [\partial(\delta u_\alpha)/\partial x^0_\beta] = [\,\delta\mathbf{d}_1, \delta\mathbf{d}_2, \delta\mathbf{d}_3\,]. \tag{6.85}$$

It is seen that the summation in the definition of the virtual strain increment $\delta E_{\alpha\beta}$ is on the subscript denoting the row number, and thus the summation can be expressed directly by the boldface notation introduced by the columns in (6.84) and (6.85). The array format (6.82) of the virtual strain then takes the form

$$\delta\mathbf{E} = \begin{bmatrix} \mathbf{f}_1^T\delta\mathbf{d}_1 \\ \mathbf{f}_2^T\delta\mathbf{d}_2 \\ \mathbf{f}_3^T\delta\mathbf{d}_3 \\ \mathbf{f}_2^T\delta\mathbf{d}_3 + \mathbf{f}_3^T\delta\mathbf{d}_2 \\ \mathbf{f}_3^T\delta\mathbf{d}_1 + \mathbf{f}_1^T\delta\mathbf{d}_3 \\ \mathbf{f}_1^T\delta\mathbf{d}_2 + \mathbf{f}_2^T\delta\mathbf{d}_1 \end{bmatrix} = \begin{bmatrix} \mathbf{f}_1^T & \mathbf{0} & \mathbf{0} \\ \mathbf{0} & \mathbf{f}_2^T & \mathbf{0} \\ \mathbf{0} & \mathbf{0} & \mathbf{f}_3^T \\ \mathbf{0} & \mathbf{f}_3^T & \mathbf{f}_2^T \\ \mathbf{f}_3^T & \mathbf{0} & \mathbf{f}_1^T \\ \mathbf{f}_2^T & \mathbf{f}_1^T & \mathbf{0} \end{bmatrix} \begin{bmatrix} \delta\mathbf{d}_1 \\ \delta\mathbf{d}_2 \\ \delta\mathbf{d}_3 \end{bmatrix}. \tag{6.86}$$

The column vectors $\delta\mathbf{d}$ in the last factor are expressed in terms of the node displacements and the shape functions by use of (6.78):

$$\delta\mathbf{d}_\beta = \frac{\partial(\delta\mathbf{u})}{\partial x_\beta^0} = \sum_n \frac{\partial h_n}{\partial x_\beta^0}\,\delta\mathbf{u}_n = \sum_n h_{n,\beta}\,\delta\mathbf{u}_n, \tag{6.87}$$

where the notation $h_{n,\beta}$ has been used for the material derivatives of the shape functions. The relation (6.86) for the virtual strain can now be expressed in the compact form

$$\delta\mathbf{E}(\mathbf{x}_0) = \sum_n \hat{\mathbf{F}}\,\mathbf{B}_n\,\delta\mathbf{u}_n = \sum_n \mathbf{B}_n^0(\mathbf{x}_0)\,\delta\mathbf{u}_n, \tag{6.88}$$

where the matrix $\hat{\mathbf{F}}$ is defined by the first factor in (6.86), and the material derivatives of the shape functions are contained in the matrices

$$\mathbf{B}_n = \begin{bmatrix} h_{n,1}\,\mathbf{I} \\ h_{n,2}\,\mathbf{I} \\ h_{n,3}\,\mathbf{I} \end{bmatrix}. \tag{6.89}$$

Most often the factors $\partial x_\alpha/\partial x_\beta^0$ and $\partial h_n/\partial x_\gamma^0$ are given implicitly and the evaluation of the matrix $\mathbf{B}_n^0(\mathbf{x}_0)$ is carried out numerically at selected points, see e.g. Hughes (1987) and Zienkiewicz and Taylor (2000) for computational details.

The internal virtual work is now used to define internal nodal forces $\mathbf{f}_n^{\text{int}}$ by substitution of the virtual strain field into (6.64):

$$\delta V_{\text{int}} = \delta\mathbf{u}_n^T\,\mathbf{f}_n^{\text{int}} = \delta\mathbf{u}_n^T\Big\{\int_{V_0} \mathbf{B}_n^0(\mathbf{x}_0)^T\mathbf{S}\,dV_0\Big\}, \tag{6.90}$$

where summation over the nodes $n = 1, 2, \dots$ is implied. By selecting non-vanishing nodal displacements one node at a time, the following definition of the internal nodal forces is obtained:

$$\mathbf{f}_n^{\text{int}} = \int_{V_0} \mathbf{B}_n^0(\mathbf{x}_0)^T\,\mathbf{S}\,dV_0. \tag{6.91}$$

The calculation of the internal nodal forces requires evaluation of a volume integral of the product of the current Piola–Kirchhoff stress and the matrix $\mathbf{B}_n^0(\mathbf{x}_0)$.

The integral form of the residual force relation (6.65) can now be written in discretized form in terms of nodal residual forces $\mathbf{r}_n$,

$$\mathbf{r}_n = \mathbf{f}_n^{\text{ext}} - \mathbf{f}_n^{\text{int}}. \tag{6.92}$$

For a given configuration $\mathbf{x}(\mathbf{x}_0)$ with Piola–Kirchhoff stress distribution $\mathbf{S}(\mathbf{x}_0)$ the external and internal forces are calculated from (6.80) and (6.91), respectively. Any deviation from equilibrium is then expressed by the nodal residual forces $\mathbf{r}_n$, determined by (6.92). In the absence of equilibrium the residual forces are eliminated by iterations using the tangent stiffness matrix.

The tangent stiffness is evaluated from the increment of the internal virtual work (6.70) by substitution of the displacement and the incremental strain representations (6.78) and (6.88), whereby

$$d(\delta V_{\text{int}}) = \delta\mathbf{u}_n^T \Big\{ \mathbf{I} \int_{V_0} \frac{\partial h_n}{\partial x_\alpha^0} S_{\alpha\beta} \frac{\partial h_m}{\partial x_\beta^0}\, dV_0 + \int_{V_0} \mathbf{B}_n^{0T}\, \mathbb{C}_0\, \mathbf{B}_m^0\, dV_0 \Big\}\, d\mathbf{u}_m, \tag{6.93}$$

where $\mathbf{I}$ is the 3×3 unit tensor with components $\delta_{\alpha\beta}$. The 6×6 matrix $\mathbb{C}_0$ is the incremental stiffness corresponding to the relation (6.68), when written in terms of the array notation (6.82) and (6.83) for strains and stresses,

$$d\mathbf{S} = \mathbb{C}_0\, d\mathbf{E}. \tag{6.94}$$

The incremental virtual work relation (6.93) is of the form

$$d(\delta V_{\text{int}}) = \delta\mathbf{u}_n^T\, \mathbf{K}_{nm}\, d\mathbf{u}_m, \tag{6.95}$$

where summation over n and m is implied, and $\mathbf{K}_{nm}$ is the contribution to the tangent stiffness matrix from the combination of nodes n and m. It follows from (6.93) and (6.95) that the stiffness matrix contribution $\mathbf{K}_{nm}$ is given by

$$\mathbf{K}_{nm} = \mathbf{I} \int_{V_0} \frac{\partial h_n}{\partial x_\alpha^0} S_{\alpha\beta} \frac{\partial h_m}{\partial x_\beta^0}\, dV_0 + \int_{V_0} \mathbf{B}_n^{0T}\, \mathbb{C}_0\, \mathbf{B}_m^0\, dV_0. \tag{6.96}$$

The global stiffness matrix is assembled as discussed e.g. in Section 2.5.

The total Lagrangian formulation is complicated by the fact that the matrix $\mathbf{B}_n^0(\mathbf{x}_0)$ depends on the components of the deformation gradient $\mathbf{F} = \partial\mathbf{x}/\partial\mathbf{x}_0$. Simplifications arise if the current configuration $\mathbf{x}$ coincides with the initial configuration $\mathbf{x}_0$. In that case $\partial x_\alpha/\partial x_\beta^0 = \delta_{\alpha\beta}$, and the components of the matrix $\mathbf{F}$ are either unity or zero. Hereby the formulation becomes similar to that of a linear finite element analysis, in which an initial stress term has been included. This simplification of the computation of internal forces and the tangent stiffness matrix is a characteristic feature of the updated Lagrangian formulation discussed in the next section.

6.4 Updated Lagrangian formulation

In the previous section it was demonstrated how the finite deformation primarily enters into the total Lagrangian formulation via the non-linear form of the strain increments. In this section it is described how the non-linearities may be transferred into the constitutive behavior by use of an updated Lagrangian formulation. In the updated Lagrangian formulation each increment uses the current configuration as reference. This gives some simplifications in the kinematic description, but on the other side the change of stress must be formulated in an objective format to justify evaluation by a constitutive relation for the material properties. First, a direct transformation of the total format into updated form is demonstrated. Then, formulations derived directly from the principle of virtual work in the current state are discussed, and finally a brief description of implementation issues is given.

6.4.1 Transformation from total to updated format

One way to obtain an updated Lagrangian formulation is by transformation of the total format presented above. The quantities needed are: the external virtual work, the internal virtual work, and the incremental virtual work representing the tangent stiffness.

The external virtual work follows immediately from (6.63) by introducing $\mathbf{t}$ as the traction vector with reference to the current bounding surface element dS and $\mathbf{p}$ as the distributed load intensity with reference to the current volume element dV. The external virtual work then follows immediately from (6.63) in the form

$$\delta V_{\text{ext}} = \int_S \delta u_\gamma \, t_\gamma \, dS + \int_V \delta u_\gamma \, p_\gamma \, dV. \tag{6.97}$$

In a similar way the internal virtual work follows from (6.64) by using the relation (6.34), changing the format to linearized current strain, Cauchy stress and current volume:

$$\delta V_{\text{int}} = \int_V \delta \varepsilon_{\alpha\beta} \, \sigma_{\alpha\beta} \, dV. \tag{6.98}$$

These relations enable evaluation of residual forces representing any deviations from equilibrium.

The incremental internal virtual work was given by (6.67). The first term represents geometric effects, while the second term is the constitutive relation representing the material behavior. The geometric contribution can be transformed into current configuration by differentiation 'through' the current configuration via the chain rule of differentiation. The result is

$$\int_{V_0} \frac{\partial(\delta u_\gamma)}{\partial x_\alpha^0} S_{\alpha\beta} \frac{\partial(du_\gamma)}{\partial x_\beta^0}\, dV_0 = \int_V \frac{\partial(\delta u_\gamma)}{\partial x_\kappa} \Big(\frac{\partial x_\kappa}{\partial x_\alpha^0} \frac{S_{\alpha\beta}}{J} \frac{\partial x_\lambda}{\partial x_\beta^0} \Big) \frac{\partial(du_\gamma)}{\partial x_\lambda}\, dV, \tag{6.99}$$

where the volume element has also been changed to current volume by use of the relation $dV = JdV_0$. By using the relation (6.38) it is seen that the factor in parentheses is simply the Cauchy stress component matrix $\sigma_{\kappa\lambda}$.

The second term involves the stress increment $dS_{\alpha\beta}$. The defining property of the Truesdell rate of the Cauchy stress is that objective increments of Cauchy and Piola–Kirchhoff stress are related by the same transformation as their total components. Thus, the incremental form of the virtual work relation (6.34) is

$$\int_{V_0} \delta E_{\alpha\beta}\, dS_{\alpha\beta}\, dV_0 = \int_V \delta\varepsilon_{\alpha\beta}\, \overset{\circ}{D}\sigma_{\alpha\beta}\, dV, \tag{6.100}$$

where the notation $\overset{\circ}{D}$ has been introduced to denote the increment corresponding to the Truesdell rate. The precise form follows from (6.52):

$$\overset{\circ}{D}\sigma_{\alpha\beta} = d(\sigma_{\alpha\beta}J)\, J^{-1} - \frac{\partial(dx_\alpha)}{\partial x_\gamma}\sigma_{\gamma\beta} - \sigma_{\alpha\gamma}\frac{\partial(dx_\beta)}{\partial x_\gamma}, \tag{6.101}$$

where $d\sigma_{\alpha\beta}$ represents the increments of the components $\sigma_{\alpha\beta}$.

As discussed in Section 6.2.3, constitutive relations must be formulated in terms of objective stress rates. In the present context this means a relation between the Truesdell stress increment and the increment of the linearized stress of the form

$$\overset{\circ}{D}\sigma_{\alpha\beta} = C_{\alpha\beta\gamma\delta}\, d\varepsilon_{\gamma\delta}. \tag{6.102}$$

When using the strain and stress transformation rules (6.33) and (6.37) it can be demonstrated that the new stiffness component matrix $C_{\alpha\beta\gamma\delta}$ is connected to the previous stiffness component matrix $C^0_{\alpha\beta\gamma\delta}$ via the relation

$$C_{\xi\eta\kappa\lambda} = \frac{\partial x_\xi}{\partial x_\alpha^0}\frac{\partial x_\eta}{\partial x_\beta^0}\frac{C^0_{\alpha\beta\gamma\delta}}{J}\frac{\partial x_\kappa}{\partial x_\gamma^0}\frac{\partial x_\lambda}{\partial x_\delta^0}. \tag{6.103}$$

This relation can also be established by differentiating the first and last factors of the last integral in (6.70) 'through' the current state variable x_κ. Note that the initial coordinates $\mathbf{x}_0$ with index 0 are contracted by the indices of the stiffness coefficients of $\mathbb{C}_0$, also with index 0, while the stiffness coefficients of the current state $\mathbb{C}$ correspond to the indices of the current configuration $\mathbf{x}$. The relation (6.103) corresponds to a transformation of the material properties from the initial to the current configuration.

Combination of the geometric contribution (6.99) and the constitutive contribution (6.100) leads to the incremental virtual work in the form

$$d(\delta V_{\text{int}}) = \int_V \frac{\partial(\delta u_\gamma)}{\partial x_\alpha} \sigma_{\alpha\beta} \frac{\partial(du_\gamma)}{\partial x_\beta}\, dV + \int_V \delta\varepsilon_{\alpha\beta}\, C_{\alpha\beta\gamma\delta}\, d\varepsilon_{\gamma\delta}\, dV. \qquad (6.104)$$

It is seen that the external and internal force relations (6.97) and (6.98) and the incremental virtual work relation (6.104) are obtained from their total Lagrangian equivalents (6.63), (6.64) and (6.70) by changing to Cauchy stress and linearized current strain and using current surface and volume elements. In this formulation the complications associated with the non-linear Green strain interpolation matrix $\mathbf{B}_n^0$ have been absorbed into the constitutive relations for the material. However, this does not imply that the complications arising from the non-linear kinematics of the problem have vanished. As described in Chapter 7, the classical definition of (hyper)elastic materials as derivable from an elastic energy function implies that the elastic coefficients $C^0_{\alpha\beta\gamma\delta}$ from the material description can be constants, while the corresponding coefficients $C_{\alpha\beta\gamma\delta}$ must depend on the deformation gradient $F_{\xi\eta}$.

It is of interest to note that the present reformulation of the total Lagrangian equations into updated Lagrangian format establishes exact correspondence between the use of Green strain and Piola–Kirchhoff stress in the total formulation with the use of linearized current strain and Cauchy stress combined with its Truesdell rate in the updated formulation. Alternatively, other stress rates may be introduced, when the updated Lagrangian formulation is derived directly from the equilibrium equation in the current configuration.

6.4.2 Virtual work in the current configuration

The updated Lagrangian formulation may be obtained directly from the equation of virtual work formulated in the current configuration. The starting point is the equilibrium equation

$$\frac{\partial \sigma_{\alpha\beta}}{\partial x_\beta} + p_\alpha = 0 \quad \text{in } V \tag{6.105}$$

expressed in terms of the components $\sigma_{\alpha\beta}(\mathbf{x})$ of the Cauchy stress and the distributed force $p_\alpha(\mathbf{x})$ per unit current volume.

The virtual work equation is obtained by multiplying the equilibrium equation (6.105) with the virtual displacement field $\delta u_\alpha(\mathbf{x})$, followed by integration over the current volume. After use of the divergence theorem, the equation of virtual work takes the form

$$\int_V \delta\varepsilon_{\alpha\beta}\, \sigma_{\alpha\beta}\, dV = \int_S \delta u_\alpha\, t_\alpha\, dS + \int_V \delta u_\alpha\, p_\alpha\, dV, \tag{6.106}$$

where the virtual strain components are

$$\delta\varepsilon_{\alpha\beta} = \frac{1}{2}\left(\frac{\partial(\delta u_\alpha)}{\partial x_\beta} + \frac{\partial(\delta u_\beta)}{\partial x_\alpha} \right) \tag{6.107}$$

and the surface traction components t_α are related to the stress components by

$$t_\alpha = \sigma_{\alpha\beta}\, n_\beta. \tag{6.108}$$

In this formulation the distributed surface loads as well as the surface normal components n_β refer to the current surface area.

It follows immediately from the virtual work equation (6.106) that the external virtual work in the current configuration is

$$\delta V_{\text{ext}} = \int_S \delta u_\alpha\, t_\alpha\, dS + \int_V \delta u_\alpha\, p_\alpha\, dV, \tag{6.109}$$

where the loads are normalized with respect to the current surface and volume, respectively. The internal virtual work is

$$\delta V_{\text{int}} = \int_V \delta\varepsilon_{\alpha\beta}\, \sigma_{\alpha\beta}\, dV. \tag{6.110}$$

These relations are identical to (6.97) and (6.98) derived from transformation of the total Lagrangian formulation.

The tangent stiffness follows from the increment of the internal virtual work,

$$d(\delta V_{\text{int}}) = \int_V d(\delta\varepsilon_{\alpha\beta})\, \sigma_{\alpha\beta}\, dV + \int_V \delta\varepsilon_{\alpha\beta}\, d(\sigma_{\alpha\beta} J)\, J^{-1}\, dV. \tag{6.111}$$

The Jacobi determinant is included in the stress factor, because $J^{-1}dV = dV_0$ is then invariant with respect to the increment. In order to derive the stress increment from a constitutive material relation it must be expressed in objective form. It turns out to be convenient first to use the Truesdell stress rate to establish the basic result. The corresponding results for other stress rates can then be obtained from this result by simple substitution.

When the Truesdell stress increment $\overset{\circ}{D}\sigma_{\alpha\beta}$ defined in (6.101) and the linearized virtual strain from (6.107) are introduced into the second integral, the following result is obtained:

$$\begin{aligned} \int_V \delta\varepsilon_{\alpha\beta}\, d(\sigma_{\alpha\beta} J)\, J^{-1}\, dV &= \int_V \delta\varepsilon_{\alpha\beta}\, \overset{\circ}{D}\sigma_{\alpha\beta}\, dV \\ + \tfrac{1}{2}\int_V \Big[\frac{\partial(\delta u_\alpha)}{\partial x_\gamma}\sigma_{\gamma\beta}\frac{\partial(du_\alpha)}{\partial x_\beta} &+ \frac{\partial(\delta u_\beta)}{\partial x_\alpha}\sigma_{\alpha\gamma}\frac{\partial(du_\beta)}{\partial x_\gamma}\Big]\, dV \\ + \tfrac{1}{2}\int_V \Big[\frac{\partial(\delta u_\alpha)}{\partial x_\beta}\sigma_{\alpha\gamma}\frac{\partial(du_\beta)}{\partial x_\gamma} &+ \frac{\partial(\delta u_\beta)}{\partial x_\alpha}\sigma_{\gamma\beta}\frac{\partial(du_\alpha)}{\partial x_\gamma}\Big]\, dV. \end{aligned} \tag{6.112}$$

The stress component matrix $\sigma_{\alpha\beta}$ is symmetric, and therefore the two terms in each of the square brackets are identical. Thus, the relation may be written in the more compact form

$$\begin{aligned} \int_V \delta\varepsilon_{\alpha\beta}\, d(\sigma_{\alpha\beta} J)\, J^{-1}\, dV &= \int_V \delta\varepsilon_{\alpha\beta}\, \overset{\circ}{D}\sigma_{\alpha\beta}\, dV \\ + \int_V \frac{\partial(\delta u_\gamma)}{\partial x_\alpha}\sigma_{\alpha\beta}\frac{\partial(du_\gamma)}{\partial x_\beta}\, dV &+ \int_V \frac{\partial(\delta u_\alpha)}{\partial x_\gamma}\sigma_{\alpha\beta}\frac{\partial(du_\gamma)}{\partial x_\beta}\, dV. \end{aligned} \tag{6.113}$$

The first two integrals on the right side correspond to the incremental virtual work as given in (6.104). It is now demonstrated that in the full form (6.111) of the incremental virtual work the first integral containing the contribution from the increment of virtual strain exactly cancels the last integral in (6.113).

The contribution to the incremental virtual work from the increment of the virtual strain is given by

$$\int_V d(\delta\varepsilon_{\alpha\beta})\, \sigma_{\alpha\beta}\, dV = \int_V d\Big(\frac{\partial(\delta u_\alpha)}{\partial x_\beta}\Big)\sigma_{\alpha\beta}\, dV, \tag{6.114}$$

where the definition (6.107) of the linearized virtual strain and the symmetry of the strain component matrix have been used. The variation δu_α is kept constant, and the increment arises because the position of a material point currently at $\mathbf{x} = \mathbf{x}_1$ changes position to $\mathbf{x} = \mathbf{x}_1 + d\mathbf{u}$. The influence of this on the partial derivatives in (6.114) is found by first differentiating through the current state variables x^1_γ by the chain rule of differentiation,

$$d\Big(\frac{\partial(\delta u_\alpha)}{\partial x_\beta}\Big) = \frac{\partial(\delta u_\alpha)}{\partial x^1_\gamma} d\Big(\frac{\partial x^1_\gamma}{\partial x_\beta}\Big). \tag{6.115}$$

The increment has now been isolated to the last factor, which is independent of any particular variational displacement δu_α. It is evaluated by considering the identity

$$\frac{\partial x^1_\gamma}{\partial x_\lambda}\frac{\partial x_\lambda}{\partial x^1_\beta} = \delta_{\gamma\beta}. \tag{6.116}$$

This relation is simply a statement of the definition of partial differentiation of the variable x^1_γ with respect to the variable x^1_λ. For identical variables $\gamma = \lambda$ the result is unity, while for different variables $\gamma \neq \lambda$ it is zero. When the variables x_γ are fixed, the increment of this relation is

$$d\Big(\frac{\partial x^1_\gamma}{\partial x_\lambda}\frac{\partial x_\lambda}{\partial x^1_\beta}\Big) = d\Big(\frac{\partial x^1_\gamma}{\partial x_\lambda}\Big)\frac{\partial x_\lambda}{\partial x^1_\beta} + \frac{\partial x^1_\gamma}{\partial x_\lambda}\frac{\partial(dx_\lambda)}{\partial x^1_\beta} = 0. \tag{6.117}$$

At the current state $x_\gamma = x^1_\gamma$, and thus $\partial x^1_\alpha/\partial x_\beta = \partial x_\alpha/\partial x^1_\beta = \delta_{\alpha\beta}$ in this state. When using this

$$d\Big(\frac{\partial x^1_\gamma}{\partial x_\beta}\Big) = -\frac{\partial(dx_\gamma)}{\partial x^1_\beta}. \tag{6.118}$$

By use of this result in (6.115), the integral (6.114) takes the form

$$\int_V d(\delta\varepsilon_{\alpha\beta})\,\sigma_{\alpha\beta}\,dV = -\int_V \frac{\partial(\delta u_\alpha)}{\partial x^1_\gamma}\sigma_{\alpha\beta}\frac{\partial(du_\gamma)}{\partial x^1_\beta}\,dV. \tag{6.119}$$

In the current state $x^1_\alpha = x_\alpha$, and it is seen that this integral is exactly equal to the last term in (6.113) with opposite sign.

Thus, the final form of the incremental virtual work derived directly from the virtual work equation in the current configuration is

$$d(\delta V_{\text{int}}) = \int_V \frac{\partial(\delta u_\gamma)}{\partial x_\alpha}\sigma_{\alpha\beta}\frac{\partial(du_\gamma)}{\partial x_\beta}\,dV + \int_V \delta\varepsilon_{\alpha\beta}\,\overset{\circ}{D}\sigma_{\alpha\beta}\,dV. \tag{6.120}$$

This form of the incremental virtual work is identical to (6.104), when the linear relation (6.102) is introduced between the Truesdell stress increment and the incremental strains $d\varepsilon_{\gamma\delta}$. It is seen that this simple form of the incremental virtual work is connected with the use of the Truesdell stress increment. In this formulation the Truesdell stress increment is assumed to be provided by a constitutive relation representing the material behavior. A different objective stress rate or increment can be used, but then it must be introduced into the incremental virtual work equation in a consistent manner. In essence this amounts to substitution of the new objective stress increment into the incremental virtual work equation (6.120) via its relation to the Truesdell stress increment. As a result, additional terms will appear in the first integral on the right side of (6.120), representing a modification of the geometric stiffness following from a modified interpretation of convection of the current state of stress. In general this will lead to a non-symmetric tangent stiffness, as seen in the case of the Jaumann stress increment by substitution of the incremental form of (6.58). The lack of symmetry of the tangent stiffness for alternative stress rates is discussed in detail in connection with the rotated Green–Naghdi stress rate in section 7.3.2.2 of Simo and Hughes (1998).

6.4.3 Finite element implementation

The finite element implementation of each step of the updated Lagrangian formulation follows that of the total Lagrangian formulation closely, but contains three essential differences: the current geometry with current surfaces and volumes is used, the virtual and incremental strains are simplified because they refer to current geometry, and finally the stress components must be updated from the objective increment provided by the constitutive relation in order to refer to fixed spatial coordinates.

In the updated Lagrangian formulation the virtual and incremental displacements $\delta\mathbf{u}(\mathbf{x})$ and $d\mathbf{u}(\mathbf{x})$ are considered as functions of the current configuration $\mathbf{x}$. This implies a representation of the virtual displacements in the form

$$\delta\mathbf{u}(\mathbf{x}) = \sum_n h_n(\mathbf{x})\, \delta\mathbf{u}_n \tag{6.121}$$

in terms of shape functions $h_n(\mathbf{x})$ and nodal values $\delta\mathbf{u}_n$, $n = 1, 2, \dots$ A similar representation is used for the incremental displacements in terms of their nodal values $d\mathbf{u}_n$.

The external virtual work is used to define equivalent external nodal forces $\mathbf{f}_n^{\text{ext}}$ via the relation

$$\delta V_{\text{ext}} = \delta\mathbf{u}_n^T\, \mathbf{f}_n^{\text{ext}} = \delta\mathbf{u}_n^T \Big\{ \int_S h_n(\mathbf{x})\, \mathbf{t}\; dS + \int_V h_n(\mathbf{x})\, \mathbf{p}\; dV \Big\}, \tag{6.122}$$

where summation over the nodes $n = 1, 2, \dots$ is implied, and loads refer to current surface and volume. By the standard procedure, in which each nodal component is selected as the only non-vanishing component, this scalar equation generates the external nodal forces

$$\mathbf{f}_n^{\rm ext} = \int_S h_n(\mathbf{x})\,\mathbf{t}\,dS + \int_V h_n(\mathbf{x})\,\mathbf{p}\,dV. \tag{6.123}$$

Thus, in the updated Lagrangian formulation the external nodal forces are obtained by integrating the surface traction and the volume force, weighted by the corresponding shape function $h_n(\mathbf{x})$.

Also the Cauchy stress and the linear strain components satisfy component symmetry conditions that permit replacement of the nine-component tensor index format with a six-component format. The six-component form of the Cauchy stress is

$$\boldsymbol{\sigma} = [\,\sigma_{11}\,,\,\sigma_{22}\,,\,\sigma_{33}\,,\,\sigma_{23}\,,\,\sigma_{31}\,,\,\sigma_{12}\,]^T, \tag{6.124}$$

while the corresponding virtual strain components are

$$\delta\boldsymbol{\varepsilon} = [\,\delta\varepsilon_{11}, \delta\varepsilon_{22}, \delta\varepsilon_{33}, 2\delta\varepsilon_{23}, 2\delta\varepsilon_{31}, 2\delta\varepsilon_{12}\,]^T. \tag{6.125}$$

A similar formula holds for the strain increment $d\boldsymbol{\varepsilon}$.

In the updated Lagrangian formulation the internal virtual work involves the virtual strain components $\delta\varepsilon_{\alpha\beta}$, given by (6.107). When substituting the virtual displacement representation (6.121) into the strain definition, the virtual strain is represented as

$$\delta\boldsymbol{\varepsilon}(\mathbf{x}) = \sum_n \mathbf{B}_n(\mathbf{x})\,\delta\mathbf{u}_n, \tag{6.126}$$

where the matrix strain interpolation matrix $\mathbf{B}_n(\mathbf{x})$ corresponding to node n is given by

$$\mathbf{B}_n = \begin{bmatrix} \dfrac{\partial h_n}{\partial x_1} & 0 & 0 \\ 0 & \dfrac{\partial h_n}{\partial x_2} & 0 \\ 0 & 0 & \dfrac{\partial h_n}{\partial x_3} \\ 0 & \dfrac{\partial h_n}{\partial x_3} & \dfrac{\partial h_n}{\partial x_2} \\ \dfrac{\partial h_n}{\partial x_3} & 0 & \dfrac{\partial h_n}{\partial x_1} \\ \dfrac{\partial h_n}{\partial x_2} & \dfrac{\partial h_n}{\partial x_1} & 0 \end{bmatrix}. \tag{6.127}$$

The strain interpolation matrix $\mathbf{B}_n(\mathbf{x})$ for the updated Lagrangian formulation is seen to be a special case of the strain interpolation matrix $\mathbf{B}_n^0(\mathbf{x}_0)$ for the total Lagrangian formulation. It corresponds to the choice of the current configuration $\mathbf{x}$ as initial configuration $\mathbf{x}_0$. This introduces the simplification $\partial x_\alpha / \partial x^0_\beta = \delta_{\alpha\beta}$ and removes the need for the superscript 0 in the formula (6.88) for $\mathbf{B}^0_n(\mathbf{x}_0)$. In spite of the apparent simplicity of the formula (6.127) for $\mathbf{B}_n(\mathbf{x})$, the shape functions $h_n(\mathbf{x})$ are usually only given as implicit functions of the current coordinates $\mathbf{x}$ and thus the matrix $\mathbf{B}_n(\mathbf{x})$ must be evaluated numerically at specific points.

The internal virtual work defines internal nodal forces $\mathbf{f}^{\text{int}}_n$ by substitution of the virtual strain field into (6.110):

$$\delta V_{\text{int}} = \delta \mathbf{u}_n^T \, \mathbf{f}^{\text{int}}_n = \delta \mathbf{u}_n^T \Big\{ \int_V \mathbf{B}_n(\mathbf{x})^T \boldsymbol{\sigma} \, dV \Big\}, \tag{6.128}$$

where summation over the nodes $n = 1, 2, \ldots$ is implied. This scalar work equation generates the internal nodal forces

$$\mathbf{f}^{\text{int}}_n = \int_V \mathbf{B}_n(\mathbf{x})^T \boldsymbol{\sigma} \, dV. \tag{6.129}$$

The calculation of the internal nodal forces requires evaluation of a volume integral of the product of the current Cauchy stress and the strain representation matrix $\mathbf{B}_n(\mathbf{x})$.

In the discretized problem equilibrium, or deviation from equilibrium, is expressed in terms of the nodal residual forces

$$\mathbf{r}_n = \mathbf{f}^{\text{ext}}_n - \mathbf{f}^{\text{int}}_n. \tag{6.130}$$

The residual forces are calculated via the external and internal nodal forces by integration of the current loads and estimate of the Cauchy stresses. If the residual forces are not negligible, they are eliminated by iterations using the tangent stiffness matrix.

The tangent stiffness is evaluated from the increment of the internal virtual work (6.120) using the constitutive incremental relation (6.102) for the Truesdell stress increment. When substituting the virtual and incremental strains from representations of the form (6.121), the incremental internal virtual work is obtained in the form

$$d(\delta V_{\text{int}}) = \delta \mathbf{u}_n^T \Big\{ \mathbf{I} \int_V \frac{\partial h_n}{\partial x_\alpha} \sigma_{\alpha\beta} \frac{\partial h_m}{\partial x_\beta} \, dV + \int_V \mathbf{B}_n^T \mathbb{C} \, \mathbf{B}_m \, dV \Big\} \, d\mathbf{u}_m, \tag{6.131}$$

where $\mathbf{I}$ is the two-dimensional unit tensor. The 6×6 matrix $\mathbb{C}$ is the incremental stiffness (6.102) for the Truesdell increment of the Cauchy stress,

$$\overset{\circ}{D}\boldsymbol{\sigma} = \mathbb{C} \, d\boldsymbol{\varepsilon}. \tag{6.132}$$

As for the total Lagrangian formulation, the incremental internal virtual work relation is of the form

$$d(\delta V_{\text{int}}) = \delta\mathbf{u}_n^T\,\mathbf{K}_{nm}\,d\mathbf{u}_m, \tag{6.133}$$

where summation over n and m is implied, and $\mathbf{K}_{nm}$ is the contribution to the tangent stiffness matrix from the combination of nodes n and m. In the updated Lagrangian format the tangent stiffness matrix contribution $\mathbf{K}_{nm}$ is given by

$$\mathbf{K}_{nm} = \mathbf{I}\int_V \frac{\partial h_n}{\partial x_\alpha}\,\sigma_{\alpha\beta}\,\frac{\partial h_m}{\partial x_\beta}\,dV + \int_V \mathbf{B}_n^T\,\mathbb{C}\,\mathbf{B}_m\,dV. \tag{6.134}$$

When comparing this expression to the equivalent total Lagrangian formula (6.96) it is seen that in the first integral, representing the geometric stiffness, the deformation gradient changes the gradient factors, leaving the Cauchy stress instead of the Piola–Kirchhoff stress, and in the second integral, representing the constitutive relation of the material, the deformation gradient also changes the strain representation matrix to $\mathbf{B}$ and the constitutive stiffness to $\mathbb{C}$ for the Truesdell increment of the Cauchy stress. While the first of these changes is a simple substitution, the second involves basic considerations if the coefficient matrix $\mathbb{C}$ is to be derived directly from material behavior.

The iterative procedure provides an estimate of the displacement increment $d\mathbf{u}_n$ at each of the nodes. This estimate gives the strain increment $d\boldsymbol{\varepsilon}$ by the representation (6.126). In the updated Lagrangian formulation the strain increment is related to an objective stress increment, e.g. the Truesdell increment of the Cauchy stress shown in (6.132). The update of the stress components must then be made by use of the defining relation for that particular objective stress increment. In the case of the Truesdell stress increment the component formula follows from (6.101) as

$$d\sigma_{\alpha\beta} = \overset{\circ}{D}\sigma_{\alpha\beta} - d\varepsilon_{\gamma\gamma}\,\sigma_{\alpha\beta} + \frac{\partial(du_\alpha)}{\partial x_\gamma}\,\sigma_{\gamma\beta} + \sigma_{\alpha\gamma}\frac{\partial(du_\beta)}{\partial x_\gamma}. \tag{6.135}$$

This formula can be expressed in a computationally more convenient form using the six-component stress format,

$$d\boldsymbol{\sigma} = \overset{\circ}{D}\boldsymbol{\sigma} - \boldsymbol{\sigma}\,d\varepsilon_{\gamma\gamma} + \mathbf{T}_n\,d\mathbf{u}_n. \tag{6.136}$$

The increment of the volume strain is calculated from

$$d\varepsilon_{\gamma\gamma} = \left[\frac{\partial h_n}{\partial x_1}\;\frac{\partial h_n}{\partial x_2}\;\frac{\partial h_n}{\partial x_3}\right]_n d\mathbf{u}_n \tag{6.137}$$

while the matrix $\mathbf{T}_n$ is given by a formula similar to that of the strain representation matrix $\mathbf{B}_n$, but with the terms weighted by the stress components,

$$\mathbf{T}_n = \begin{bmatrix} 2\sigma_{1\gamma}\frac{\partial h_n}{\partial x_\gamma} & 0 & 0 \\ 0 & 2\sigma_{2\gamma}\frac{\partial h_n}{\partial x_\gamma} & 0 \\ 0 & 0 & 2\sigma_{3\gamma}\frac{\partial h_n}{\partial x_\gamma} \\ 0 & \sigma_{3\gamma}\frac{\partial h_n}{\partial x_\gamma} & \sigma_{2\gamma}\frac{\partial h_n}{\partial x_\gamma} \\ \sigma_{3\gamma}\frac{\partial h_n}{\partial x_\gamma} & 0 & \sigma_{1\gamma}\frac{\partial h_n}{\partial x_\gamma} \\ \sigma_{2\gamma}\frac{\partial h_n}{\partial x_\gamma} & \sigma_{1\gamma}\frac{\partial h_n}{\partial x_\gamma} & 0 \end{bmatrix}. \tag{6.138}$$

In all these formulae, summation over the contributing nodes n is implied.

The formula (6.135) is for infinitesimal increments. Actual calculations work with finite increments. This suggests a reinterpretation of the stress update in which the strain increment $\Delta\boldsymbol{\varepsilon}$ is considered as a linearized Green strain increment with the current state as reference state. This strain increment produces a Piola–Kirchhoff stress increment $\Delta\mathbf{S}$, also with the current state as reference state. The current configuration is denoted $\mathbf{x}_1$, and the displacement increment $\Delta\mathbf{u}$ leads to the new configuration $\mathbf{x}_2 = \mathbf{x}_1 + \Delta\mathbf{u}$. The updated Cauchy stress then follows from the transformation formula (6.38) in the form

$$\sigma^2_{\alpha\beta} = \frac{J_1}{J_2}\frac{\partial x^2_\alpha}{\partial x^1_\gamma}\left(\sigma^1_{\gamma\delta} + \Delta S_{\gamma\delta}\right)\frac{\partial x^2_\beta}{\partial x^1_\delta}, \tag{6.139}$$

where the terms in parentheses represent the updated Piola–Kirchhoff stress with $\mathbf{x}_1$ as basis configuration.

In summary, the implementation of the updated Lagrangian formulation using the Truesdell increment of the Cauchy stress is similar in structure to the total Lagrangian formulation but leads to a simpler strain representation matrix $\mathbf{B}_n$. However, some of this simplification is offset by the need to update the stress components from their objective increments. This leads to a matrix fairly similar to the strain representation matrix $\mathbf{B}^0_n$ of the total Lagrangian formulation. The use of an alternative objective stress increment will introduce additional terms in the tangent stiffness matrix, that will then generally lose its symmetry, even for a symmetric constitutive matrix.

6.5 Summary of non-linear motion of solids

Motion of a solid body is described either with reference to the individual particles of the body, the material or Lagrangian description, or with reference to the current positions in space, the spatial or Eulerian description. In the material formulation the properties of the local deformed state around a particle are described via the deformation gradient, which can be used to define several different strain measures. A special role is taken by the Green strain, which is a quadratic function of the displacement gradient. While the Green strain may not be ideal for the formulation of constitutive relations for large deformation, its quadratic form leads to a particularly simple formulation of problems with large displacement and moderate strain, used e.g. in the formulation of momentum- and energy-conserving algorithms for dynamics in Chapter 9.

The principle of virtual work serves as an important tool for the formulation of consistent theories of continuous deformable bodies. In its basic form it is a statement of equality between the external virtual work, performed by the external loads through a virtual displacement field, and the internal virtual work, performed by the internal stresses through the corresponding virtual strain field. In the case of the Green strain the postulated form of the internal virtual work leads to a clear interpretation of the associated Piola–Kirchhoff stress, simply by reformulation of the involved integrals. The principle of virtual work also serves as a convenient means of connecting a material description in terms of e.g. Green strain and Piola–Kirchhoff stress with an equivalent spatial formulation in terms of linearized small strain and Cauchy stress. When the kinematic relation between the strain increments has been established, the relation between the corresponding conjugate stresses again follows from simple reformulation of the integrals.

The solution of specific problems relies on the constitutive relation between the stresses and strains in the body. The material properties, including the state of deformation, reside in the material frame of reference, and thus a spatial formulation must be based on proper representation of these material properties in the spatial frame. An important aspect is the relation between the change of stress as observed in the spatial frame, and as observed relative to a local material frame. This issue is dealt with by the so-called objective stress rate, i.e. a formulation in the spatial frame, that accounts for the convection effects introduced by the motion of the material body. The basic form, corresponding to the Lie derivative from tensor analysis, is derived for the Cauchy stress. It is observed that the use of current volume in the definition of the Cauchy stress leads to an additional term in the objective stress rate accounting for material dilation. A simpler formulation is obtained by re-normalizing the stress with respect to a constant amount of material, and this leads to the introduction of the Kirchhoff stress. The Kirchhoff stress plays a more direct role when accounting for internal properties such as internal energy, because it refers to a specific amount of material.

The chapter concludes with a discussion of the formulation of finite deformation problems for solids by the total and the updated Lagrangian methods. The total Lagrangian formulation is obtained from the virtual work equation in terms of Green strain and Piola–Kirchhoff stress and leads to the definition of residual forces and tangent stiffness in terms of a shape function representation of the displacement field. In the updated Lagrangian formulation the current state is used as reference state. The corresponding formulation is obtained from the total formulation by transforming the stresses and strains. This transformation leads directly to the objective stress rate in terms of the Lie derivative. Specific constitutive relations for elastic and elasto-plastic materials are discussed in the following chapter.

6.6 Exercises

Exercise 6.1 A uniform state of deformation can be expressed by the linear relation $\mathbf{x} = \mathbf{a} + \mathbf{A}\mathbf{x}_0$. Consider the special two-dimensional case given by

$$\begin{bmatrix} x_1 \\ x_2 \end{bmatrix} = \begin{bmatrix} 3.0 \\ 2.0 \end{bmatrix} + \begin{bmatrix} 0.8 & -0.2 \\ 0.6 & 1.2 \end{bmatrix} \begin{bmatrix} x_1^0 \\ x_2^0 \end{bmatrix}.$$

(a) Plot the initial and current base vectors $\mathbf{e}_1, \mathbf{e}_2$ and the current location of the initial square with corner coordinates $\mathbf{x}_0 = [\pm 1, \pm 1]^T$.
(b) Find the deformation gradient component matrix $F_{\alpha\beta}$.
(c) Find the Green strain component matrix $E_{\alpha\beta}$.
(d) Find the ratio of current area to original area via the determinant J.

Exercise 6.2 In a rectangular two-dimensional bi-linear element the displacement field is of the form

$$\begin{aligned} u_1 &= a_0 + a_1 x_1^0 + a_2 x_2^0 + a_3 x_1^0 x_2^0, \\ u_2 &= b_0 + b_1 x_1^0 + b_2 x_2^0 + b_3 x_1^0 x_2^0. \end{aligned}$$

(a) Find the deformation gradient component matrix $F_{\alpha\beta}$.
(b) Find the Green strain component matrix $E_{\alpha\beta}$.
(c) Determine the polynomial degree of the components $E_{11}^2, E_{22}^2, E_{12}^2$ and the necessary order of Gauss quadrature for exact integration of all terms in a linear elastic material.

Exercise 6.3 The purpose of the objective stress rates discussed in this chapter is to isolate changes associated with the material frame, and thus they must be independent of any superimposed rigid body rotation. A rigid body rotation transforms the current coordinates $\mathbf{x}$ to $\mathbf{x}_R = \mathbf{R}\mathbf{x}$. The corresponding transformed deformation gradient is $\mathbf{F}_R = \mathbf{R}\mathbf{F}$, while the transformed Kirchhoff stress is $\boldsymbol{\tau}_R = \mathbf{R}\boldsymbol{\tau}\mathbf{R}^T$.

(a) Express the transformed velocity gradient $\mathbf{L}_R$ in terms of the original velocity gradient $\mathbf{L}$, the rigid body rotation $\mathbf{R}$ and its time derivative $\dot{\mathbf{R}}$.

(b) Use the expression (6.48) for the Truesdell rate of Kirchhoff stress to prove the transformation formula $\overset{\circ}{\boldsymbol{\tau}}_R = \mathbf{R}\overset{\circ}{\boldsymbol{\tau}}\mathbf{R}^T$.

Exercise 6.4 The symmetric matrix $\mathbf{U}$ defined by the polar decomposition theorem (6.17) is called the (right) stretch tensor. Its principal directions define the directions of zero shear, and the corresponding eigenvalues $\lambda_1, \lambda_2, \lambda_3$ are called the stretches.

(a) Find the stretches λ_1, λ_2 for the shear problem of Example 6.2 in terms of the parameter γ and illustrate the result graphically.

(b) Introduce η by the variable transformation $\sinh(\eta) = \frac{1}{2}\gamma$, and express the stretches λ_1 and λ_2 in terms of η.

(c) Give an interpretation of η in terms of logarithmic strain, introduced in Chapter 2.

Exercise 6.5 The polar decomposition theorem (6.17) provides a factorization of the deformation gradient tensor, $\mathbf{F} = \mathbf{R}\mathbf{U}$.

(a) Obtain an expression for the velocity gradient tensor $\mathbf{L}$ in terms of $\mathbf{R}$ and $\mathbf{U}$, and identify the angular velocity tensor $\boldsymbol{\Omega} = \dot{\mathbf{R}}\mathbf{R}^T$.

(b) Express the angular velocity tensor $\boldsymbol{\Omega}$ for the shear problem of Example 6.2 in terms of $\dot{\varphi}$.

Exercise 6.6 In the shear problem considered in the examples of this chapter there is no volume change, and for a linear isotropic elastic material the stress rate is then proportional to the strain rate. In terms of the Jaumann stress rate this amounts to the relation $\overset{\nabla}{\boldsymbol{\tau}} = 2\mu\dot{\boldsymbol{\varepsilon}}$, where μ is the shear modulus.

(a) Show, on the basis of the expressions in Example 6.5, that the mean stress vanishes identically, $\tau_{11} + \tau_{22} = 0$.

(b) Show that the remaining elasticity equations can be expressed as

$$\dot{\tau}_{11} - \dot{\gamma}\,\tau_{12} = 0, \qquad \dot{\tau}_{12} + \dot{\gamma}\,\tau_{11} = \mu\,\dot{\gamma}.$$

(c) Integrate these equations for $\dot{\gamma} = \text{const.}$ to find the solution

$$\tau_{12} = \mu\sin\gamma, \qquad \tau_{11} = -\tau_{22} = \mu\,(1 - \cos\gamma).$$

(d) Plot and discuss the solution.

This solution was given by Dienes (1979) as part of a general discussion of stress rates.

7

Elasto-plastic solids

A central feature of the mechanical behavior of solids and structures is the constitutive relation connecting stresses and strains. This chapter presents the basic theory of elastic and plastic solids, including the techniques needed to implement the models in the numerical framework described in previous chapters. The theoretical basis for the material models described here has developed over the last 50 years, from the early work of Ziegler (1963), over potentials with internal variables (Rice, 1971; Hill and Rice, 1973), to a fully developed theory including conjugate variables and potentials (Halphen and Son, 1975; Germain *et al.*, 1983). A full account of this development is outside the present scope, and the chapter is limited to elastic and elasto-plastic solids. The presentation is deliberately simplified to purely mechanical effects, leaving out e.g. thermal effects, but retaining the general structure of the formulation.

First the notion of reversible elastic deformation is introduced in Section 7.1 in connection with an internal energy potential, and the concept of conjugate variables and potentials is described. The role of stress and strain invariants and some of their properties is described and used in the formulation of isotropic elasticity. The general theory of rate-independent elasto-plasticity with internal variables is then developed in Section 7.2. The key ingredients are the internal energy potential, expressing the recoverable energy in terms of elastic strains and the primary internal variables, and a yield potential, describing the limit of the region of reversible behavior. The equations of plasticity theory are then developed from the postulate of maximum rate of dissipation during plastic deformation. It is demonstrated by elementary arguments that this requirement leads to evolution conditions formulated in terms of the gradients with respect to the conjugate variables. The natural appearance of the conjugate variables has fundamental implications, explaining the difference between the classic notions of associated and non-associated plastic flow in terms of properties of the internal energy function.

The following two sections describe basic properties of elasto-plastic models and their numerical implementation on two levels. First, the basic properties like hardening and numerical integration are treated on a simple basis in connection with von Mises plasticity, and then the corresponding problems are treated in a general form. Section 7.5 illustrates the formulation of a non-associated plasticity theory for granular materials by using simple arguments on special stress states, and then 'unfolding' the theory for general stress states by more qualitative arguments. For simplicity the main presentation is based on small displacement theory. Extension of the results to finite displacements is treated in Section 7.6, where two different definitions of finite elastic strain are identified and discussed.

Some elastic materials, such as e.g. wood and fiber composites, have anisotropic properties, i.e. properties that depend on the orientation of loading relative to directions fixed in the material, see e.g. Lekhnitskii (1963) and Cowin and Mehrabadi (1995). However, in the following the theory will be specialized to isotropic materials, i.e. materials whose properties are independent of the orientation of the material. This enables a fairly general, yet simple, identification of two basic deformation mechanisms: change of volume and change of shape.

7.1 Elastic solids

A material is called elastic when the current state of stress is a function of the current state of the conjugate strain. Here the basic considerations will be based on linear kinematics using the Cauchy stress $\boldsymbol{\sigma}$ with components $\sigma_{\alpha\beta}$ and the conjugate linear strain $\boldsymbol{\varepsilon}$ with components $\varepsilon_{\alpha\beta}$. Thus, the property of elasticity is expressed as

$$\sigma_{\alpha\beta} = \sigma_{\alpha\beta}(\varepsilon_{\gamma\delta}). \tag{7.1}$$

This relation is illustrated in Fig. 7.1 as a single curve, as reversing the strain history leads to reversal of the stress history.

During straining of the material the stresses perform work as defined via the principle of virtual work. The accumulated work per unit volume is

$$\varphi(\varepsilon_{\alpha\beta}) = \int_0^{\varepsilon_{\alpha\beta}} \sigma_{\alpha\beta}(\bar{\varepsilon}_{\gamma\delta})\, d\bar{\varepsilon}_{\alpha\beta}. \tag{7.2}$$

The integral is illustrated schematically in Fig. 7.1. In order for the accumulated work to be a unique function of the current state of strain, the

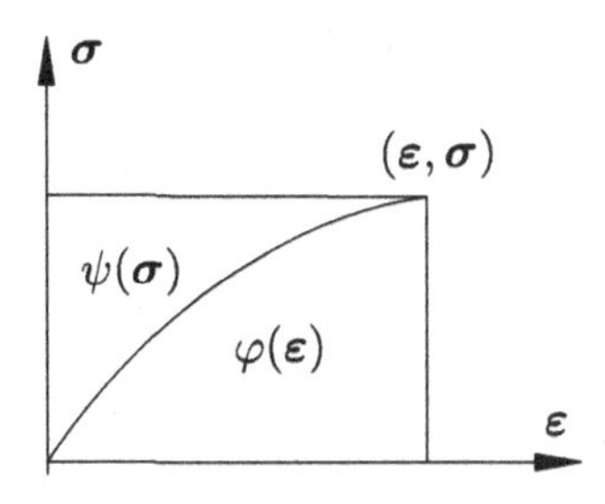

Fig. 7.1. Elastic stress–strain curve with internal energy density $\varphi(\boldsymbol{\varepsilon})$ and complementary energy density $\psi(\boldsymbol{\sigma})$.

integrand must satisfy the conditions

$$\frac{\partial \sigma_{\alpha\beta}}{\partial \varepsilon_{\gamma\delta}} = \frac{\partial \sigma_{\gamma\delta}}{\partial \varepsilon_{\alpha\beta}}. \tag{7.3}$$

Conversely, if these conditions are satisfied, the integral $\varphi(\varepsilon)$ is only a function of the current state of strain and the stresses are given in terms of the internal energy density function as

$$\sigma_{\alpha\beta} = \frac{\partial \varphi(\varepsilon)}{\partial \varepsilon_{\alpha\beta}}. \tag{7.4}$$

Such a material is called hyper-elastic or Green-elastic. Hyper-elasticity implies that reversal to a previous state of strain will imply reversal to the energy at this previous state. Thus, the deformation of a hyper-elastic material is fully reversible. An elastic material without the integrability relations (7.3) is called Cauchy-elastic. The additional freedom in the formulation of Cauchy-elastic models is obtained at the cost of lack of energy reversibility for closed strain cycles, see e.g. Ottosen and Ristinmaa (2005).

It is sometimes more convenient to use the complementary energy density $\psi(\sigma_{\gamma\delta})$, defined in terms of the strain energy density as

$$\psi(\boldsymbol{\sigma}) = \sigma_{\alpha\beta}\, \varepsilon_{\alpha\beta} - \varphi(\boldsymbol{\varepsilon}). \tag{7.5}$$

The role of ψ as the complement of φ is illustrated schematically in Fig. 7.1. The complementary energy is considered as a function of current stress. The differential increment of (7.5) therefore takes the form

$$\frac{\partial \psi}{\partial \sigma_{\alpha\beta}} d\sigma_{\alpha\beta} = \varepsilon_{\alpha\beta}\, d\sigma_{\alpha\beta} + \Big(\sigma_{\alpha\beta} - \frac{\partial \varphi}{\partial \varepsilon_{\alpha\beta}}\Big) d\varepsilon_{\alpha\beta}. \tag{7.6}$$

The terms in parentheses cancel due to (7.4), leaving the relation between the strain and the complementary energy density,

$$\varepsilon_{\alpha\beta} = \frac{\partial \psi(\boldsymbol{\sigma})}{\partial \sigma_{\alpha\beta}}. \tag{7.7}$$

This relation is a clear analogue to the relation (7.4) between stress and the strain energy density. The use of a relation like (7.5) to define a complementary potential is called a Legendre transformation and is a procedure widely used in mechanics, see e.g. Washizu (1974) and Sewell (1987). The purpose here is to interchange the roles of stress and strain relative to the original potential.

In the following it is convenient to use boldface notation for the stress and strain tensors, and also to introduce the notion of a compact array notation, in which only the independent components appear. In this notation $\boldsymbol{\sigma}$ and $\boldsymbol{\varepsilon}$ are represented by the column arrays

$$\boldsymbol{\sigma} = [\sigma_{11}, \sigma_{22}, \sigma_{33}, \sigma_{23}, \sigma_{31}, \sigma_{12}]^T, \tag{7.8}$$

$$\boldsymbol{\varepsilon} = [\varepsilon_{11}, \varepsilon_{22}, \varepsilon_{33}, 2\varepsilon_{23}, 2\varepsilon_{31}, 2\varepsilon_{12}]^T. \tag{7.9}$$

As discussed in the previous chapter, this implies that the contribution to the internal virtual work is represented as $\boldsymbol{\sigma}^T d\boldsymbol{\varepsilon}$. This establishes a precise analogy with the boldface tensor notation for conjugate tensor quantities, when the transpose is indicated explicitly. A similar analogy applies to the energy potentials $\varphi(\boldsymbol{\varepsilon})$ and $\psi(\boldsymbol{\sigma})$, where the increments can be expressed as

$$d\varphi = \frac{\partial \varphi}{\partial \boldsymbol{\varepsilon}} d\boldsymbol{\varepsilon} = \boldsymbol{\sigma}^T d\boldsymbol{\varepsilon} \tag{7.10}$$

and

$$d\psi = \frac{\partial \psi}{\partial \boldsymbol{\sigma}} d\boldsymbol{\sigma} = \boldsymbol{\varepsilon}^T d\boldsymbol{\sigma}. \tag{7.11}$$

It is seen that derivatives correspond to the transpose of the array. In the following the boldface notation will be used with the possibility of interpretation as tensors or as component arrays. In the few cases where ambiguities of interpretation can arise, these are resolved by indicating e.g. products in a more explicit form.

7.1.1 Stress invariants

The discussion of material models is here limited to isotropic materials, i.e. materials whose properties are independent of the orientation in space. This gives the possibility of formulating the theory in terms of the so-called stress and strain invariants, i.e. special combinations of tensor components that are independent of the orientation. While a similar approach is also possible for anisotropic materials, the number of invariants increases considerably because for anisotropic materials additional invariants must be included to account for the orientation of the stress and strain components relative to special directions in the material.

The properties of the stress tensor $\boldsymbol{\sigma}$, and thereby the state of stress at a point, can be described via the eigenvalue problem

$$\boldsymbol{\sigma}\,\mathbf{n}_j = \sigma_j\,\mathbf{n}_j \qquad \text{(no sum)}. \tag{7.12}$$

In this equation $\mathbf{n}_j$ is a unit vector, defining a section through the point with stress tensor $\boldsymbol{\sigma}$. The equation defines sections with stress vector σ_j in the direction of the normal $\mathbf{n}_j$. These stresses are called the principal stresses and the directions are similarly called the principal directions of the stress tensor. The principal stresses and directions are found by writing the eigenvalue problem in matrix form,

$$(\boldsymbol{\sigma} - \sigma_j\mathbf{I})\,\mathbf{n}_j = \mathbf{0} \qquad \text{(no sum)}. \tag{7.13}$$

A non-trivial solution for the direction vector $\mathbf{n}_j$ requires the matrix to be singular, corresponding to the component determinant equation

$$\begin{vmatrix} \sigma_{11}-\sigma & \sigma_{12} & \sigma_{13} \\ \sigma_{21} & \sigma_{22}-\sigma & \sigma_{23} \\ \sigma_{31} & \sigma_{32} & \sigma_{33}-\sigma \end{vmatrix} = 0. \tag{7.14}$$

The principal stresses σ_j are the solution to the corresponding cubic characteristic equation

$$\sigma^3 - I_1\sigma^2 + I_2\sigma - I_3 = 0, \tag{7.15}$$

in which the coefficients I_1, I_2 and I_3 are the so-called principal stress invariants. The principal stresses σ_1, σ_2 and σ_3 are the roots of the characteristic equation, and it can therefore be expressed in the form

$$\sigma^3 - (\sigma_1+\sigma_2+\sigma_3)\sigma^2 + (\sigma_2\sigma_3+\sigma_3\sigma_1+\sigma_1\sigma_2)\sigma - \sigma_1\sigma_2\sigma_3 = 0. \tag{7.16}$$

The invariance of the parameters I_1, I_2 and I_3 follows from their identification with the coefficients in terms of principal stresses. They can be expressed in terms of the general stress components via the determinant equation (7.14) as

$$I_1 = \operatorname{tr}(\boldsymbol{\sigma}) = \sigma_{\gamma\gamma} = \sigma_1 + \sigma_2 + \sigma_3, \tag{7.17}$$

$$\begin{aligned} I_2 &= \tfrac{1}{2}[\operatorname{tr}(\boldsymbol{\sigma})^2 - \operatorname{tr}(\boldsymbol{\sigma}^2)] = \tfrac{1}{2}(\sigma_{\alpha\alpha}\sigma_{\beta\beta} - \sigma_{\alpha\beta}\sigma_{\beta\alpha}) \\ &= \sigma_2\sigma_3 + \sigma_3\sigma_1 + \sigma_1\sigma_2, \end{aligned} \tag{7.18}$$

$$I_3 = \det(\boldsymbol{\sigma}) = \varepsilon_{\alpha\beta\gamma}\sigma_{1\alpha}\sigma_{2\beta}\sigma_{3\gamma} = \sigma_1\sigma_2\sigma_3. \tag{7.19}$$

Here $\operatorname{tr}(\boldsymbol{\sigma})$ denotes the trace defined as the sum of the diagonal terms and $\det(\boldsymbol{\sigma})$ denotes the determinant of the component matrix.

It is an interesting fact that the stress tensor $\boldsymbol{\sigma}$ satisfies its own characteristic equation, when interpreted in tensor format. The proof is quite simple and consists in pre-multiplication of the eigenvalue equation (7.13) with factors containing the two remaining principal stress components. The order of the factors can be interchanged, and thus the resulting equation can be written in the form

$$(\boldsymbol{\sigma} - \sigma_1\mathbf{I})(\boldsymbol{\sigma} - \sigma_2\mathbf{I})(\boldsymbol{\sigma} - \sigma_3\mathbf{I})\,\mathbf{n}_j = \mathbf{0}. \tag{7.20}$$

By changing the order of the factors to place the factor with subscript j last, the equation is seen to be satisfied for all three directions $\mathbf{n}_1$, $\mathbf{n}_2$ and $\mathbf{n}_3$. Thus, the product of the three factors must be the zero tensor, and multiplication gives the cubic tensor equation

$$\boldsymbol{\sigma}^3 - (\sigma_1 + \sigma_2 + \sigma_3)\boldsymbol{\sigma}^2 + (\sigma_2\sigma_3 + \sigma_3\sigma_1 + \sigma_1\sigma_2)\boldsymbol{\sigma} - \sigma_1\sigma_2\sigma_3\mathbf{I} = \mathbf{0}. \tag{7.21}$$

This is the Cayley–Hamilton equation, stating that the stress tensor satisfies its own characteristic equation. The equation implies that $\boldsymbol{\sigma}^3$ can be expressed as a linear combination of $\boldsymbol{\sigma}^2$, $\boldsymbol{\sigma}$ and $\mathbf{I}$ with coefficients in terms of the stress invariants I_1, I_2 and I_3. If the equation is multiplied with $\boldsymbol{\sigma}$ the fourth power $\boldsymbol{\sigma}^4$ can be reduced to lower powers, and so on. By continuing this process it can be concluded that a power series in the stress tensor $\boldsymbol{\sigma}$ can be expressed in terms of a basis consisting of the three members $\mathbf{I}$, $\boldsymbol{\sigma}$ and $\boldsymbol{\sigma}^2$ with coefficients in terms of the stress invariants I_1, I_2 and I_3. Scalar functions of the stress tensor $\boldsymbol{\sigma}$ must be formed by invariants of the series expansions, and thus must be functions of the three stress invariants.

The implications may be illustrated with reference to the complementary energy $\psi(\boldsymbol{\sigma})$. It follows from the argument above that it can be expressed as a function of the three stress invariants, i.e. in the form $\psi(I_1, I_2, I_3)$. The strain was found by the stress derivatives as shown in (7.7). This can be expressed using the invariant format as

$$\varepsilon = \frac{\partial \psi}{\partial \boldsymbol{\sigma}^T} = \frac{\partial \psi}{\partial I_1}\frac{\partial I_1}{\partial \boldsymbol{\sigma}^T} + \frac{\partial \psi}{\partial I_2}\frac{\partial I_2}{\partial \boldsymbol{\sigma}^T} + \frac{\partial \psi}{\partial I_3}\frac{\partial I_3}{\partial \boldsymbol{\sigma}^T}, \tag{7.22}$$

where the transpose has been introduced to make the notation consistent with the condensed component format. It is seen from this relation that the derivatives of the invariants with respect to $\boldsymbol{\sigma}$ serve as a three-component basis for the strain tensor.

The derivatives of the invariants follow from their defining relations (7.17)–(7.19). They are most easily derived in component form. The derivative of I_1 follows directly from the definition as the trace,

$$\frac{\partial I_1}{\partial \sigma_{\alpha\beta}} = \delta_{\alpha\beta}. \tag{7.23}$$

Similarly, the derivative of the second invariant I_2 is found as

$$\frac{\partial I_2}{\partial \sigma_{\alpha\beta}} = \sigma_{\gamma\gamma}\delta_{\alpha\beta} - \sigma_{\alpha\beta}. \tag{7.24}$$

Finally, the derivatives of the invariant I_3 follow from the observation that the definition as the stress tensor determinant can be expanded as products of components and co-factors along the row or column containing the component $\sigma_{\alpha\beta}$. In this expansion the selected component $\sigma_{\alpha\beta}$ only appears once, namely as factor to the corresponding co-factor. Thus, the derivative is the co-factor corresponding to $\sigma_{\alpha\beta}$. The co-factor can be expressed in component form as

$$\frac{\partial I_3}{\partial \sigma_{\alpha\beta}} = \tfrac{1}{2}\varepsilon_{\alpha\gamma\delta}\varepsilon_{\beta\kappa\lambda}\sigma_{\gamma\kappa}\sigma_{\delta\lambda}. \tag{7.25}$$

The factor $\frac{1}{2}$ compensates for the inclusion of two permutations that each contribute to the full result.

It is often convenient to use a different set of stress invariants that correspond more directly to the properties of materials. In this connection the mean stress

$$\sigma_m = \tfrac{1}{3}\mathrm{tr}(\boldsymbol{\sigma}) = \tfrac{1}{3}\sigma_{\gamma\gamma} = \tfrac{1}{3}I_1 \tag{7.26}$$

plays a central role. The idea is to decompose the total stress into an isotropic stress state defined by the mean stress and the remaining stress representing the deviation from an isotropic state of stress, and therefore called the deviatoric stress. The deviatoric stress is defined as

$$\boldsymbol{\sigma}' = \boldsymbol{\sigma} - \tfrac{1}{3}\mathrm{tr}(\boldsymbol{\sigma})\mathbf{I} = \boldsymbol{\sigma} - \sigma_m\mathbf{I} \tag{7.27}$$

corresponding to the component form

$$\sigma'_{\alpha\beta} = \sigma_{\alpha\beta} - \tfrac{1}{3}\sigma_{\gamma\gamma}\,\delta_{\alpha\beta} = \sigma_{\alpha\beta} - \sigma_m\,\delta_{\alpha\beta}. \tag{7.28}$$

The mean stress state is described by the principal stress invariant I_1, and two new invariants that do not contain the mean stress are now introduced to characterize the deviatoric stress state.

The deviatoric stress invariants are defined by the corresponding powers of the deviatoric stress tensor as

$$J_1 = \mathrm{tr}(\boldsymbol{\sigma}') = \sigma'_{\gamma\gamma} = 0, \tag{7.29}$$

$$J_2 = \tfrac{1}{2}\mathrm{tr}(\boldsymbol{\sigma}'^2) = \tfrac{1}{2}\sigma'_{\alpha\beta}\sigma'_{\beta\alpha}, \tag{7.30}$$

$$J_3 = \tfrac{1}{3}\mathrm{tr}(\boldsymbol{\sigma}'^3) = \tfrac{1}{3}\sigma'_{\alpha\beta}\sigma'_{\beta\gamma}\sigma'_{\gamma\alpha}. \tag{7.31}$$

The fact that the mean deviatoric stress vanishes provides a relation between the components. This leads to two alternative formats for the expressions of the deviatoric invariants in terms of principal stresses:

$$J_2 = \tfrac{1}{2}(\sigma_1'^2 + \sigma_2'^2 + \sigma_3'^2) = -(\sigma_2'\sigma_3' + \sigma_3'\sigma_1' + \sigma_1'\sigma_2'), \tag{7.32}$$

$$J_3 = \tfrac{1}{3}(\sigma_1'^3 + \sigma_2'^3 + \sigma_3'^3) = \sigma_1'\sigma_2'\sigma_3', \tag{7.33}$$

where the first expression follows directly from the definition, while the second can be demonstrated by use of the zero mean deviatoric stress property.

An important use of the deviatoric invariants is in the expression of the complementary energy density in the form $\psi(I_1, J_2, J_3)$. When expressed in terms of these variables, the strain relation (7.7) takes the form

$$\boldsymbol{\varepsilon} = \frac{\partial\psi}{\partial\boldsymbol{\sigma}^T} = \frac{\partial\psi}{\partial I_1}\frac{\partial I_1}{\partial\boldsymbol{\sigma}^T} + \frac{\partial\psi}{\partial J_2}\frac{\partial J_2}{\partial\boldsymbol{\sigma}^T} + \frac{\partial\psi}{\partial J_3}\frac{\partial J_3}{\partial\boldsymbol{\sigma}^T}. \tag{7.34}$$

In this format the basis consists of the derivatives $\partial I_1/\partial\boldsymbol{\sigma}^T$, $J_2/\partial\boldsymbol{\sigma}^T$ and $\partial J_3/\partial\boldsymbol{\sigma}^T$. The first of these is the unit tensor $\mathbf{I}$ as shown in (7.23). The two remaining terms are evaluated by differentiation through the deviatoric stress tensor $\boldsymbol{\sigma}'$. The components of this differentiation follow from (7.27) as

$$\frac{\partial\sigma'_{\gamma\delta}}{\partial\sigma_{\alpha\beta}} = \delta_{\alpha\gamma}\delta_{\beta\delta} - \tfrac{1}{3}\delta_{\gamma\delta}\delta_{\alpha\beta}. \tag{7.35}$$

The derivative of the second deviatoric stress invariant J_2 then follows from (7.30) as

$$\frac{\partial J_2}{\partial\sigma_{\alpha\beta}} = \sigma'_{\alpha\beta}. \tag{7.36}$$

Clearly, this component is independent of the mean stress σ_m, in contrast to the derivative of I_2 given in (7.24). This also follows from the property $\partial J_2/\partial\sigma_{\gamma\gamma} = 0$. The derivative of the third deviatoric stress invariant J_3 follows from (7.31) as

$$\frac{\partial J_3}{\partial \sigma_{\alpha\beta}} = \sigma'_{\alpha\gamma}\sigma'_{\gamma\beta} - \tfrac{2}{3} J_2 \delta_{\alpha\beta}. \tag{7.37}$$

Also for this derivative the contracted form vanishes, $\partial J_3/\partial\sigma_{\gamma\gamma} = 0$, indicating the independence of the mean stress.

Example 7.1. Graphic representation of stresses. A state of stress $\boldsymbol{\sigma}$ is conveniently represented graphically in terms of its principal components $\sigma_1, \sigma_2, \sigma_3$. In particular, this representation permits an illustration of the decomposition of the stress state into the sum of the mean stress state $\mathbf{I}\sigma_m$ and the deviatoric stress state $\boldsymbol{\sigma}'$, as shown in Fig. 7.2. The geometry of this stress space is central to the development of elasto-plastic material models and is discussed in some detail, e.g. by Chen and Han (1988).

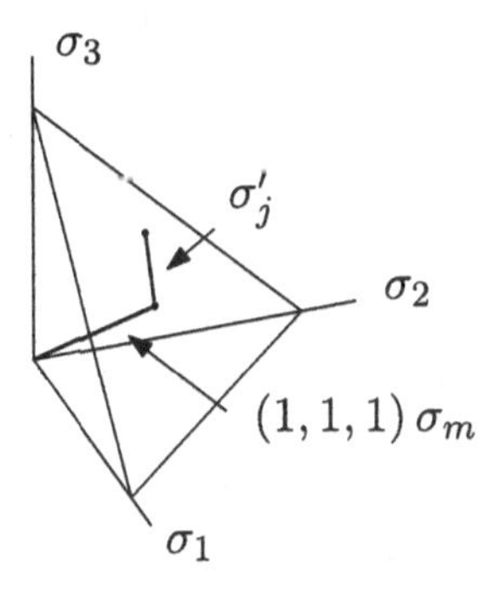

Fig. 7.2. Principal stress space with deviatoric plane.

The mean stress is represented by the vector $(1,1,1)\sigma_m$ in the direction of the principal diagonal. The deviatoric components are represented by the vector $(\sigma'_1, \sigma'_2, \sigma'_3)$ that is orthogonal to the mean state vector. Thus, the deviatoric stress lies in the deviatoric plane illustrated by the triangle that intersects the axes at $3\sigma_m$. The location of the stress in the principal stress space is conveniently determined by the stress invariants I_1, J_2 and J_3. The location of the deviatoric plane is described by its distance from the origin $\sqrt{3}\sigma_m = I_1/\sqrt{3}$. The length of the deviatoric stress vector $(\sigma'_1, \sigma'_2, \sigma'_3)$ is $\sqrt{2J_2}$, and the direction in the deviatoric plane is determined by a combination of J_3 and J_2 as discussed in Section 7.5.

7.1.2 Strain invariants and small strain elasticity

The properties relating to principal components, invariants, etc. have been described relating to the stress tensor $\boldsymbol{\sigma}$. Similar results apply to the strain tensor $\boldsymbol{\varepsilon}$. Within small strain analysis it is customary to use the linear volumetric strain

$$\varepsilon_v = \mathrm{tr}(\boldsymbol{\varepsilon}) = \varepsilon_{\gamma\gamma} \tag{7.38}$$

instead of the mean strain. This gives the deviatoric strain components in the form

$$\varepsilon'_{\alpha\beta} = \varepsilon_{\alpha\beta} - \tfrac{1}{3}\varepsilon_{\gamma\gamma}\,\delta_{\alpha\beta} = \varepsilon_{\alpha\beta} - \tfrac{1}{3}\varepsilon_v\,\delta_{\alpha\beta}. \tag{7.39}$$

The principal strain invariants $\bar{I}_1$, $\bar{I}_2$ and $\bar{I}_3$ follow from (7.17)–(7.19), when replacing the stress symbol σ with the strain symbol ε. As in the case of stresses, the first principal strain invariant

$$\bar{I}_1 = \mathrm{tr}(\boldsymbol{\varepsilon}) = \varepsilon_{\gamma\gamma} = \varepsilon_1 + \varepsilon_2 + \varepsilon_3 \tag{7.40}$$

can be used together with the second and third deviatoric strain invariants, defined in the same way as the deviatoric stress invariants,

$$\bar{J}_2 = \tfrac{1}{2}\mathrm{tr}(\boldsymbol{\varepsilon}'^2) = \tfrac{1}{2}\varepsilon'_{\alpha\beta}\varepsilon'_{\beta\alpha}, \tag{7.41}$$

$$\bar{J}_3 = \tfrac{1}{3}\mathrm{tr}(\boldsymbol{\varepsilon}'^3) = \tfrac{1}{3}\varepsilon'_{\alpha\beta}\varepsilon'_{\beta\gamma}\varepsilon'_{\gamma\alpha}. \tag{7.42}$$

The property of zero mean value of the deviatoric strain leads to two alternative forms of the formulae for the principal component forms, exactly like (7.32) and (7.33) for the deviatoric stress invariants.

For isotropic materials the internal energy can be expressed in terms of the deviatoric strain invariants in the form $\varphi(\bar{I}_1, \bar{J}_2, \bar{J}_3)$. The stress relation (7.10) then takes the form

$$\boldsymbol{\sigma} = \frac{\partial\varphi}{\partial\boldsymbol{\varepsilon}^T} = \frac{\partial\varphi}{\partial\bar{I}_1}\frac{\partial\bar{I}_1}{\partial\boldsymbol{\varepsilon}^T} + \frac{\partial\varphi}{\partial\bar{J}_2}\frac{\partial\bar{J}_2}{\partial\boldsymbol{\varepsilon}^T} + \frac{\partial\varphi}{\partial\bar{J}_3}\frac{\partial\bar{J}_3}{\partial\boldsymbol{\varepsilon}^T}. \tag{7.43}$$

In this format the basis consists of the derivatives $\partial\bar{I}_1/\partial\boldsymbol{\varepsilon}^T$, $\bar{J}_2/\partial\boldsymbol{\varepsilon}^T$ and $\partial\bar{J}_3/\partial\boldsymbol{\varepsilon}^T$. The components follow from the similar stress basis given in (7.23), (7.36) and (7.37) as

$$\boldsymbol{\sigma} = \Big(\frac{\partial\varphi}{\partial\bar{I}_1} - \tfrac{2}{3}\bar{J}_2\frac{\partial\varphi}{\partial\bar{J}_3}\Big)\mathbf{I} + \frac{\partial\varphi}{\partial\bar{J}_2}\boldsymbol{\varepsilon}' + \frac{\partial\varphi}{\partial\bar{J}_3}\boldsymbol{\varepsilon}'\boldsymbol{\varepsilon}'. \tag{7.44}$$

It follows from this relation that a linear relation between conjugate stress and strain tensors implies an internal energy function of the form $\varphi(\bar{I}_1, \bar{J}_2)$, because dependence on $\bar{J}_3$ would lead to a term with the quadratic tensor product $\boldsymbol{\varepsilon}'\boldsymbol{\varepsilon}'$.

The linear isotropic elastic model follows from (7.44), when the internal energy is a homogeneous function of degree two in the strain components as

$$\varphi(\bar{I}_1, \bar{J}_2) = \tfrac{1}{2}k\bar{I}_1^2 + 2\mu\bar{J}_2 = \tfrac{1}{2}k(\varepsilon_{\gamma\gamma})^2 + \mu\varepsilon'_{\alpha\beta}\varepsilon'_{\beta\alpha}. \tag{7.45}$$

This energy function is described by two additive terms: a volumetric term multiplied by the bulk modulus k, and a deviatoric term multiplied by the shear modulus μ. The corresponding stresses follow by differentiation with respect to the strain components as

$$\sigma_{\alpha\beta} = \frac{\partial\varphi}{\partial\varepsilon_{\alpha\beta}} = k\varepsilon_{\gamma\gamma}\delta_{\alpha\beta} + 2\mu\varepsilon'_{\alpha\beta} = (k - \tfrac{2}{3}\mu)\varepsilon_{\gamma\gamma}\delta_{\alpha\beta} + 2\mu\varepsilon_{\alpha\beta}. \tag{7.46}$$

The first form shows the split into mean stress and deviatoric stress:

$$\sigma_m = \tfrac{1}{3}\sigma_{\gamma\gamma} = k\varepsilon_{\gamma\gamma}, \qquad \sigma'_{\alpha\beta} = 2\mu\varepsilon'_{\alpha\beta}. \tag{7.47}$$

These two equations describe the two basic deformation mechanisms of linear isotropic materials – change of volume and change of shape. The change of shape is associated with zero-mean stress and strain states and is sometimes called generalized shear. The relations are illustrated in Fig. 7.3.

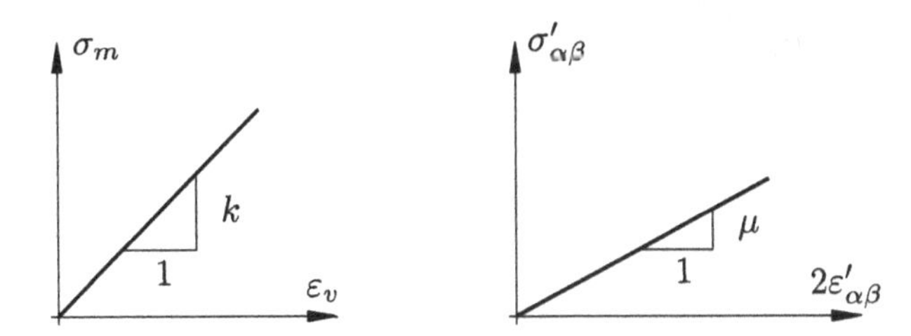

Fig. 7.3. Linear isotropic elasticity in dilation and generalized shear.

The linear elastic relation (7.46) can be written in boldface format as

$$\boldsymbol{\sigma} = \mathbb{C}\,\boldsymbol{\varepsilon}, \qquad \boldsymbol{\varepsilon} = \mathbb{C}^{-1}\boldsymbol{\sigma}, \tag{7.48}$$

where the relations can be interpreted either in full tensor format or in the reduced vector format as need may be. The stiffness tensor components follow from (7.46) as

$$C_{\alpha\beta\gamma\delta} = (k - \tfrac{2}{3}\mu)\delta_{\alpha\beta}\delta_{\gamma\delta} + \mu(\delta_{\alpha\gamma}\delta_{\beta\delta} + \delta_{\alpha\delta}\delta_{\beta\gamma}). \tag{7.49}$$

In the vector format $\mathbb{C}$ is a 6×6 stiffness matrix, and its inverse $\mathbb{C}^{-1}$ a 6×6 flexibility matrix. For hyper-elastic materials the existence of an energy function implies the symmetry relations

$$C_{\alpha\beta\gamma\delta} = C_{\gamma\delta\alpha\beta}. \tag{7.50}$$

These relations are equivalent to symmetry of the 6×6 stiffness matrix $\mathbb{C}$ and its inverse. For linear isotropic materials the elastic parameters are often represented by Young's modulus E and Poisson's ratio ν, describing the uniaxial tension stiffness and transverse contraction, respectively. The somewhat more basic parameters k and μ are expressed in terms of these as

$$k = \frac{E}{3(1-2\nu)}, \qquad \mu = \frac{E}{2(1+\nu)}. \tag{7.51}$$

The elastic flexibility matrix

$$\mathbb{C}^{-1} = \frac{1}{E}\begin{bmatrix} 1 & -\nu & -\nu & & & \\ -\nu & 1 & -\nu & & & \\ -\nu & -\nu & 1 & & & \\ & & & 2(1+\nu) & & \\ & & & & 2(1+\nu) & \\ & & & & & 2(1+\nu) \end{bmatrix} \tag{7.52}$$

and the corresponding stiffness matrix

$$\mathbb{C} = \frac{E}{(1+\nu)(1-2\nu)}\begin{bmatrix} 1-\nu & \nu & \nu & & & \\ \nu & 1-\nu & \nu & & & \\ \nu & \nu & 1-\nu & & & \\ & & & \frac{1}{2}-\nu & & \\ & & & & \frac{1}{2}-\nu & \\ & & & & & \frac{1}{2}-\nu \end{bmatrix} \tag{7.53}$$

follow from (7.49), when using the vector representation (7.8) for stresses and (7.9) for strains.

7.1.3 Isotropic elasticity at finite strain

In practice the isotropic linear elastic relation developed in the previous section is limited to small strains. The theory demonstrates the important distinction between change of volume and change of shape. In the small strain theory the change of volume is represented by the linear volume strain, defined in (7.38) as $\varepsilon_v = \varepsilon_{\gamma\gamma}$. However, the interpretation of ε_v as a volume strain only holds within linearized displacement theory. In the case of finite motion and finite strain, volume changes must be described by a precise measure of the change of volume, e.g. the Jacobian J, introduced in (6.36) as the ratio between a volume element in the current state and in the reference state, $J = dV/dV_0$. A general discussion of finite strain models in elasticity can be found in e.g. Ogden (1984) and Holzapfel (2000).

In the context of finite deformation it is often convenient to describe the current configuration relative to the original in a factored format without explicit use of strains. For this purpose the right Green deformation tensor is introduced by the definition

$$\mathbf{C} = \mathbf{F}^T\mathbf{F}. \tag{7.54}$$

In terms of this tensor the Green strain tensor definition (6.12) takes the form

$$\mathbf{E} = \tfrac{1}{2}(\mathbf{C} - \mathbf{I}). \tag{7.55}$$

In the development of constitutive relations, volume changes play an important role and will often be represented directly in the model, e.g. via the Jacobian determinant of the deformation gradient tensor $\mathbf{F}$,

$$J = \det(\mathbf{F}) = \det(\mathbf{C})^{1/2}. \tag{7.56}$$

The dilatational part of the motion, described by J, must be supplemented by a tensor measure of the change of shape, the so-called isochoric or volume-preserving deformation. A split of the infinitesimal motion into a volumetric and an isochoric part can be accomplished by introducing a factored format, in which the local dilation appears as a separate factor (Simo *et al.*, 1985; Simo, 1988a,b). This is expressed by the factored format

$$\mathbf{F} = \mathbf{F}_{\mathrm{vol}}\mathbf{F}_{\mathrm{iso}} = \mathbf{F}_{\mathrm{iso}}\mathbf{F}_{\mathrm{vol}}. \tag{7.57}$$

The volumetric part $\mathbf{F}_{\mathrm{vol}}$ represents a simple isotropic scaling with the appropriate power of the Jacobian determinant J:

$$\mathbf{F}_{\mathrm{vol}} = J^{1/3}\mathbf{I}, \qquad \mathbf{F}_{\mathrm{iso}} = J^{-1/3}\mathbf{F}. \tag{7.58}$$

Hereby the motion represented by $\mathbf{F}_{\mathrm{iso}}$ is volume-preserving, as demonstrated by the property $\det(\mathbf{F}_{\mathrm{iso}}) = 1$. The factored format of the deformation gradient leads to the following factored format for the corresponding right Green deformation tensor:

$$\mathbf{C} = J^{2/3}\mathbf{C}_{\mathrm{iso}}, \qquad \mathbf{C}_{\mathrm{iso}} = \mathbf{F}_{\mathrm{iso}}^T\mathbf{F}_{\mathrm{iso}}. \tag{7.59}$$

The deformation tensor $\mathbf{C}_{\mathrm{iso}}$ is then used to describe the isochoric part of the motion by assuming an internal energy density composed of two additive parts, whereby

$$\varphi(\mathbf{C}) = \varphi_{\mathrm{vol}}(J) + \varphi_{\mathrm{iso}}(\mathbf{C}_{\mathrm{iso}}), \tag{7.60}$$

representing the dilatational and the isochoric part of the motion, respectively.

The Piola–Kirchhoff components of the stress follow from differentiation of the energy density function with respect to the finite Green strain $\mathbf{E}$. When using the relation (7.55) the differentiation is changed to the deformation tensor $\mathbf{C}$,

$$\mathbf{S} = \frac{\partial \varphi}{\partial \mathbf{E}} = 2\frac{\partial \varphi}{\partial \mathbf{C}} = 2\frac{\partial \varphi_{\mathrm{vol}}}{\partial J}\frac{\partial J}{\partial \mathbf{C}} + 2\frac{\partial \varphi_{\mathrm{iso}}}{\partial \mathbf{C}_{\mathrm{iso}}}\frac{\partial \mathbf{C}_{\mathrm{iso}}}{\partial \mathbf{C}}. \tag{7.61}$$

The first term describes the dilatational behavior. The potential $\varphi_{\mathrm{vol}}(J)$ should exhibit monotonic dependence on J on both sides of a minimum at $J = 1$. The potential

$$\varphi_{\mathrm{vol}}(J) = \tfrac{1}{2}k\,[\,\tfrac{1}{2}(J^2 - 1) - \ln(J)\,] \tag{7.62}$$

was introduced by Simo (1988a,b) and Simo and Miehe (1992). Other specific forms can be used, e.g. $\varphi_{\mathrm{vol}}(J) = \frac{1}{2}k(\ln J)^2$ from Belytschko *et al.* (2000). The derivative with respect to the deformation tensor $\mathbf{C}$ is evaluated in two steps. First, the derivative of the particular function $\varphi_{\mathrm{vol}}(J)$ with respect to J,

$$\frac{\partial \varphi_{\mathrm{vol}}}{\partial J} = \tfrac{1}{2}k\,(\,J - J^{-1}\,), \tag{7.63}$$

followed by the derivative of J.

The derivative of J follows from a general formula for the derivative of a determinant with respect to its components. The contribution to a determinant from any particular component can be expressed as the product of this component with the corresponding co-factor. The co-factor is defined by the determinant of the matrix, where the row and column containing the component have been deleted. Thus, differentiation with respect to the component gives the co-factor. This result can be expressed for the matrix $\mathbf{C}$ in the compact notation

$$\frac{\partial \det(\mathbf{C})}{\partial \mathbf{C}} = \mathrm{cof}(\mathbf{C}) = \det(\mathbf{C})\,\mathbf{C}^{-T}. \tag{7.64}$$

The last result follows from the expression of the inverse in terms of co-factors. When using the expression (7.56) for J and the symmetry of $\mathbf{C}$, the formula takes the form

$$\frac{\partial J}{\partial \mathbf{C}} = \tfrac{1}{2}J\,\mathbf{C}^{-1}. \tag{7.65}$$

This completes the evaluation of the dilatational part,

$$\mathbf{S}_{\mathrm{vol}} = 2\frac{\partial \varphi_{\mathrm{vol}}}{\partial J}\frac{\partial J}{\partial \mathbf{C}} = \tfrac{1}{2}k\,(J^2 - 1)\,\mathbf{C}^{-1}. \tag{7.66}$$

For small displacements the leading term of this relation is

$$\mathbf{S}_{\mathrm{vol}} \simeq \tfrac{1}{2}k\,(J^2 - 1)\,\mathbf{I} = \tfrac{1}{2}k\,[\,\det(\mathbf{C}) - 1\,]\,\mathbf{I} \simeq k\,E_{\gamma\gamma}\,\mathbf{I}. \tag{7.67}$$

This corresponds to the dilation relation (7.47a) for the linearized theory.

The isochoric motion is described in terms of the scaled deformation tensor $\mathbf{C}_{\mathrm{iso}}$ defined in (7.59),

$$\varphi_{\mathrm{iso}}(\mathbf{C}_{\mathrm{iso}}) = \tfrac{1}{2}\mu\,[\,\mathrm{tr}(\mathbf{C}_{\mathrm{iso}}) - 3\,]. \tag{7.68}$$

Differentiation of this energy potential gives

$$\mathbf{S}_{\mathrm{iso}} = 2\frac{\partial\varphi_{\mathrm{iso}}}{\partial\mathbf{C}_{\mathrm{iso}}}\frac{\partial\mathbf{C}_{\mathrm{iso}}}{\partial\mathbf{C}} = \mu\, J^{-2/3}\,[\,\mathbf{I} - \tfrac{1}{3}\mathrm{tr}(\mathbf{C})\mathbf{C}^{-1}\,], \tag{7.69}$$

where $\mathbf{C}_{\mathrm{iso}}$ is represented by (7.59).

In total the two contributions give the Piola–Kirchhoff stress

$$\mathbf{S} = \tfrac{1}{2}k\,(J^2 - 1)\,\mathbf{C}^{-1} + \mu\, J^{-2/3}\,[\,\mathbf{I} - \tfrac{1}{3}\mathrm{tr}(\mathbf{C})\mathbf{C}^{-1}\,]. \tag{7.70}$$

The physical content of this relation is more easily seen when expressed in terms of the spatial Kirchhoff stress. The Kirchhoff stress is obtained by the transformation (6.40). The transformation cancels the factors $\mathbf{C}^{-1}$, and leads to the left Green deformation tensor $\mathbf{B}$ together with its scaled isochoric counterpart $\mathbf{B}_{\mathrm{iso}}$,

$$\mathbf{B} = \mathbf{F}\,\mathbf{F}^T, \qquad \mathbf{B}_{\mathrm{iso}} = \mathbf{F}_{\mathrm{iso}}\mathbf{F}_{\mathrm{iso}}^T = J^{-2/3}\mathbf{B}. \tag{7.71}$$

In terms of these tensors the Kirchhoff stress is found as

$$\boldsymbol{\tau} = \tfrac{1}{2}k\,(J^2 - 1)\,\mathbf{I} + \mu[\,\mathbf{B}_{\mathrm{iso}} - \tfrac{1}{3}\mathrm{tr}(\mathbf{B}_{\mathrm{iso}})\,\mathbf{I}\,]. \tag{7.72}$$

This formula brings out the similarity with the linear stress–strain relations (7.47) for dilatational and isochoric motion.

7.2 General plasticity theory

Plasticity theory is used to denote constitutive relations in which irreversible processes take part, when the stress or strain state exceeds a certain limit, described by the yield condition. Thus, plasticity theory is concerned with constitutive relations which combine an elastic state of deformation with additional irreversible effects, typically represented by additional straining. The discussion in this chapter is limited to rate independent forms of plasticity, i.e. forms of plasticity theory in which the additional straining is independent of time. Theories in which the irreversible deformation is associated with a time scale are called visco-plasticity, see e.g. Lemaitre and Chaboche (1990) or Ottosen and Ristinmaa (2005). In this section the structure of rate-independent plasticity theory is presented by using the modern approach of potentials, but without including non-mechanical effects like temperature. Thus, the theory takes the simplest form that illustrates the interplay between internal energy and the limit of reversible deformation, described by a yield surface. This section presents the general framework and proceeds to discuss the special case of associated plasticity theory, often used e.g. for metals. The implications of the more general non-associated plasticity theory are discussed in Section 7.4, and application of the general theory to the area of geotechnical problems is presented in Section 7.5.

7.2.1 Reversible deformation

Consider a material point in a state of deformation described by the elastic strain $\boldsymbol{\varepsilon}_e$ and a set of internal variables $\boldsymbol{\kappa}$. The subscript e denotes elastic and refers to reversible elastic deformation. The internal variables $\boldsymbol{\kappa}$ are parameters describing irreversible processes in the material. They are chosen such that they do not change during reversible elastic deformation. An example is the specific pore volume in a porous material that will contain a part due to the elastic deformation, but may also develop an irreversible contribution due to constitutive changes, e.g. in a granular material. The internal parameters are here selected such that they represent only the irreversible part, while any reversible contribution is included in the elastic properties.

The material is assumed to have an internal energy density described by the function $\varphi(\boldsymbol{\varepsilon}_e, \boldsymbol{\kappa})$. This form includes two extensions relative to the elastic deformation described in Section 7.1: the elastic strain $\boldsymbol{\varepsilon}_e$ may be only part of the total straining of the material, and the internal variables $\boldsymbol{\kappa}$ may contribute to the energy density. Conjugate variables are defined via the partial derivatives of the energy function,

$$\boldsymbol{\sigma} = \frac{\partial \varphi}{\partial \boldsymbol{\varepsilon}_e^T} = \boldsymbol{\nabla}_{\varepsilon} \varphi(\boldsymbol{\varepsilon}_e, \boldsymbol{\kappa}), \qquad \boldsymbol{\zeta} = \frac{\partial \varphi}{\partial \boldsymbol{\kappa}^T} = \boldsymbol{\nabla}_{\kappa} \varphi(\boldsymbol{\varepsilon}_e, \boldsymbol{\kappa}). \tag{7.73}$$

Here $\boldsymbol{\sigma}$ is the stress tensor, and $\boldsymbol{\zeta}$ is the set of internal variables, conjugate to $\boldsymbol{\kappa}$. It is here convenient to introduce the special notation for partial derivatives in terms of the symbol $\boldsymbol{\nabla}$ with an appropriate subscript indicating the variable with respect to which the function is differentiated.

The system is assumed to change with time, and the time derivatives are denoted by a dot over the symbol. The theory to be developed is rate independent, and thus time appears merely as a parameter, and can be considered as 'pseudo time'. The time derivatives of the conjugate variables in (7.73) follow by differentiation through $\boldsymbol{\varepsilon}_e$ and $\boldsymbol{\kappa}$,

$$\dot{\boldsymbol{\sigma}} = (\boldsymbol{\nabla}_{\varepsilon\varepsilon}\varphi)\,\dot{\boldsymbol{\varepsilon}}_e + (\boldsymbol{\nabla}_{\varepsilon\kappa}\varphi)\,\dot{\boldsymbol{\kappa}}, \tag{7.74}$$

$$\dot{\boldsymbol{\zeta}} = (\boldsymbol{\nabla}_{\kappa\varepsilon}\varphi)\,\dot{\boldsymbol{\varepsilon}}_e + (\boldsymbol{\nabla}_{\kappa\kappa}\varphi)\,\dot{\boldsymbol{\kappa}}. \tag{7.75}$$

The role of the internal parameters $\boldsymbol{\kappa}$ as describing irreversible changes of material configuration implies that $\dot{\boldsymbol{\kappa}} = \mathbf{0}$ in the elastic state. Thus, the reversible development of the conjugate external variables $\boldsymbol{\sigma}$ and internal variables $\boldsymbol{\zeta}$ is governed by the relations

$$\dot{\boldsymbol{\sigma}} = (\boldsymbol{\nabla}_{\varepsilon\varepsilon}\varphi)\,\dot{\boldsymbol{\varepsilon}}_e, \qquad \dot{\boldsymbol{\zeta}} = (\boldsymbol{\nabla}_{\kappa\varepsilon}\varphi)\,\dot{\boldsymbol{\varepsilon}}_e. \tag{7.76}$$

The first relation identifies $\boldsymbol{\nabla}_{\varepsilon\varepsilon}\varphi$ as the reversible elastic incremental stiffness tensor,

$$\mathbb{C}(\boldsymbol{\varepsilon}_e, \boldsymbol{\kappa}) = \boldsymbol{\nabla}_{\varepsilon\varepsilon}\varphi(\boldsymbol{\varepsilon}_e, \boldsymbol{\kappa}). \tag{7.77}$$

The second relation demonstrates that for a material in which the external and internal variables are coupled, the conjugate internal variables $\boldsymbol{\zeta}$ will change under reversible elastic deformation, while the primary internal variables $\boldsymbol{\kappa}$ remain constant.

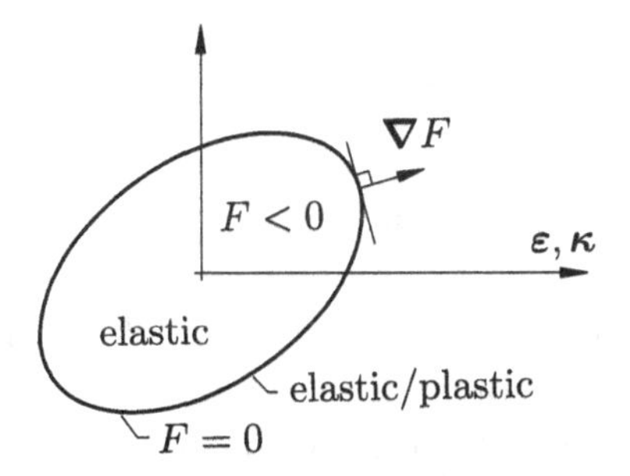

Fig. 7.4. Domain of reversible elastic deformation.

Reversible deformation only takes place when the primary variables $\boldsymbol{\varepsilon}_e$ and $\boldsymbol{\kappa}$ remain within a limited domain, described by a condition of the form

$$F(\boldsymbol{\varepsilon}_e, \boldsymbol{\kappa}) \leq 0. \tag{7.78}$$

This relation and the role of the yield function $F(\boldsymbol{\varepsilon}_e, \boldsymbol{\kappa})$ defining the domain of reversible deformation are illustrated in Fig. 7.4. While the material point is inside the domain of reversible deformation the strains $\boldsymbol{\varepsilon}_e$ and thereby the stresses $\boldsymbol{\sigma}$ may change, while the internal variables $\boldsymbol{\kappa}$ remain constant. Thus, the yield function can be expressed in terms of the stress tensor $\boldsymbol{\sigma}$ and $\boldsymbol{\kappa}$. For convenience this is expressed here without change of symbol as

$$F(\boldsymbol{\sigma}, \boldsymbol{\kappa}) \leq 0. \tag{7.79}$$

The stress gradient of the yield function $F(\boldsymbol{\sigma}, \boldsymbol{\kappa})$ is obtained by using the incremental relation between the stress $\boldsymbol{\sigma}$ and the elastic strain $\boldsymbol{\varepsilon}_e$ for constant value of the internal parameters $\boldsymbol{\kappa}$:

$$\boldsymbol{\nabla}_\sigma F = (\boldsymbol{\nabla}_{\varepsilon\varepsilon}\varphi)^{-1}\boldsymbol{\nabla}_\varepsilon F = \mathbb{C}^{-1}\boldsymbol{\nabla}_\varepsilon F. \tag{7.80}$$

In any reversible elastic process the internal variables $\boldsymbol{\kappa}$ remain constant, and thus in this particular state the limiting surface, also called the yield surface, can be described as a function of stress $F(\boldsymbol{\sigma}, \boldsymbol{\kappa})$ with the internal variables $\boldsymbol{\kappa}$ acting as internal parameters. This, the most commonly used form, is illustrated in Fig. 7.5.

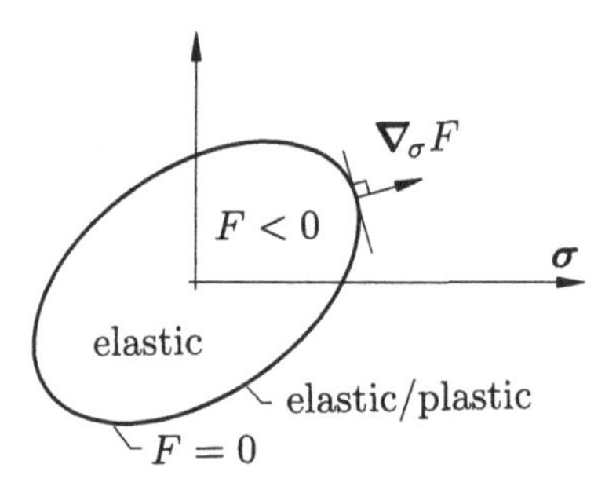

Fig. 7.5. Domain of reversible elastic deformation in stress space.

It is important to note that while the yield surface can be expressed either by using the reversible elastic strain $\boldsymbol{\varepsilon}_e$ or the stress $\boldsymbol{\sigma}$, there may be restrictions on the representation in terms of the internal variables. The original internal variables $\boldsymbol{\kappa}$ were introduced in such a way that they remain constant during reversible elastic deformation. Thus, the internal variables $\boldsymbol{\kappa}$ appear as constant parameters for processes inside the current yield surface. In contrast, the conjugate internal variables ζ will change during reversible processes according to (7.76b), if not all the mixed derivative components $\boldsymbol{\nabla}_{\kappa\varepsilon}\varphi$ vanish. When all mixed derivatives vanish, the conjugate variables $\boldsymbol{\zeta}$ remain constant during reversible processes. In that case the yield condition can be expressed in terms of the conjugate internal variables $\boldsymbol{\zeta}$ instead of the original internal variables $\boldsymbol{\kappa}$. However, in the general case the original internal variables $\boldsymbol{\kappa}$, that remain constant during reversible deformation, must be used.

In the following it will be assumed that the yield function F is convex in its arguments. This corresponds to the statement that for any two states within the yield surface any weighted linear average of the state variables will also be inside the yield surface. This condition is important for the uniqueness of the irreversible processes to be described next.

7.2.2 Maximum plastic dissipation rate

In the case of reversible deformation the internal variables $\boldsymbol{\kappa}$ remain constant. Thus, it follows from (7.73a) that the work of the stresses $\boldsymbol{\sigma}$ through the strain rate $\dot{\boldsymbol{\varepsilon}}$ is equal to the rate of change of the internal energy φ. It is noted that in reversible deformation the elastic strain constitutes the total strain, and therefore (7.73a) can be written in time rate form as

$$\boldsymbol{\sigma}^T\dot{\boldsymbol{\varepsilon}} = \dot{\boldsymbol{\varepsilon}}_e^T\boldsymbol{\nabla}_\varepsilon\varphi(\boldsymbol{\varepsilon}_e,\boldsymbol{\kappa}) = \dot{\varphi}(\boldsymbol{\varepsilon}_e,\boldsymbol{\kappa}), \tag{7.81}$$

where the superscript T denotes transpose, and has been introduced to make the notation cover tensor and matrix notation. In essence this is a statement that the total work of the stresses is stored in the internal energy φ, and thus is recoverable.

In the case of irreversible deformation dissipation will occur, and the rate of dissipation is defined by

$$\mathcal{D} = \boldsymbol{\sigma}^T\dot{\boldsymbol{\varepsilon}} - \dot{\varphi}(\boldsymbol{\varepsilon}_e,\boldsymbol{\kappa}) \geq 0. \tag{7.82}$$

The first term is the work of the stresses $\boldsymbol{\sigma}$ through the strain rate $\dot{\boldsymbol{\varepsilon}}$, while the second term is the rate of recoverable work stored in the internal energy function $\varphi(\boldsymbol{\varepsilon}_e,\boldsymbol{\kappa})$. The energy function φ is a function of the current state of the recoverable strain $\boldsymbol{\varepsilon}_e$ and the internal variables $\boldsymbol{\kappa}$, but does not depend explicitly on time. Thus, the rate of change of the internal energy is given by

$$\dot{\varphi}(\boldsymbol{\varepsilon}_e,\boldsymbol{\kappa}) = \dot{\boldsymbol{\varepsilon}}_e^T\,\boldsymbol{\nabla}_\varepsilon\varphi + \dot{\boldsymbol{\kappa}}^T\,\boldsymbol{\nabla}_\kappa\varphi. \tag{7.83}$$

The gradients in this expression are the conjugate variables $\boldsymbol{\sigma}$ and $\boldsymbol{\zeta}$ defined in (7.73). Thus, the internal rate of energy can be written in terms of the rates of the primary variables, multiplied by the corresponding conjugate variables:

$$\dot{\varphi}(\boldsymbol{\varepsilon}_e,\boldsymbol{\kappa}) = \boldsymbol{\sigma}^T\dot{\boldsymbol{\varepsilon}}_e + \boldsymbol{\zeta}^T\dot{\boldsymbol{\kappa}}. \tag{7.84}$$

When this is introduced into (7.82), the rate of dissipation takes the form

$$\mathcal{D} = \boldsymbol{\sigma}^T(\dot{\boldsymbol{\varepsilon}} - \dot{\boldsymbol{\varepsilon}}_e) - \boldsymbol{\zeta}^T\dot{\boldsymbol{\kappa}}. \tag{7.85}$$

The plastic strain rate $\dot{\boldsymbol{\varepsilon}}_p$ is now defined as the difference of the total and the elastic strain rates, appearing in the first term. Hereby the total strain rate is the sum of the elastic and the plastic strain rates, i.e.

$$\dot{\boldsymbol{\varepsilon}} = \dot{\boldsymbol{\varepsilon}}_e + \dot{\boldsymbol{\varepsilon}}_p. \tag{7.86}$$

The dissipation relation then takes the simpler form

$$\mathcal{D} = \boldsymbol{\sigma}^T \dot{\boldsymbol{\varepsilon}}_p - \boldsymbol{\zeta}^T \dot{\boldsymbol{\kappa}}, \tag{7.87}$$

where it is clear that only the plastic part of the strain rate contributes to the dissipation, together with a contribution from the internal variables.

The constitutive relations of the material follow from a postulate of maximum dissipation rate under the condition that the variables are bounded by the yield condition defined by the inequality (7.78). The basic idea goes back to von Mises (1928) and Hill (1948), and has since been used by many authors, typically in relation to convex analysis, see e.g. Eve *et al.* (1990) for a survey. Here, a simple and direct formulation is given, following the approach of optimization theory. The maximum dissipation problem can be stated as a constrained extremum problem of the form

$$\begin{aligned} &\underset{\boldsymbol{\sigma},\boldsymbol{\zeta}}{\text{maximize}} \qquad && \mathcal{D} = \boldsymbol{\sigma}^T \dot{\boldsymbol{\varepsilon}}_p - \boldsymbol{\zeta}^T \dot{\boldsymbol{\kappa}} \\ &\text{subject to} && F(\boldsymbol{\varepsilon}_e, \boldsymbol{\kappa}) \le 0. \end{aligned} \tag{7.88}$$

Here the original form of the yield function $F(\boldsymbol{\varepsilon}_e, \boldsymbol{\kappa})$ has been retained to stress its role as the bounding surface of the domain of reversible deformation, described by the variables $\boldsymbol{\varepsilon}_e$ and $\boldsymbol{\kappa}$. This form helps clarify the role of the internal variables in determining the rules for the development of plastic strains.

The maximum rate of dissipation is found by considering the rates $\dot{\boldsymbol{\varepsilon}}_p$ and $\dot{\boldsymbol{\kappa}}$ as fixed, and then searching for the maximum by considering the conjugate variables $\boldsymbol{\sigma}$ and $\boldsymbol{\zeta}$ as subject to variation. The yield function is given in terms of the original variables $\boldsymbol{\varepsilon}_e$ and $\boldsymbol{\kappa}$, and therefore derivatives with respect to the conjugate variables must be obtained by using the defining relations (7.73). If the maximum is attained at a point inside the domain defined by the yield condition, the derivatives of $\mathcal{D}$ with respect to the independent variables $\boldsymbol{\sigma}, \boldsymbol{\zeta}$ must vanish. This gives the equations

$$\dot{\boldsymbol{\varepsilon}}_p = \mathbf{0}, \qquad \dot{\boldsymbol{\kappa}} = \mathbf{0}. \tag{7.89}$$

These equations state that there is no development of plastic strain and no development of the internal variables $\boldsymbol{\kappa}$. These are precisely the conditions characterizing reversible deformation, as described in Section 7.2.1.

Now, assume that the maximum rate of dissipation is attained for a set of variables $\boldsymbol{\sigma}, \boldsymbol{\zeta}$ corresponding to a point on the yield surface. According to (7.88a), surfaces corresponding to equal dissipation rate are hyperplanes with normal vector $(\dot{\boldsymbol{\varepsilon}}_p, -\dot{\boldsymbol{\kappa}})$. The point of maximum dissipation rate must therefore be located on the yield surface and have the same normal vector in $\boldsymbol{\sigma}, \boldsymbol{\zeta}$ space. If this were not the case, a larger value of the dissipation rate $\mathcal{D}$ could be found in the neighborhood of the assumed intersection of the dissipation contour and the yield surface. It is convenient to formulate this condition in terms of a potential function $G(\boldsymbol{\sigma}, \boldsymbol{\zeta})$ that describes the yield surface in terms of the variables $\boldsymbol{\sigma}, \boldsymbol{\zeta}$:

$$G(\boldsymbol{\sigma}, \boldsymbol{\zeta}) = F(\boldsymbol{\varepsilon}_e, \boldsymbol{\kappa}), \tag{7.90}$$

where the variables are transformed according to (7.73). The partial derivatives of the two potentials $F(\boldsymbol{\varepsilon}, \boldsymbol{\kappa})$ and $G(\boldsymbol{\sigma}, \boldsymbol{\zeta})$ are related via the chain rule of differentiation:

$$\begin{aligned} \boldsymbol{\nabla}_{\varepsilon} F &= \frac{\partial \boldsymbol{\sigma}}{\partial \boldsymbol{\varepsilon}_e}\Big|_{\kappa}^{T} \boldsymbol{\nabla}_{\sigma} G + \frac{\partial \boldsymbol{\zeta}}{\partial \boldsymbol{\varepsilon}_e}\Big|_{\kappa}^{T} \boldsymbol{\nabla}_{\zeta} G, \\ \boldsymbol{\nabla}_{\kappa} F &= \frac{\partial \boldsymbol{\sigma}}{\partial \boldsymbol{\kappa}}\Big|_{\varepsilon}^{T} \boldsymbol{\nabla}_{\sigma} G + \frac{\partial \boldsymbol{\zeta}}{\partial \boldsymbol{\kappa}}\Big|_{\varepsilon}^{T} \boldsymbol{\nabla}_{\zeta} G. \end{aligned} \tag{7.91}$$

When the conjugate variables $\boldsymbol{\sigma}, \boldsymbol{\zeta}$ are introduced by their definition (7.73), the relation between the partial derivatives of the potentials takes the form

$$\begin{bmatrix} \boldsymbol{\nabla}_{\varepsilon} F \\ \boldsymbol{\nabla}_{\kappa} F \end{bmatrix} = \begin{bmatrix} \boldsymbol{\nabla}_{\varepsilon\varepsilon}\varphi & \boldsymbol{\nabla}_{\varepsilon\kappa}\varphi \\ \boldsymbol{\nabla}_{\kappa\varepsilon}\varphi & \boldsymbol{\nabla}_{\kappa\kappa}\varphi \end{bmatrix} \begin{bmatrix} \boldsymbol{\nabla}_{\sigma} G \\ \boldsymbol{\nabla}_{\zeta} G \end{bmatrix}. \tag{7.92}$$

This provides the relation of the partial derivatives of the yield function F in terms of the original variables $\boldsymbol{\varepsilon}_e, \boldsymbol{\kappa}$ and the yield function G in terms of the conjugate variables $\boldsymbol{\sigma}, \boldsymbol{\zeta}$.

The solution to the constrained maximization problem (7.88) can now be stated in concise form. In the $\boldsymbol{\sigma}, \boldsymbol{\zeta}$ space the condition of co-directional normals of the dissipation rate function $\mathcal{D}$ and the yield potential G takes the form

$$\begin{bmatrix} \boldsymbol{\nabla}_{\sigma} \mathcal{D} \\ \boldsymbol{\nabla}_{\zeta} \mathcal{D} \end{bmatrix} = \dot{\lambda} \begin{bmatrix} \boldsymbol{\nabla}_{\sigma} G \\ \boldsymbol{\nabla}_{\zeta} G \end{bmatrix}, \tag{7.93}$$

where $\dot{\lambda} \geq 0$ is a factor of proportionality. This factor is conveniently combined with the yield potential in the product condition

$$\dot{\lambda}\, F(\boldsymbol{\varepsilon}_e, \boldsymbol{\kappa}) = 0. \tag{7.94}$$

This condition covers both of the following two cases. Inside the yield surface

$\dot{\lambda} = 0$, and the processes are fully reversible with $\mathcal{D} = 0$, as discussed in connection with (7.89). Conversely, points on the yield surface have $F = 0$ and thus permit $\dot{\lambda} > 0$, corresponding to dissipative processes. These are the Kuhn–Tucker conditions, here expressed in terms of conjugate variables (Kuhn and Tucker, 1951).

When substituting the rate of dissipation from (7.85), the evolution equations are obtained as

$$\begin{aligned} \dot{\boldsymbol{\varepsilon}} &= \dot{\boldsymbol{\varepsilon}}_e + \dot{\lambda}\, \boldsymbol{\nabla}_\sigma G, \\ \dot{\boldsymbol{\kappa}} &= \quad - \dot{\lambda}\, \boldsymbol{\nabla}_\zeta G. \end{aligned} \tag{7.95}$$

For $\dot{\lambda} = 0$ the elastic equations (7.89) corresponding to reversible deformation are recovered. The internal variables $\boldsymbol{\kappa}$ have been chosen to represent the deviation from the quasi-stationary equilibrium in the elastic state, and as a consequence changes are limited to the plastic state. The plastic strain rate and the rate of the internal variables $\dot{\boldsymbol{\kappa}}$ are given by the gradients of the potential function G, and this is therefore often called the flow potential.

A similar set of equations was obtained by Lemaitre and Chaboche (1990) by a different approach, using special properties of the Legendre transformation of homogeneous functions of degree one instead of the maximum dissipation condition, see e.g. Sewell (1987) for the mathematical background. The plastic strain rate and the gradient of the potential G are illustrated in Fig. 7.6. It is seen that in the case of coupling between external and internal variables the plastic strain increment does not appear as orthogonal to the yield surface in stress space.

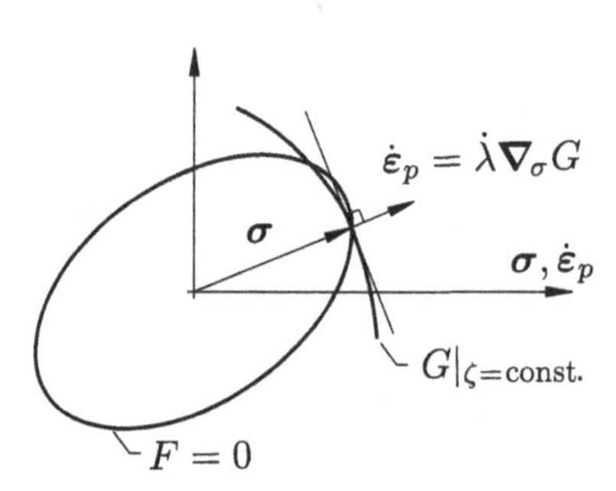

Fig. 7.6. Yield surface with coupled plastic flow.

The solution (7.95) for the rates of plastic strain and the internal variables leads to a simple form of the rate of dissipation $\mathcal{D}$ in terms of gradients of the flow potential G. Substitution of $\dot{\boldsymbol{\varepsilon}}_p$ and $\dot{\boldsymbol{\kappa}}_e$ into (7.85) gives

$$\mathcal{D} = \dot{\lambda}\big(\boldsymbol{\sigma}^T \boldsymbol{\nabla}_\sigma G + \boldsymbol{\zeta}^T \boldsymbol{\nabla}_\zeta G\big). \tag{7.96}$$

The dissipation rate is here given entirely as a function of the stress $\boldsymbol{\sigma}$ and the conjugate internal variable $\boldsymbol{\zeta}$. Thus, this expression may be used as a theoretical basis for derivation of the flow potential $G(\boldsymbol{\sigma}, \boldsymbol{\zeta})$ from the physics of the dissipation process, see e.g. Schofield and Wroth (1968) and Krenk (1998) for the case of granular friction materials, discussed in Section 7.5.

Much work on plasticity theory and many specific plasticity models have been based on the uncoupled case, in which $\boldsymbol{\nabla}_{\varepsilon\kappa}\varphi = \boldsymbol{\nabla}_{\kappa\varepsilon}\varphi = 0$, corresponding to an additive format of the internal energy:

$$\varphi(\boldsymbol{\varepsilon}, \boldsymbol{\kappa}) = \varphi_\varepsilon(\boldsymbol{\varepsilon}) + \varphi_\kappa(\boldsymbol{\kappa}). \tag{7.97}$$

For this class of plasticity problems the conjugate variables are defined by the uncoupled relations

$$\boldsymbol{\sigma} = \boldsymbol{\nabla}_\varepsilon \varphi_\varepsilon(\boldsymbol{\varepsilon}_e), \qquad \boldsymbol{\zeta} = \boldsymbol{\nabla}_\kappa \varphi_\kappa(\boldsymbol{\kappa}). \tag{7.98}$$

This implies that the transformation between the variable $\boldsymbol{\varepsilon}_e$ and its conjugate variable $\boldsymbol{\sigma}$ is independent of the transformation between the internal variable $\boldsymbol{\kappa}$ and its conjugate internal variable $\boldsymbol{\zeta}$. As a consequence, the gradients of the two yield function representations F and G are given by the uncoupled form of (7.92) as

$$\boldsymbol{\nabla}_\varepsilon F = (\boldsymbol{\nabla}_{\varepsilon\varepsilon}\varphi)\boldsymbol{\nabla}_\sigma G, \qquad \boldsymbol{\nabla}_\kappa F = (\boldsymbol{\nabla}_{\kappa\kappa}\varphi)\boldsymbol{\nabla}_\zeta G \tag{7.99}$$

for the class of uncoupled materials. For this class of plastic materials the plastic strain rate is proportional to the stress gradient of the yield function $F(\boldsymbol{\sigma}, \boldsymbol{\kappa})$ as illustrated in Fig. 7.7.

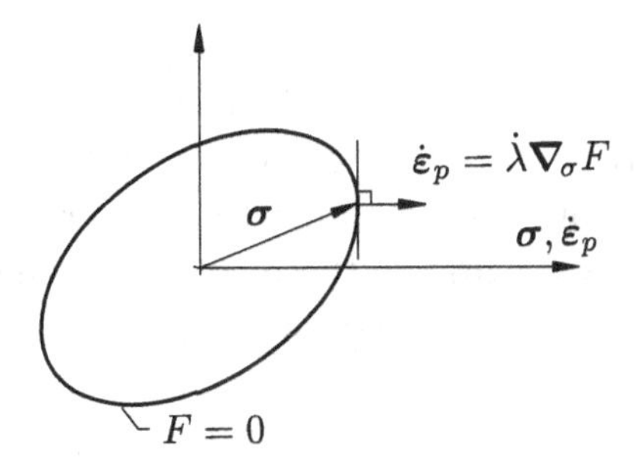

Fig. 7.7. Yield surface with uncoupled plastic flow.

In the literature, plasticity theory with internal variables that are uncoupled from the elastic strain (or the stress) is often called associated plasticity, because the gradient property used in the flow rule follows directly from the yield function F. In contrast, the general coupled theory is called non-associated plasticity theory, because the direction of plastic flow is not directly given by the original yield function $F(\boldsymbol{\sigma},\boldsymbol{\kappa})$. In the present context it is probably more descriptive to classify these two types of models as uncoupled and coupled, respectively.

7.2.3 Evolution equations

Plastic flow requires $\dot{\lambda} > 0$, and according to the Kuhn–Tucker criterion (7.94) this leads to the condition $F = G = 0$. This identity must remain satisfied during plastic flow, a property often called the consistency condition. This provides the extra equation needed to determine the multiplier $\dot{\lambda}$, thereby closing the system of evolution equations. The most elegant general formulation is obtained by expressing the consistency condition in terms of the yield function $F(\boldsymbol{\varepsilon}_e,\boldsymbol{\kappa})$. Time differentiation of the yield condition in terms of this potential gives the consistency condition in the form

$$\dot{F}(\boldsymbol{\varepsilon}_e,\boldsymbol{\kappa}) \;=\; \dot{\boldsymbol{\varepsilon}}_e^T\boldsymbol{\nabla}_{\varepsilon}F \,+\, \dot{\boldsymbol{\kappa}}^T\boldsymbol{\nabla}_{\kappa}F \;=\; 0. \tag{7.100}$$

The time derivatives $\dot{\boldsymbol{\varepsilon}}_e$ and $\dot{\boldsymbol{\kappa}}$ are substituted from the flow equations (7.95), whereby

$$\dot{F} \;=\; \dot{\boldsymbol{\varepsilon}}^T\boldsymbol{\nabla}_{\varepsilon}F \;-\; \dot{\lambda}\big[(\boldsymbol{\nabla}_{\sigma}G)^T(\boldsymbol{\nabla}_{\varepsilon}F) + (\boldsymbol{\nabla}_{\varsigma}G)^T(\boldsymbol{\nabla}_{\kappa}F)\big] \;=\; 0. \tag{7.101}$$

This gives an expression for the plastic multiplier $\dot{\lambda}$ in terms of the total strain rate $\dot{\boldsymbol{\varepsilon}}$,

$$\dot{\lambda} \;=\; \frac{(\boldsymbol{\nabla}_{\varepsilon}F)^T\dot{\boldsymbol{\varepsilon}}}{(\boldsymbol{\nabla}_{\sigma}G)^T(\boldsymbol{\nabla}_{\varepsilon}F) \,+\, (\boldsymbol{\nabla}_{\varsigma}G)^T(\boldsymbol{\nabla}_{\kappa}F)}. \tag{7.102}$$

It is to be noted that the products of the gradients of the potentials F and G are formed directly by products of their components. This property is a consequence of the gradients being formed in terms of conjugate variables.

The evolution equations for elasto-plastic flow are now found from (7.95) by substitution of $\dot{\lambda}$. In the first of these equations the rate of elastic strain $\dot{\boldsymbol{\varepsilon}}_e$ is replaced by the stress rate $\boldsymbol{\sigma}$ by multiplication with the elastic stiffness according to (7.76a):

$$\dot{\sigma} \;=\; (\boldsymbol{\nabla}_{\varepsilon\varepsilon}\varphi)\Big[\,\mathbf{I} \,-\, \frac{(\boldsymbol{\nabla}_{\sigma}G)\otimes(\boldsymbol{\nabla}_{\varepsilon}F)^T}{(\boldsymbol{\nabla}_{\sigma}G)^T(\boldsymbol{\nabla}_{\varepsilon}F) \,+\, (\boldsymbol{\nabla}_{\varsigma}G)^T(\boldsymbol{\nabla}_{\kappa}F)}\,\Big]\,\dot{\varepsilon}, \tag{7.103}$$

where $\mathbf{I}$ is the appropriate unit tensor, and the symbol $\otimes$ is introduced to indicate that the components of two factors are not contracted. The corresponding evolution law for the internal variables $\boldsymbol{\kappa}$ follows from the second equation (7.95b) as

$$\dot{\kappa} = -\frac{(\boldsymbol{\nabla}_\sigma G) \otimes (\boldsymbol{\nabla}_\varepsilon F)^T}{(\boldsymbol{\nabla}_\sigma G)^T(\boldsymbol{\nabla}_\varepsilon F) + (\boldsymbol{\nabla}_\varsigma G)^T(\boldsymbol{\nabla}_\kappa F)}\,\dot{\boldsymbol{\varepsilon}}. \tag{7.104}$$

This is the most compact explicit formulation of the evolution equations, making use of the dual yield and flow potentials $F(\boldsymbol{\varepsilon}_e, \boldsymbol{\kappa})$ and $G(\boldsymbol{\sigma}, \boldsymbol{\zeta})$.

In practice, plasticity problems are integrated by forming finite increments, and thus the gradients appearing in the evolution equations typically change value over the time increment. The format of the evolution equations (7.103)–(7.104) is well suited for explicit integration schemes, in which the non-linear gradients are evaluated using the current value of the variables. However, explicit integration schemes have limitations regarding accuracy and numerical stability, and therefore implicit schemes are often preferred, see e.g. the analyses of Krieg and Krieg (1977), Ortiz and Popov (1985), Ortiz and Simo (1986). In implicit schemes the functions occurring in the evolution formulae are evaluated at an intermediate or the final time. This requires iterations, and the evolution equations must therefore be given a format that is suitable for this. In the context of plasticity the incremental step will most often start with a predicted value of the total strain increment $\Delta\boldsymbol{\varepsilon}$, and the objective of the numerical scheme will then be to find corresponding increments of the variables $\Delta\boldsymbol{\sigma}$, $\Delta\boldsymbol{\kappa}$ and $\Delta\lambda$. Here the evolution equations will be reformulated using the time derivatives, while the finite increment form will be discussed in connection with numerical integration schemes later.

The idea is to cast the set of the two evolution equations (7.95a,b) and the consistency equation (7.100) in the form of three equations for the rates $\dot{\boldsymbol{\sigma}}$, $\dot{\boldsymbol{\kappa}}$ and $\dot{\lambda}$. First the strain increment equation (7.95a) is written in the form

$$\dot{\boldsymbol{\varepsilon}} = \mathbb{C}^{-1}\dot{\boldsymbol{\sigma}} + \dot{\lambda}\,\boldsymbol{\nabla}_\sigma G, \tag{7.105}$$

where the elastic strain rate has been replaced by the stress rate by use of the elastic stiffness tensor $\mathbb{C} = \boldsymbol{\nabla}_{\varepsilon\varepsilon}\varphi$ from (7.77). The equation (7.95b) for the internal variable $\boldsymbol{\kappa}$ is written in homogeneous form as

$$\mathbf{0} = \dot{\boldsymbol{\kappa}} + \dot{\lambda}\,\boldsymbol{\nabla}_\varsigma G. \tag{7.106}$$

In the final equation, the consistency equation (7.100), the elastic strain is replaced as independent variable by the stress, yielding the equation

$$\dot{F} = \dot{\boldsymbol{\sigma}}^T\boldsymbol{\nabla}_\sigma F + \dot{\boldsymbol{\kappa}}^T\boldsymbol{\nabla}_\kappa F = 0. \tag{7.107}$$

These three equations are conveniently collected in block matrix form as

$$\begin{bmatrix} \dot{\boldsymbol{\varepsilon}} \\ \mathbf{0} \\ \dot{F} \end{bmatrix} = \begin{bmatrix} \mathbb{C}^{-1} & \mathbf{0} & (\boldsymbol{\nabla}_\sigma G) \\ \mathbf{0} & \mathbf{I} & (\boldsymbol{\nabla}_\varsigma G) \\ (\boldsymbol{\nabla}_\sigma F)^T & (\boldsymbol{\nabla}_\kappa F)^T & 0 \end{bmatrix} \begin{bmatrix} \dot{\boldsymbol{\sigma}} \\ \dot{\boldsymbol{\kappa}} \\ \dot{\lambda} \end{bmatrix}, \tag{7.108}$$

where $\dot{F}$ has been indicated explicitly for clarity. It is noted that for materials, where the internal variables are coupled with the elastic strains, these equations are non-symmetric. However, for materials with uncoupled internal parameters symmetry can be obtained by multiplication of the second row with the symmetric matrix $(\boldsymbol{\nabla}_{\kappa\kappa}\varphi)$. By (7.99b) this replaces the gradient $\boldsymbol{\nabla}_\varsigma G$ with $\boldsymbol{\nabla}_\kappa F$, and the equations then take the symmetric form

$$\begin{bmatrix} \dot{\boldsymbol{\varepsilon}} \\ \mathbf{0} \\ \dot{F} \end{bmatrix} = \begin{bmatrix} \mathbb{C}^{-1} & \mathbf{0} & (\boldsymbol{\nabla}_\sigma F) \\ \mathbf{0} & (\boldsymbol{\nabla}_{\kappa\kappa}\varphi) & (\boldsymbol{\nabla}_\kappa F) \\ (\boldsymbol{\nabla}_\sigma F)^T & (\boldsymbol{\nabla}_\kappa F)^T & 0 \end{bmatrix} \begin{bmatrix} \dot{\boldsymbol{\sigma}} \\ \dot{\boldsymbol{\kappa}} \\ \dot{\lambda} \end{bmatrix}. \tag{7.109}$$

The integrated forms, in which the time derivatives are replaced by finite increments, are suitable for numerical plasticity algorithms, when the entries in the matrix are recognized as representative mean values, typically evaluated iteratively to correspond to the final time.

It is sometimes convenient to use a reduced format, in which the effect of the internal parameters is combined into a single parameter H, and the update of the internal parameters $\boldsymbol{\kappa}$ is considered separately. This format follows, when the time rate of internal variables $\dot{\boldsymbol{\kappa}}$ is eliminated from the consistency condition (7.107) by use of the evolution equation (7.106). This gives the consistency equation in the form

$$\dot{F} = \dot{\boldsymbol{\sigma}}^T \boldsymbol{\nabla}_\sigma F - H\dot{\lambda} = 0, \tag{7.110}$$

where the contribution from the internal parameters has been combined in the hardening modulus

$$H = (\boldsymbol{\nabla}_\varsigma G)^T (\boldsymbol{\nabla}_\kappa F). \tag{7.111}$$

In some cases the hardening modulus may be prescribed directly in the model, and the equations (7.108) can then be used in the reduced form

$$\begin{bmatrix} \dot{\boldsymbol{\varepsilon}} \\ \dot{F} \end{bmatrix} = \begin{bmatrix} \mathbb{C}^{-1} & (\boldsymbol{\nabla}_\sigma G) \\ (\boldsymbol{\nabla}_\sigma F)^T & -H \end{bmatrix} \begin{bmatrix} \dot{\boldsymbol{\sigma}} \\ \dot{\lambda} \end{bmatrix} \tag{7.112}$$

supplemented with the evolution equation for the internal parameters,

$$\dot{\boldsymbol{\kappa}} = -\dot{\lambda}\, \boldsymbol{\nabla}_\varsigma G. \tag{7.113}$$

This format can be used for a combined implicit–explicit iteration scheme for the finite increments corresponding to $\dot{\boldsymbol{\sigma}}$, $\dot{\boldsymbol{\kappa}}$ and $\dot{\lambda}$.

If the internal variables are uncoupled from the elastic strain, these equations can be recast in a symmetric form, where the potential G is eliminated. In the definition of the hardening modulus the gradient $\boldsymbol{\nabla}_{\zeta} G$ is replaced by the gradient $\boldsymbol{\nabla}_{\kappa} F$ by use of (7.99b),

$$H = (\boldsymbol{\nabla}_{\kappa} F)^T (\boldsymbol{\nabla}_{\kappa\kappa}\varphi)^{-1} (\boldsymbol{\nabla}_{\kappa} F). \tag{7.114}$$

It follows from (7.99a) that $\boldsymbol{\nabla}_{\sigma} G = \boldsymbol{\nabla}_{\sigma} F$, and thus the block matrix equation takes the symmetric form

$$\begin{bmatrix} \dot{\varepsilon} \\ \dot{F} \end{bmatrix} = \begin{bmatrix} \mathbb{C}^{-1} & (\boldsymbol{\nabla}_{\sigma} F) \\ (\boldsymbol{\nabla}_{\sigma} F)^T & -H \end{bmatrix} \begin{bmatrix} \dot{\sigma} \\ \dot{\lambda} \end{bmatrix}. \tag{7.115}$$

Finally, the evolution equation for the internal parameters is expressed in terms of the gradient $\boldsymbol{\nabla}_{\kappa} F$ by use of (7.99b) as

$$\dot{\boldsymbol{\kappa}} = -\dot{\lambda}\, (\boldsymbol{\nabla}_{\kappa\kappa}\varphi)^{-1} \boldsymbol{\nabla}_{\kappa} F. \tag{7.116}$$

This simple format is often found in the literature, but in many cases without the identification of the coefficient matrix $(\boldsymbol{\nabla}_{\kappa\kappa}\varphi)$ as part of the internal energy function.

To round off the general discussion of evolution equations the rate equations (7.102)–(7.104) are given in terms of stress derivatives, the elastic stiffness tensor $\mathbb{C}$ and the hardening modulus H:

$$\dot{\lambda} = \frac{(\boldsymbol{\nabla}_{\sigma} F)^T \mathbb{C}\, \dot{\varepsilon}}{(\boldsymbol{\nabla}_{\sigma} G)^T \mathbb{C}\, (\boldsymbol{\nabla}_{\sigma} F) + H}, \tag{7.117}$$

$$\dot{\sigma} = \left[\mathbb{C} - \frac{\mathbb{C}\, (\boldsymbol{\nabla}_{\sigma} G) \otimes (\boldsymbol{\nabla}_{\sigma} F)^T \mathbb{C}}{(\boldsymbol{\nabla}_{\sigma} G)^T \mathbb{C}\, (\boldsymbol{\nabla}_{\sigma} F) + H} \right] \dot{\varepsilon}, \tag{7.118}$$

$$\dot{\kappa} = -\frac{(\boldsymbol{\nabla}_{\sigma} G) \otimes (\boldsymbol{\nabla}_{\sigma} F)^T}{(\boldsymbol{\nabla}_{\sigma} G)^T \mathbb{C}\, (\boldsymbol{\nabla}_{\sigma} F) + H}\, \mathbb{C}\, \dot{\varepsilon}. \tag{7.119}$$

For materials with uncoupled internal variables, i.e. for $\boldsymbol{\nabla}_{\varepsilon\kappa}\varphi = 0$, the stress gradients are identical, $\boldsymbol{\nabla}_{\sigma} G = \boldsymbol{\nabla}_{\sigma} F$, and the hardening modulus is given by the symmetric relation (7.114).

For materials with internal variables that are coupled to the elastic strains, i.e. for $\boldsymbol{\nabla}_{\varepsilon\kappa}\varphi \neq 0$, the relation between the gradients of the potentials $F(\boldsymbol{\varepsilon}_e, \boldsymbol{\kappa})$ and $G(\boldsymbol{\sigma}_e, \boldsymbol{\zeta})$ is non-trivial, and the directions of the two corresponding stress gradients are different. This is generally called non-associated flow. It should be noted that, according to the present formulation based on potentials and maximum energy dissipation, the stress gradients of the yield potential $F(\boldsymbol{\varepsilon}_e, \boldsymbol{\kappa})$ and its dual, the flow potential $G(\boldsymbol{\sigma}, \boldsymbol{\zeta})$, are not independent, but related through the energy function. In the literature the two potentials are often considered as independent potentials. These theories may be consistent in the sense of dual potentials, if they can be related via a complete internal energy function containing the internal variables. This issue will be considered further in Sections 7.4 and 7.5.

7.2.4 Isotropic and kinematic hardening

As seen in the previous section, the plastic deformation of a material is governed by the internal energy function $\varphi(\boldsymbol{\varepsilon}, \boldsymbol{\kappa})$ and the yield function $F(\boldsymbol{\sigma}, \boldsymbol{\kappa})$. Thus, the initial shape of the yield surface and its possible development during plastic deformation are important for the representation of elasto-plastic material models. Elasto-plastic materials whose yield surface does not change with plastic deformation are called perfectly plastic. For these materials the yield function $F(\boldsymbol{\sigma})$ does not depend on internal parameters, and the hardening modulus H is identically zero, as follows from (7.110). This group has been intensively studied by analytical methods, see e.g. Hill (1950), Chen and Hahn (1988) and Chakrabarty (2000). In the present context perfect plasticity arises as a special case, in which the yield function $F(\boldsymbol{\sigma})$ is independent of internal variables. Thus, perfect plasticity will not be given special attention here, but merely be considered as a limiting case.

At any particular instant the current yield surface can be described by a relation of the form

$$f(\boldsymbol{\sigma}) = \sigma_0, \tag{7.120}$$

where σ_0 represents the yield stress under specific conditions, e.g. uniaxial loading. This corresponds to a yield function $F(\boldsymbol{\sigma})$ of the form

$$F(\boldsymbol{\sigma}) = f(\boldsymbol{\sigma}) - \sigma_0. \tag{7.121}$$

In general the yield function F depends on internal variables $\boldsymbol{\kappa}$. Often the details of this dependence are not known in terms of first principles, and two simple assumptions are often used, either separately or in combination. The first is the so-called isotropic hardening, in which only the yield stress $\sigma_0 = \sigma_0(\boldsymbol{\kappa})$ is assumed to depend on the internal variables. This gives the functional form

$$F(\boldsymbol{\sigma}, \boldsymbol{\kappa}) = f(\boldsymbol{\sigma}) - \sigma_0(\boldsymbol{\kappa}). \tag{7.122}$$

Due to the particular functional form, the yield surfaces constitute a set of surfaces typically enclosing each other as illustrated in Fig. 7.8. The name arises from the fact that the expansion is independent of precisely which stress path generated the change of size of the yield surface. As a consequence, there are infinitely many stress paths that create the same sequence of yield surfaces, and thus the final state gives only limited information on the process leading to the final state.

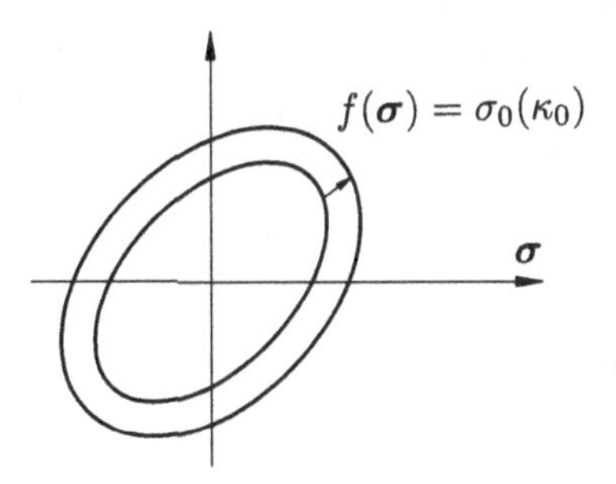

Fig. 7.8. Isotropic hardening: expanding yield surface.

In the isotropic hardening model the yield surface expands in all directions. Thus, the yield stress experienced after a load reversal will also be increased. This is not in accordance with the so-called Bauschinger effect by which an increase in the yield stress for loading in one direction is typically accompanied by a change of the reverse yield limit in the same direction. A simple hardening model that includes the Bauschinger effect is the so-called kinematic hardening, in which the yield surface retains its shape but translates during plastic loading, (Fig. 7.9).

A translating yield surface with initial yield stress σ_0 can be written in

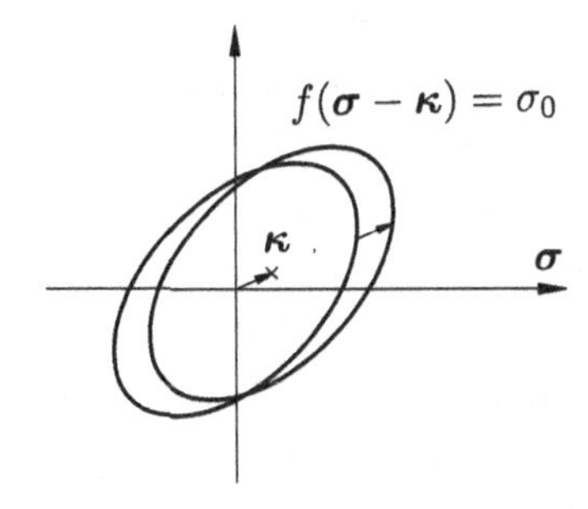

Fig. 7.9. Kinematic hardening: translating yield surface.

terms of the yield function

$$F(\boldsymbol{\sigma}, \boldsymbol{\kappa}) = f(\boldsymbol{\sigma} - \boldsymbol{\kappa}) - \sigma_0. \tag{7.123}$$

The parameters $\boldsymbol{\kappa}$ represent the current position of the initial center of the yield surface. The direct interpretation of the parameters $\boldsymbol{\sigma}$ in terms of stresses has led to names like 'pseudo stress' and 'back stress' for these parameters.

In the kinematic hardening model the translation is governed by a set of relations of the form (7.112) for the rate of the internal parameters $\dot{\boldsymbol{\kappa}}$. Commonly used forms are the rule of Melan (1938) and Prager (1955), defining the translation by the plastic strain rate

$$\dot{\boldsymbol{\kappa}} = a(\boldsymbol{\kappa})\, \dot{\boldsymbol{\varepsilon}}_p \tag{7.124}$$

and the rule of Ziegler (1959), using a radial direction from an assumed center

$$\dot{\boldsymbol{\kappa}} = \dot{\lambda}\, a(\boldsymbol{\kappa})\, (\boldsymbol{\sigma} - \boldsymbol{\kappa}) \tag{7.125}$$

corresponding to radial motion controlled by the single scalar function $a(\boldsymbol{\kappa})$ for calibration. These rules coincide and follow from the potential formulation if the yield function is a quadratic function of the shifted stress variables $\boldsymbol{\sigma} - \boldsymbol{\kappa}$, as discussed in Section 7.4.1.

7.3 Von Mises plasticity models

It has been found that the theory of plasticity can provide accurate representation of deformation of materials like metals. These materials typically exhibit plastic strains that are orthogonal to the yield surface described by the yield function F, and thereby suggest the use of uncoupled plasticity theory, without need for the extra flow potential G. It also appears that in the absence of voids in the material, the plastic strains are purely deviatoric, i.e. the plastic contribution to the volumetric strain vanishes. It then follows from the flow rule (7.95a) for plastic strains that the yield function is independent of the mean stress and only depends on the deviatoric stress components. For isotropic materials this implies that the yield surface may be generated via its contour in the deviatoric stress plane illustrated in Fig. 7.2. The central model for this class of materials is the model of von Mises (1928), in which the yield surface is a circular cylinder. This model leads to several particularly simple theoretical relations between stresses, strains and hardening parameters, and also has a number of very attractive features with respect to numerical integration algorithms. This section contains a brief presentation of some of these features.

7.3.1 Yield surface and flow potential

The von Mises yield surface constitutes a circular cylinder in principal stress space with the principal diagonal as center axis. The radius is expressed in terms of the so-called equivalent stress σ_e, defined via the second deviatoric stress invariant as

$$\sigma_e^2 = 3J_2 = \tfrac{3}{2}\sigma'_{\alpha\beta}\sigma'_{\alpha\beta} = \tfrac{3}{2}\left(\sigma'_1\sigma'_1 + \sigma'_2\sigma'_2 + \sigma'_3\sigma'_3\right). \tag{7.126}$$

The numerical factor has been chosen such that a uniaxial stress σ corresponds to the equivalent stress $\sigma_e = |\sigma|$. The cylindrical von Mises yield surface is illustrated in principal stress space in Fig. 7.10.

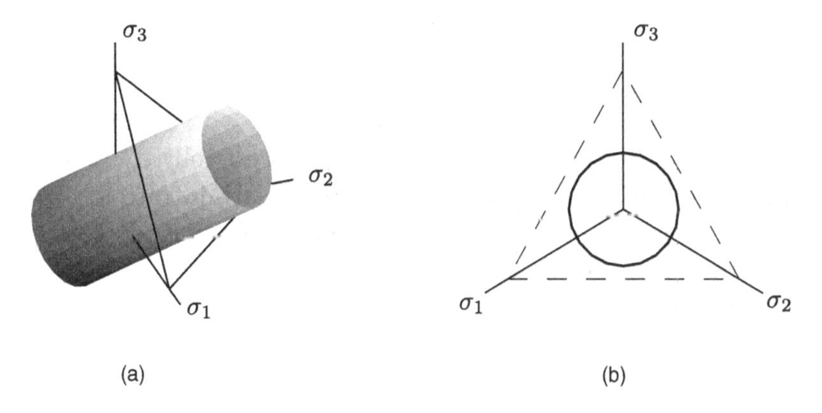

Fig. 7.10. Von Mises yield surface in principal stress space.

The yield function is conveniently expressed in terms of the equivalent stress as

$$F(\boldsymbol{\sigma}, \sigma_0) = \sigma_e(\boldsymbol{\sigma}) - \sigma_0. \tag{7.127}$$

In this format σ_0 represents the uniaxial yield stress. Note that the yield function is formulated with the dimension of stress. This leads to considerable algorithmic advantages, as discussed later. The yield stress σ_0 is considered as an internal parameter that may change during plastic deformation. Extension to kinematic hardening is discussed later. The strains are derived from the yield function $F(\boldsymbol{\sigma}, \sigma_0)$ with respect to the stresses. This requires the derivatives of the equivalent stress σ_e, conveniently calculated by the component relations

$$\frac{\partial \sigma_e}{\partial \sigma_{\alpha\beta}} = \frac{1}{2\sigma_e}\frac{\partial \sigma_e^2}{\partial \sigma_{\alpha\beta}} = \frac{3}{2}\frac{\sigma'_{\alpha\beta}}{\sigma_e}. \tag{7.128}$$

This result may be expressed in the more compact form

$$\boldsymbol{\nabla}_\sigma \sigma_e = \tfrac{3}{2}\,\sigma_e^{-1}\,\boldsymbol{\sigma}'. \tag{7.129}$$

It is noted that the stress gradient of the equivalent stress σ_e is purely deviatoric and therefore has mean value zero. In geometric terms this corresponds to the normals of the cylinder shown in Fig. 7.10(a) being orthogonal to the axis of the cylinder.

The plastic strain increments are evaluated according to the rate equation (7.95a),

$$\dot{\boldsymbol{\varepsilon}}_p = \dot{\lambda}\,\boldsymbol{\nabla}_\sigma \sigma_e = \tfrac{3}{2}\,\dot{\lambda}\,\sigma_e^{-1}\,\boldsymbol{\sigma}'. \tag{7.130}$$

It follows from the definition of the deviatoric stress that the volumetric plastic strain rate vanishes, $\dot{\varepsilon}_v^p = 0$. It is convenient to define an equivalent plastic strain rate $\dot{\varepsilon}_p$, similar to the equivalent stress σ_e. For incompressible strain increments the definition is

$$\dot{\varepsilon}_p^2 = \tfrac{2}{3}\dot{\varepsilon}_{\alpha\beta}^p \dot{\varepsilon}_{\alpha\beta}^p. \tag{7.131}$$

In pure extension in the principal direction 1 with $\dot{\varepsilon}_1^p \neq 0$ the incompressibility condition gives $\dot{\varepsilon}_2^p = \dot{\varepsilon}_3^p = -\tfrac{1}{2}\dot{\varepsilon}_1^p$ and thereby $\dot{\varepsilon}_p = |\dot{\varepsilon}_1^p|$. Thus, the equivalent strain rate $\dot{\varepsilon}_p$ corresponds to the strain rate for uniaxial stress loading. For the von Mises model the equivalent strain rate follows from substitution of the plastic strain rate tensor (7.130) into the definition of the equivalent plastic strain rate (7.131):

$$\dot{\varepsilon}_p^2 = \tfrac{2}{3}\big(\tfrac{3}{2}\,\dot{\lambda}\,\sigma_e^{-1}\big)^2 \sigma'_{\alpha\beta}\sigma'_{\alpha\beta} = \dot{\lambda}^2, \tag{7.132}$$

and thus the rate multiplier $\dot{\lambda}$ is equal to the equivalent strain rate in this model:

$$\dot{\lambda} = \dot{\varepsilon}_p. \tag{7.133}$$

This leads to some simple illustrative interpretations of e.g. the multiplier $\dot{\lambda}$.

EXAMPLE 7.2. HARDENING IN VON MISES PLASTICITY. Let the hardening be described in terms of the accumulated plastic strain ε_p. The consistency condition is now expressed as the time derivative of the yield function in the form

$$\dot{F}(\boldsymbol{\sigma}, \sigma_0) = \dot{\boldsymbol{\sigma}}\boldsymbol{\nabla}_\sigma \sigma_e - \frac{d\sigma_0}{d\varepsilon_p}\dot{\varepsilon}_p = 0. \tag{7.134}$$

Comparison with the form (7.110) identifies the hardening modulus in the present model as

$$H = \frac{d\sigma_0}{d\varepsilon_p}, \tag{7.135}$$

i.e. the slope of the uniaxial stress–strain relation corresponding to yielding of the material.

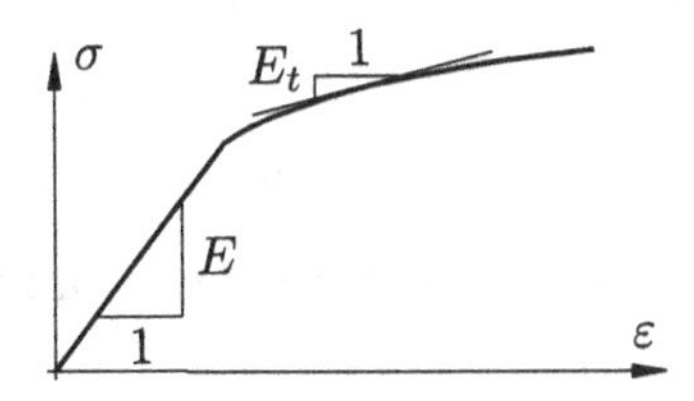

Fig. 7.11. Uniaxial stress–strain curve.

Figure 7.11 shows a uniaxial stress strain curve and identifies the initial modulus of elasticity E and the tangent modulus E_t. The representation of the total strain rate as the sum of the elastic and plastic strain rates can be written as

$$\frac{\dot{\sigma}}{E_t} = \frac{\dot{\sigma}}{E} + \frac{\dot{\sigma}}{H}, \tag{7.136}$$

from which the plastic hardening modulus H is obtained as

$$\frac{1}{H} = \frac{1}{E_t} - \frac{1}{E}. \tag{7.137}$$

H has a range from zero for perfect plasticity with $E_t = 0$ to infinity for vanishing plastic strain. If the tangent modulus E_t is negative the material is said to be softening, and the corresponding value of the hardening modulus H is negative. This phenomenon typically leads to localization of the strains in smaller regions and requires special computational measures.

The tangent stiffness $\mathbb{C}_{ep}$ of elasto-plastic deformation is defined via the stress–strain relation

$$\dot{\boldsymbol{\sigma}} = \mathbb{C}_{ep}\, \dot{\boldsymbol{\varepsilon}}, \tag{7.138}$$

where the specific form of the elasto-plastic tangent stiffness follows from (7.118), when specialized to associated form:

$$\mathbb{C}_{ep} = \mathbb{C} - \frac{\mathbb{C}\,(\boldsymbol{\nabla}_\sigma F) \otimes (\boldsymbol{\nabla}_\sigma F)^T \mathbb{C}}{(\boldsymbol{\nabla}_\sigma F)^T \mathbb{C}\,(\boldsymbol{\nabla}_\sigma F) + H}. \tag{7.139}$$

In the present case the elastic properties are assumed to be isotropic linear elastic, and the stiffness tensor is then given in component form by (7.49). In von Mises plasticity with yield function (7.127) the stress gradient of the yield function F is equal to the gradient of the equivalent stress σ_e,

$\nabla F = \nabla \sigma_e$. This gradient has already been calculated in (7.129), where it was demonstrated to be purely deviatoric. Therefore, multiplication with the isotropic elastic stiffness tensor $\mathbb{C}$ simply corresponds to multiplication with the scalar factor 2μ:

$$\mathbb{C}\,(\nabla_\sigma \sigma_e) \;=\; 3\mu\,\sigma_e^{-1}\boldsymbol{\sigma}'. \tag{7.140}$$

The product in the denominator then follows from the definition (7.126) of the equivalent stress as

$$(\nabla_\sigma \sigma_e)^T \mathbb{C}\,(\nabla_\sigma \sigma_e) \;=\; 3\mu. \tag{7.141}$$

The elasto-plastic tangent stiffness now follows directly by substitution of these expressions into (7.139) as

$$\mathbb{C}_{ep} \;=\; \mathbb{C} \;-\; \frac{9\mu^2}{H+3\mu}\,\frac{\boldsymbol{\sigma}'}{\sigma_e} \otimes \frac{\boldsymbol{\sigma}'^T}{\sigma_e}. \tag{7.142}$$

With the transpose sign the expression is valid both when interpreted as a four-dimensional tensor and when interpreted in association with the vector representation (7.8)–(7.9) of stresses and strains as a 6×6 matrix. In neither formulation are the two last factors contracted.

The problem of integrating the constitutive relations of plasticity is due to the fact that the elasto-plastic stiffness $\mathbb{C}_{ep}$ depends on the current stresses – and possibly on hardening parameters as well. Thus, generally approximate numerical methods must be used. These can be either explicit, in which case the needed parameters are evaluated at the initial time of the increment, or implicit, which leaves a number of possibilities for representation of the development of the parameters over the increment. The following two subsections give a brief presentation of explicit and implicit algorithms, illustrated specifically with reference to von Mises plasticity.

7.3.2 Explicit integration

In elasto-plastic analysis the solution is advanced by increments of the form $\boldsymbol{\sigma}_{n+1} = \boldsymbol{\sigma}_n + \Delta\boldsymbol{\sigma}$ and $\boldsymbol{\varepsilon}_{n+1} = \boldsymbol{\varepsilon}_n + \Delta\boldsymbol{\varepsilon}$. Typically, the elasto-plastic stiffness is used in connection with a specified load increment to predict the corresponding strain increment $\Delta\boldsymbol{\varepsilon}$ at the individual material points – typically element Gauss points – of the body to be analyzed. This leads to a local analysis at the individual material points, in which the strain increment $\Delta\boldsymbol{\varepsilon}$ is imposed, and the corresponding stress increment $\Delta\boldsymbol{\sigma}$ needs to be calculated. Once the stresses are known, they can be used to carry out an equilibrium check via residual forces, or in the case of explicit methods the stresses can be accepted and used as a basis for the next step. In either case it is necessary to solve the problem of determining the stress increment $\Delta\boldsymbol{\sigma}$ resulting from an imposed strain increment $\Delta\boldsymbol{\varepsilon}$. This limited task, carried out one or more times at each material point of the model within each load increment, is discussed in terms of explicit methods in this subsection and in terms of implicit methods in the next subsection.

Explicit integration, often termed the forward Euler procedure, consists in calculating the stress increment $\Delta\boldsymbol{\sigma}$ directly from the rate equation (7.138) by suitable representation of the elasto-plastic stiffness $\mathbb{C}_{ep}$ in terms of parameter values that have already been obtained. In its simplest form the stress and strain rates are replaced by finite increments, and the variables defining the elasto-plastic stiffness are evaluated corresponding to the initial state. Thus, the incremental relation (7.138) for infinitesimal increments is replaced by

$$\Delta\boldsymbol{\sigma} = \mathbb{C}_{ep}(\boldsymbol{\sigma}^0, \boldsymbol{\kappa}^0)\, \Delta\boldsymbol{\varepsilon}, \tag{7.143}$$

where the notation $\boldsymbol{\kappa}$ is retained for internal variables, and superscript 0 denotes initial state. A similar representation can be used to update λ and $\boldsymbol{\kappa}$ from (7.117) and (7.119) if necessary. In the forward Euler scheme the new stress state may not satisfy the yield condition, and the errors tend to accumulate. Thus, refined methods are needed.

The simplest refinement of the forward Euler method follows from the observation that the error in (7.143) depends on the square of the size of the increments. Therefore, the error can be reduced by dividing the original strain increment $\Delta\boldsymbol{\varepsilon}$ into sub-increments, e.g. m equal sub-increments $\Delta\boldsymbol{\varepsilon}/m$. This 'sub-incremental' method is illustrated in Fig. 7.12. The corresponding formulae are

$$\begin{aligned} \delta\boldsymbol{\sigma}^k &= \mathbb{C}_{ep}(\boldsymbol{\sigma}^{k-1}, \boldsymbol{\kappa}^{k-1})\, \Delta\boldsymbol{\varepsilon}/m, \\ \boldsymbol{\sigma}^k &= \boldsymbol{\sigma}^{k-1} + \delta\boldsymbol{\sigma}^k, \end{aligned} \tag{7.144}$$

supplemented with similar formulae for λ and $\boldsymbol{\kappa}$ if necessary. Various methods have been proposed to determine a suitable number of sub-increments, see e.g. Crisfield (1991).

Although the use of sub-increments reduces the error, there is still no check on the yield condition, and thus $f(\boldsymbol{\sigma}^k, \boldsymbol{\kappa}^k)$ may drift away from the zero value required by consistency. A special two-step procedure has been suggested by Zienkiewicz and Taylor (2000). In this procedure a mid-point stress $\boldsymbol{\sigma}^{1/2}$ is first estimated by use of the forward Euler formula

$$\Delta\boldsymbol{\sigma}^{1/2} = \mathbb{C}_{ep}(\boldsymbol{\sigma}^0, \boldsymbol{\kappa}^0)\, \tfrac{1}{2}\Delta\boldsymbol{\varepsilon}. \tag{7.145}$$

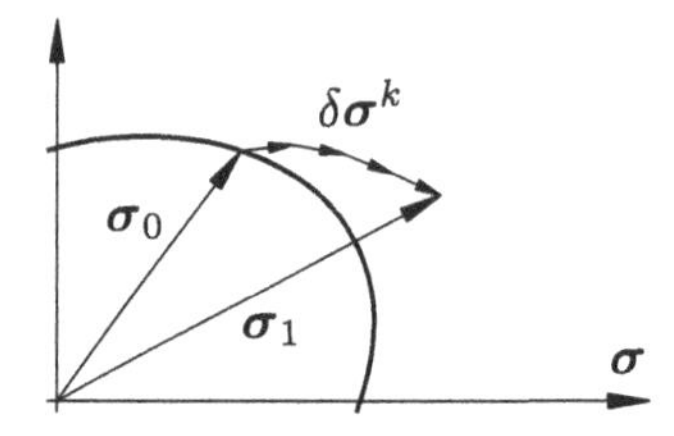

Fig. 7.12. The sub-incremental method.

If necessary $\Delta\boldsymbol{\kappa}^{1/2}$ is also determined. The increment $\Delta\boldsymbol{\sigma}^{1/2}$ is now used to determine the mid-point stress $\boldsymbol{\sigma}^{1/2}$, and the *full* increment $\Delta\boldsymbol{\sigma}$ is then calculated as

$$\Delta\boldsymbol{\sigma} \;=\; \mathbb{C}_{ep}(\boldsymbol{\sigma}^{1/2},\boldsymbol{\kappa}^{1/2})\,\Delta\boldsymbol{\varepsilon}. \tag{7.146}$$

The two-step procedure provides an error estimate by comparing stress increments based on the initial point and the mid-point, respectively. Thus, from (7.145) and (7.146) the error estimate is

$$\Delta\boldsymbol{\sigma} \;-\; 2\Delta\boldsymbol{\sigma}^{1/2} \;=\; \big(\mathbb{C}_{ep}^{1/2} - \mathbb{C}_{ep}^{0}\big)\Delta\boldsymbol{\varepsilon}. \tag{7.147}$$

This error measure may be used to control the increment size.

EXAMPLE 7.3. EXPLICIT INTEGRATION. The explicit integration methods are illustrated by a simple example with the von Mises yield criterion in a two-dimensional principal stress space $\boldsymbol{\sigma} = [\sigma_1, \sigma_2]^T$, with the corresponding strain components $\boldsymbol{\varepsilon} = [\varepsilon_1, \varepsilon_2]^T$. The elastic modulus is E and Poisson's ratio is assumed to be zero, $\nu = 0$, giving the simple diagonal elastic stiffness matrix $\mathbb{C} = E\mathbf{I} = 2\mu\mathbf{I}$. The yield function F is given by (7.127), with equivalent stress σ_e:

$$\sigma_e^2 \;=\; \sigma_1^2 + \sigma_2^2 - \sigma_1\sigma_2.$$

The material is assumed to be non-hardening, i.e. $H = 0$.

The material is loaded in uniaxial stress to first yield at the stress $\boldsymbol{\sigma} = [\sigma_0, 0]^T$ and the strain $\boldsymbol{\varepsilon} = [\varepsilon_0, 0]^T$, where $\varepsilon_0 = \sigma_0/E$ is the uniaxial yield strain. This state is represented by the point A in Fig. 7.13. After first yield at A an additional strain increment $\Delta\boldsymbol{\varepsilon} = [\varepsilon_0, 0]^T$ is imposed, i.e. an elongation with $\Delta\varepsilon_2 = 0$. The stress increment $\Delta\boldsymbol{\sigma}$ from this imposed strain increment is now determined by the three explicit methods described above: the single increment, division into two strain sub-increments, and the mid-point procedure. The final stress state is indicated with B, C and D, respectively in Fig. 7.13, and the numerical values given in Table 7.1.

Comparison of the results of the first two methods illustrates that half

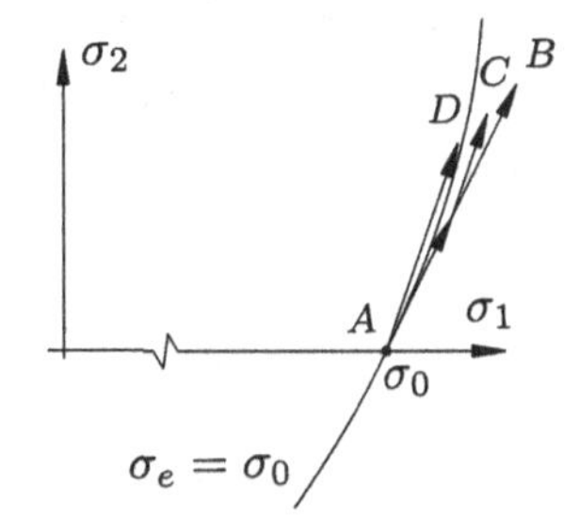

Fig. 7.13. Explicit stress increments.

Table 7.1. *Explicit plastic stress integration.*

	Single increment	Sub-increments	Mid-point
$\boldsymbol{\sigma}^T/\sigma_0$	(1.200, 0.400)	(1.155, 0.356)	(1.109, 0.312)
σ_e/σ_0	1.058	1.024	0.991

the increment gives one quarter error, but to be added from the two steps, resulting in an overall result with half the original error. Comparison between the two last methods demonstrates that given that two computations are going to be used, an extra factor of three is gained on the error by using the estimated mid-point for calculating the total step.

7.3.3 Radial return algorithm

The explicit methods have the inherent weakness that the stresses may leave the yield surface and thereby violate the consistency condition. This problem is addressed in the implicit methods, in which the consistency condition is satisfied, often by iterative procedures. The implicit methods can be developed from two different points of view: either by recasting the combined elasto-plastic strain increment into algorithmic form, or by solving the non-linear matrix evolution equations developed in Section 7.2.3 for finite increments by an iterative numerical method. The direct algorithmic formulations are often called 'return algorithms' because they typically consist of an elastic prediction of the stress, followed by a correction due to plastic effects, see e.g. Ortiz and Popov (1985), Ortiz and Simo (1986). The return algorithms have particular advantages in connection with special yield surfaces like cylinders or cones. The main idea of the radial return algorithm is illustrated in this section with reference to the von Mises plasticity model. The general integration problem is dealt with in the following subsections in connection with the discussion of internal variables and non-associated flow.

The basic idea of the return algorithms is to decompose the stress increment into two additive parts,

$$\Delta\boldsymbol{\sigma} = \mathbb{C}\,(\Delta\boldsymbol{\varepsilon} - \Delta\boldsymbol{\varepsilon}_p) = \Delta\boldsymbol{\sigma}_e - \Delta\boldsymbol{\sigma}_p. \tag{7.148}$$

Here $\Delta\boldsymbol{\sigma}_e = \mathbf{C}\Delta\boldsymbol{\varepsilon}$ is an equivalent elastic stress that would result in the absence of plastic strains, while $\Delta\boldsymbol{\sigma}_p = \mathbb{C}\Delta\boldsymbol{\varepsilon}_p$ represents a plastic correction. This decomposition is illustrated in Fig. 7.14.

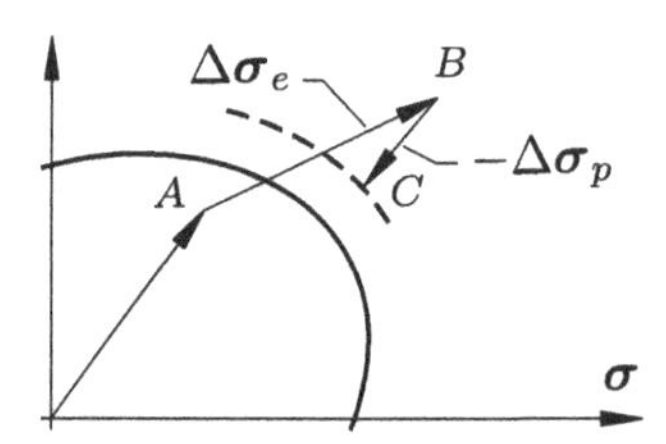

Fig. 7.14. Elastic predictor and plastic corrector.

If the point B is inside the yield surface the increment is fully elastic and $\Delta\boldsymbol{\sigma}_p = \mathbf{0}$. If B is outside the yield surface the plastic corrector $-\Delta\boldsymbol{\sigma}_p$ returns the stress to the yield surface. The plastic strain rate is given by (7.95) and a representation in finite increment form is

$$\Delta\boldsymbol{\sigma}_p = \mathbf{C}\,\Delta\boldsymbol{\varepsilon}_p = \Delta\lambda\,\mathbb{C}\,\boldsymbol{\nabla}_\sigma G. \tag{7.149}$$

Thus, the direction of the plastic corrector is determined by the gradient of the flow potential $\boldsymbol{\nabla}_\sigma G$, while the length depends on $\Delta\lambda$. The increment $\Delta\lambda$ is determined such that the point C is on the yield surface after hardening. Taylor expansion of the yield function $F(\boldsymbol{\sigma}, \boldsymbol{\kappa})$ around $(\boldsymbol{\sigma}, \boldsymbol{\kappa})_B$ gives

$$F(\boldsymbol{\sigma}_C) \simeq F(\boldsymbol{\sigma}_B) - \Delta\boldsymbol{\sigma}_p^T \boldsymbol{\nabla}_\sigma F + (\partial F/\partial\lambda)\,\Delta\lambda = 0. \tag{7.150}$$

Comparison of the last term with the differential form (7.110) of the consistency equation identifies the derivative as the hardening modulus, $\partial F/\partial\lambda = -H$. Substitution of $\Delta\boldsymbol{\sigma}^p$ from (7.149) gives

$$\Delta\lambda = \frac{F(\boldsymbol{\sigma}_B)}{(\boldsymbol{\nabla}_\sigma G)^T \mathbb{C}\,(\boldsymbol{\nabla}_\sigma F) + H}. \tag{7.151}$$

It is instructive to compare this formula with (7.117) for the infinitesimal increment $d\lambda$. The only change is that the numerator $(\nabla_\sigma F)^T \mathbb{C}\,\dot{\varepsilon}$ has been replaced with $F(\boldsymbol{\sigma}_B)$.

The derivation leaves a number of issues. In the general case the gradients $\nabla_\sigma G$ and $\nabla_\sigma F$ vary along the return path from B to C. Then an iterative scheme must be used to determine the point C. Two different approaches have been proposed: an explicit and an implicit, see Ortiz and Simo (1986), Simo and Hughes (1987). In the explicit scheme a sequence of linearized problems are solved, starting at the point B. In the implicit method the direction of the full plastic stress increment is determined by the gradient at the initially unknown point C. Each step consists of evaluating the gradient $\nabla_\sigma G$ at the most recent estimate of point C, and then solving the consistency condition for $\Delta\lambda$ by a Newton-type iterative procedure. This corresponds closely to the Newton-type iteration procedure presented in Section 7.4.3.

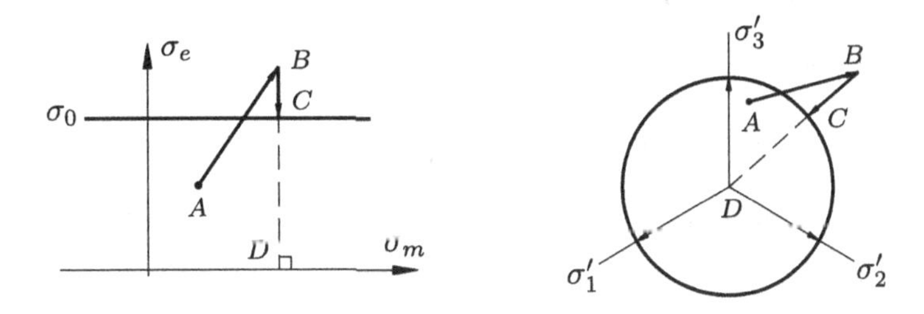

Fig. 7.15. Radial return for perfect von Mises plasticity.

In the special case of a von Mises material the return algorithm takes on a particularly simple and robust form. In this case the gradient of the flow potential is equal to the gradient of the yield function $\nabla_\sigma G = \nabla_\sigma F$. Furthermore, the particular choice (7.127) of the yield function as a linear function of the equivalent stress $\sigma_e(\boldsymbol{\sigma})$ leads to the property that the gradient of F is constant along the radial line connecting B and C. This is illustrated in Fig. 7.15 for von Mises plasticity with isotropic hardening. As seen from the explicit expression (7.129), the gradient $\nabla_\sigma \sigma_e$ represents a radial vector in the deviatoric stress plane pointing through the point D. This is the so-called radial return algorithm (Krieg and Krieg, 1977).

Example 7.4. Radial return algorithm. The radial return algorithm is illustrated for a von Mises material in three-dimensional principal stress space $\boldsymbol{\sigma} = [\sigma_1, \sigma_2, \sigma_3]^T$. As in Example 7.3 the elastic modulus is E and Poisson's ratio is assumed to be zero, $\nu = 0$, whereby the elastic stiffness tensor is $\mathbb{C} = E\,\mathbf{I} = 2\mu\,\mathbf{I}$. The yield function F is given by (7.127), with equivalent stress

$$\sigma_e^2 = \sigma_1^2 + \sigma_2^2 + \sigma_3^2 - \sigma_2\sigma_3 - \sigma_3\sigma_1 - \sigma_1\sigma_2$$

and the stress gradient from (7.129) or directly by differentiation as

$$\boldsymbol{\nabla}_\sigma \sigma_e = \tfrac{1}{2}\sigma_e^{-1}\,[2\sigma_1 - \sigma_2 - \sigma_3, 2\sigma_2 - \sigma_3 - \sigma_1, 2\sigma_3 - \sigma_1 - \sigma_2]^T.$$

The material is assumed to be non-hardening, i.e. $H = 0$, and loaded in uniaxial stress to first yield at the stress $\boldsymbol{\sigma}_A = [\sigma_0, 0, 0]^T$ with strain $\boldsymbol{\varepsilon}_A = [\varepsilon_0, 0, 0]^T$, $\varepsilon_0 = \sigma_0/E$. A strain increment $\Delta\boldsymbol{\varepsilon} = [\varepsilon_0, -\varepsilon_0, 0]^T$ is now imposed and the new stress state is to be determined.

First the equivalent elastic stress increment is determined, leading to the stress $\boldsymbol{\sigma}_B = \sigma_0[2, -1, 0]^T$. This corresponds to $\sigma_{e,B} = 2.6458\,\sigma_0$, and thereby $F(\boldsymbol{\sigma}_B) = 1.6458\,\sigma_0$. The multiplier increment $\Delta\lambda$ is determined from (7.151),

$$\Delta\lambda = F(\boldsymbol{\sigma}_B)/3\mu = 1.6458\,\sigma_0/\mu = 1.0972\,\sigma_0/E.$$

The gradient at B is

$$\boldsymbol{\nabla}_\sigma F|_B = \frac{\sigma_0}{2\sigma_{e,B}}[5, -4, -1]^T.$$

The plastic stress increment $\Delta\boldsymbol{\sigma}$ then follows from (7.149), and then the final stress state

$$\boldsymbol{\sigma}_C = \boldsymbol{\sigma}_B - \Delta\boldsymbol{\sigma}_p = \begin{bmatrix} 2 \\ -1 \\ 0 \end{bmatrix}\sigma_0 - \begin{bmatrix} 1.0367 \\ -0.8293 \\ -0.2073 \end{bmatrix}\sigma_0 = \begin{bmatrix} 0.9633 \\ -0.1707 \\ 0.2073 \end{bmatrix}\sigma_0.$$

It is easily verified that $\sigma_{e,C} = 1.000\ \sigma_0$, and thereby point C has been returned to the yield surface. In the case of isotropic hardening the yield stress is simply updated to $\sigma_{e,C}$.

It is instructive to compare the radial return algorithm with explicit integration by use of the tangent stiffness at A. Here the gradient is

$$\boldsymbol{\nabla}_\sigma F|_A = \frac{\sigma_0}{2\sigma_{e,A}}[2, -1, -1]^T = [1, -0.5, -0.5]^T.$$

The increment $\Delta\lambda$ is determined from (8.21) as

$$\Delta\lambda = \boldsymbol{\nabla}_\sigma F|_A\,\Delta\boldsymbol{\sigma}_e/3\mu = \sigma_0/E.$$

It is seen that this estimate of $\Delta\lambda$ deviates 10% from that of the radial return algorithm. The plastic strain increment is now evaluated by use of the gradient at A as

$$\boldsymbol{\sigma}_C = \boldsymbol{\sigma}_B - \Delta\boldsymbol{\sigma}_p = \begin{bmatrix} 2 \\ -1 \\ 0 \end{bmatrix}\sigma_0 - \begin{bmatrix} 1 \\ -0.5 \\ -0.5 \end{bmatrix}\sigma_0 = \begin{bmatrix} 1 \\ -0.5 \\ 0.5 \end{bmatrix}\sigma_0.$$

This stress state deviates considerably from the yield surface, $\sigma_{e,C} = 1.323\,\sigma_0$.

The example illustrates how the radial return algorithm modifies the direct tangent method in two ways: the direction of the plastic stress increment is determined from the gradient corresponding to the *final* stress state, and the length of the increment is calculated via the yield function to give exact return to the yield surface.

In the case of plane stress the von Mises yield function is constituted by a section of the cylindrical yield function shown in Fig. 7.10 in one of the principal stress planes. This leads to an elliptic surface, where the normal return procedure described above must be modified. However, the basically simple shape of the yield surface can also be used to construct efficient return mapping schemes in this case, see Jetteur (1986) and Simo and Taylor (1986).

7.4 General aspects of plasticity models

In this section three aspects of more general plasticity models are presented. The first is the use of combined isotropic and kinematic hardening, which is of central importance for the use of plasticity models for reversed and cyclic load histories. The second is a simple demonstration of the use of the dual potentials F and G in connection with internal variables and their conjugates to introduce non-associated flow. This problem is important in connection with many non-metallic materials, notably granular media, discussed in more detail in Section 7.5. Finally, the integration of the general plastic evolution equations derived in differential form in Section 7.2.3 is discussed in Section 7.4.3.

The models considered in this section are mainly intended to illustrate basic properties of the plastic deformation, and the elastic properties are therefore assumed to be linear isotropic as discussed in Section 7.1. In isotropic material models the deformation consists of two mechanisms: change of volume and change of shape. The volumetric behavior is governed by mean stress and mean strain, represented by the scalar variables

$$\sigma_m = \tfrac{1}{3}\sigma_{\gamma\gamma}, \qquad \varepsilon_v = \varepsilon_{\gamma\gamma}. \tag{7.152}$$

The change of shape is governed by the corresponding deviatoric tensor components with magnitudes

$$\sigma_e^2 = \tfrac{3}{2}\sigma'_{\alpha\beta}\sigma'_{\alpha\beta}, \qquad \varepsilon_e^2 = \tfrac{2}{3}\varepsilon'_{\alpha\beta}\varepsilon'_{\alpha\beta}. \tag{7.153}$$

The elastic energy is the sum of contributions from volumetric and deviatoric strain

$$\varphi(\varepsilon_e, \varepsilon_v) = \tfrac{1}{2}k(\varepsilon_{\gamma\gamma})^2 + \mu\varepsilon'_{\alpha\beta}\varepsilon'_{\beta\alpha} = \tfrac{1}{2}k\varepsilon_v^2 + \tfrac{3}{2}\mu\varepsilon_e^2. \tag{7.154}$$

In the models discussed below the internal energy is obtained by introducing suitable internal variable effects into the elastic strain energy, and then introducing an appropriate flow potential G in terms of the stresses and the conjugate internal variables.

7.4.1 Combined isotropic and kinematic hardening

In modeling of cyclic loading there is typically a need to include the so-called Bauschinger effect, by which an increase of the yield stress in positive loading leads to a shift in the yield stress experienced upon reversal of the loading. This involves a translation of the yield surface as discussed in general terms in the form of kinematic hardening in Section 7.2.4. Kinematic hardening requires an internal tensor variable $\boldsymbol{\kappa}$ to keep track of the translation. In addition, a scalar internal variable κ_0 is included in the model here to illustrate the combined effect. Often the translation of the yield surface in e.g. metals does not lead to noticeable deviations from normality of the plastic strain increment from the normal to the yield surface. Thus, the model will be formulated to represent associated plasticity, and therefore the internal variables will appear in the form of additive terms according to (7.97):

$$\varphi(\varepsilon_e, \varepsilon_v, \kappa) = \tfrac{1}{2}k\varepsilon_v^2 + \tfrac{3}{2}\mu\varepsilon_e^2 + \varphi_1(\kappa_e) + \varphi_0(\kappa_0). \tag{7.155}$$

The scalar invariant κ_e is here introduced in terms of the deviatoric components $\kappa'_{\alpha\beta}$ in analogy with the equivalent stress σ_e, but without the numerical factor:

$$\kappa_e^2 = \kappa'_{\alpha\beta}\kappa'_{\alpha\beta}. \tag{7.156}$$

The stress is defined as the conjugate variable to the strain, whereby

$$\sigma_{\alpha\beta} = \frac{\partial\varphi}{\partial\varepsilon_{\alpha\beta}} = k\delta_{\alpha\beta}\varepsilon_v + 2\mu\,\varepsilon'_{\alpha\beta}. \tag{7.157}$$

The conjugate internal tensor variable follows by differentiation of the potential φ_1 through the scalar argument κ_e as

$$\zeta_{\alpha\beta} = \frac{\partial\varphi_1}{\partial\kappa_e}\frac{\partial\kappa_e}{\partial\kappa_{\alpha\beta}} = \varphi_1'\,\frac{\kappa'_{\alpha\beta}}{\kappa_e}, \tag{7.158}$$

where φ_1' denotes the derivative with respect to the argument κ_e. It is noted that the conjugate internal tensor variable $\zeta_{\alpha\beta}$ is purely deviatoric. Its magnitude is characterized by the scalar invariant ζ_e, defined by a relation similar to (7.156) for κ_e. Contraction of the components on each side of (7.158) with themselves then gives the relation

$$\zeta_e = \varphi_1'(\kappa_e). \tag{7.159}$$

This relation establishes the internal variable ζ_e as conjugate to the primary internal variable κ_e. The conjugate scalar internal variable ζ_0 is defined by the similar relation

$$\zeta_0 = \varphi_0'(\kappa_0). \tag{7.160}$$

In general these relations may be non-linear, but the linear case, arising from quadratic functions φ_1 and φ_0, leads to a particularly simple formulation.

The Jacobian of the transformation between the primary internal tensor variable $\kappa'_{\alpha\beta}$ and it conjugate $\zeta'_{\alpha\beta}$ follows from differentiation of (7.158) as

$$\frac{\partial\zeta_{\alpha\beta}}{\partial\kappa_{\gamma\delta}} = \frac{\partial\varphi(\kappa_e)}{\partial\kappa_{\alpha\beta}\partial\kappa_{\gamma\delta}} = \delta_{\alpha\gamma}\delta_{\beta\delta}\Big(\frac{1}{\kappa_e}\varphi_1'\Big) + \frac{\kappa'_{\alpha\beta}}{\kappa_e}\frac{\kappa'_{\gamma\delta}}{\kappa_e}\Big(\varphi_1'' - \frac{1}{\kappa_e}\varphi_1'\Big). \tag{7.161}$$

It is seen that the first term is isotropic, while the second term is anisotropic – depending on the current direction of the internal variable tensor $\kappa'_{\alpha\beta}$. Now assume that $\kappa_e = 0$ corresponds to $\zeta_e = 0$. It then follows from (7.159) that a Taylor expansion of the potential φ_1 around this value must have the form

$$\varphi_1(\kappa_e) = \tfrac{1}{2}\alpha\,\kappa_e^2 + \mathrm{O}(\kappa_e^3). \tag{7.162}$$

It is convenient to select internal variables κ_α, κ_0 with dimension of strain, whereby the parameter α gets the dimension of stress, similar to the shear modulus μ and the bulk modulus k. Substitution of this representation into (7.161) demonstrates that if the potential φ_1 is represented by the quadratic term alone, then the second – anisotropic – term in the relation (7.161) vanishes, leaving the simple isotropic component relation

$$\zeta_{\alpha\beta} = \alpha\,\kappa_{\alpha\beta} \tag{7.163}$$

between the primary and conjugate internal variables. In contrast, any form of φ_1 that is not simply quadratic in κ_e leads to a relation with anisotropy, induced via the internal parameter components $\kappa_{\alpha\beta}$.

In the present model the energy consists of additive contributions from elastic strains and internal variables. This implies that the relation between the internal variables $\kappa_{\alpha\beta}, \kappa_0$ and their conjugate variables $\zeta_{\alpha\beta}, \zeta_0$ is independent of the state of strain. As discussed in Section 7.2.2, this implies that the yield potential F is equal to the flow potential G and can be expressed in terms of stress or strain in connection with either the primary internal variables $\kappa_{\alpha\beta}, \kappa_0$ or their conjugate internal variables $\zeta_{\alpha\beta}, \zeta_0$. In the present context it is most direct to express the potentials F and G in terms of the stress and the conjugate internal variables.

In kinematic hardening the stresses are 'centered' by the internal variable $\boldsymbol{\zeta}$. When using the von Mises yield surface this corresponds to introduction of the stresses via the translated equivalent stress

$$\bar{\sigma}_e^2(\boldsymbol{\sigma} - \boldsymbol{\zeta}) = \tfrac{3}{2}(\sigma'_{\alpha\beta} - \zeta'_{\alpha\beta})(\sigma'_{\alpha\beta} - \zeta'_{\alpha\beta}). \tag{7.164}$$

The flow potential can then be written in the form

$$G(\boldsymbol{\sigma}, \boldsymbol{\zeta}, \zeta_0) = \bar{\sigma}_e(\boldsymbol{\sigma} - \boldsymbol{\zeta}) + \tfrac{1}{2}\frac{\gamma}{\alpha}\,\boldsymbol{\zeta}' : \boldsymbol{\zeta}' - (\sigma_0 + \zeta_0). \tag{7.165}$$

The quadratic term in $\boldsymbol{\zeta}$ was introduced by Armstrong and Frederick (1966) and leads to finite translation of the yield surface, see e.g. Chaboche (1986) and Ottosen and Ristinmaa (2005).

The evolution equations follow from the general format (7.108), when it is observed that also the isotropic hardening component κ_0 must be included among the internal variables. It turns out to be advantageous to consider the effect of this internal variable first. The isotropic internal variable κ_0 has the evolution equation

$$\dot{\kappa}_0 = -\dot{\lambda}\,\frac{\partial G}{\partial \zeta_0} = \dot{\lambda}. \tag{7.166}$$

Thus, the isotropic hardening variable κ_0 can be identified with the plastic multiplier λ. In the following the plastic multiplier is replaced by the isotropic hardening parameter κ_0. Note that this representation remains valid also in the special case without isotropic hardening. In this case the internal energy is independent of κ_0, and thus $\varphi_0 \equiv 0$ and $\zeta_0 \equiv 0$. Therefore the strain-like hardening variable κ_0 is retained in the equations, and not its stress-like conjugate variable ζ_0.

When $\dot{\lambda}$ is eliminated from the general evolution equation (7.108), the following system of equations is obtained:

$$\begin{bmatrix} \dot{\boldsymbol{\varepsilon}} \\ \mathbf{0} \\ \dot{F} \end{bmatrix} = \begin{bmatrix} \mathbb{C}^{-1} & \mathbf{0} & (\boldsymbol{\nabla}_\sigma G) \\ \mathbf{0} & \mathbf{I} & (\boldsymbol{\nabla}_\zeta G) \\ (\boldsymbol{\nabla}_\sigma F)^T & (\boldsymbol{\nabla}_\kappa F)^T & \partial F/\partial \kappa_0 \end{bmatrix} \begin{bmatrix} \dot{\boldsymbol{\sigma}} \\ \dot{\boldsymbol{\kappa}} \\ \dot{\kappa}_0 \end{bmatrix}. \tag{7.167}$$

Here it is observed that $F \equiv G$, and the two last derivatives in the last column are expressed as

$$\boldsymbol{\nabla}_\zeta G = \Big(\frac{\partial \boldsymbol{\kappa}}{\partial \boldsymbol{\zeta}}\Big)^T \boldsymbol{\nabla}_\kappa F = \big(\nabla_{\kappa\kappa}\varphi_1\big)^{-1} \boldsymbol{\nabla}_\kappa F, \tag{7.168}$$

$$\frac{\partial F}{\partial \kappa_0} = \frac{\partial F}{\partial \zeta_0}\frac{\partial \zeta_0}{\partial \kappa_0} = -\varphi_0''(\kappa_0). \tag{7.169}$$

Multiplication of the middle equation in (7.167) by $\nabla_{\kappa\kappa}\varphi_1$ then gives the final form of the evolution equations,

$$\begin{bmatrix} \dot{\boldsymbol{\varepsilon}} \\ \mathbf{0} \\ \dot{F} \end{bmatrix} = \begin{bmatrix} \mathbb{C}^{-1} & \mathbf{0} & (\boldsymbol{\nabla}_\sigma F) \\ \mathbf{0} & (\nabla_{\kappa\kappa}\varphi_1) & (\boldsymbol{\nabla}_\kappa F) \\ (\boldsymbol{\nabla}_\sigma F)^T & (\boldsymbol{\nabla}_\kappa F)^T & -\varphi_0'' \end{bmatrix} \begin{bmatrix} \dot{\boldsymbol{\sigma}} \\ \dot{\boldsymbol{\kappa}} \\ \dot{\kappa}_0 \end{bmatrix}. \tag{7.170}$$

Comparison with the general form of the evolution equations (7.109), including all hardening parameters, shows that elimination of the multiplier $\dot{\lambda}$ leads to an identical system of equations apart from the isotropic hardening function $-\varphi_0''(\kappa_0)$, now appearing in the lower right position. This is the same position, and the same role, as the hardening modulus in the reduced format (7.115). The double derivative $\nabla_{\kappa\kappa}\varphi_1$ in the center term has already been expressed in component form in (7.161).

The format (7.170) of the evolution equations is general for associated plasticity theory with isotropic hardening described by an additive term ζ_0 with kinematic conjugate variable κ_0. For the particular model with modified kinematic hardening described by the flow potential (7.165), the equations are completed by introducing the gradients with respect to the stress and the kinematic hardening parameters:

$$\boldsymbol{\nabla}_\sigma G = \tfrac{3}{2}\frac{\boldsymbol{\sigma}' - \boldsymbol{\zeta}'}{\bar{\sigma}_e}, \qquad \boldsymbol{\nabla}_\zeta G = -\tfrac{3}{2}\frac{\boldsymbol{\sigma}' - \boldsymbol{\zeta}'}{\bar{\sigma}_e} + \frac{\gamma}{\alpha}\boldsymbol{\zeta}'. \tag{7.171}$$

The first of the evolution equations (7.167) gives the plastic strain rate as

$$\dot{\boldsymbol{\varepsilon}}_p = \dot{\lambda}\,\boldsymbol{\nabla}_\sigma G = \dot{\lambda}\,\tfrac{3}{2}\frac{\boldsymbol{\sigma}' - \boldsymbol{\zeta}'}{\bar{\sigma}_e}. \tag{7.172}$$

Formation of the fully contracted product of the components on each side of this equation with themselves gives the equivalent plastic strain rate as

$$\dot{\varepsilon}_e^p = \left(\tfrac{2}{3}\varepsilon_{\alpha\beta}^{p\,\prime}\varepsilon_{\alpha\beta}^{p\,\prime}\right)^{1/2} = \dot{\lambda}. \tag{7.173}$$

If the kinematic hardening potential φ_1 is assumed to be quadratic, then $\boldsymbol{\zeta} = \alpha\boldsymbol{\kappa}$ and the evolution equation for $\boldsymbol{\kappa}$ follows from the second equation in (7.167) as

$$\dot{\boldsymbol{\kappa}} = -\dot{\lambda}\,\boldsymbol{\nabla}_\zeta G = \dot{\boldsymbol{\varepsilon}}_p - \gamma\,\dot{\varepsilon}_e^p\,\boldsymbol{\kappa}. \tag{7.174}$$

For monotonic proportional loading the kinematic hardening tensor $\boldsymbol{\kappa}$ exhibits a decaying exponential approach to the limit determined by $\dot{\boldsymbol{\kappa}} = \mathbf{0}$.

The limit is conveniently expressed in terms of the scalar invariant κ_e, defined in (7.156):

$$\kappa_e^\infty = \left(\tfrac{3}{2}\right)^{1/2}\gamma^{-1}, \qquad \zeta_e^\infty = \left(\tfrac{3}{2}\right)^{1/2}\frac{\alpha}{\gamma}, \tag{7.175}$$

where the numerical factor is due to the difference in normalization between the kinematic hardening scalars and the stress/strain scalars in order to facilitate the discussion of non-linear kinematic hardening. A detailed discussion of the behavior of combined kinematic and isotropic hardening has been given by Ottosen and Ristinmaa (2005).

The numerical integration of this type of equation is discussed in Section 7.4.3. Extensions of kinematic hardening models to large deformation kinematics have been presented, e.g. by Simo (1988a,b) and Wallin *et al.* (2003). The main points of this extension are briefly indicated in Section 7.6.

7.4.2 Internal variables and non-associated flow

If the contribution of the internal variables $\boldsymbol{\kappa}$ to the internal energy φ is in the form of an additive term, the mixed derivatives $\boldsymbol{\nabla}_{\varepsilon\kappa}\varphi$ will vanish. It was demonstrated in Section 7.2.2 that if this condition is satisfied the stress gradients $\boldsymbol{\nabla}_\sigma F$ and $\boldsymbol{\nabla}_\sigma G$ are identical, and the corresponding plasticity is called associated. Conversely, if there are coupling terms in the form of non-vanishing mixed derivatives $\boldsymbol{\nabla}_{\varepsilon\kappa}\varphi$, the stress gradients of the yield potential F and the flow potential G will exhibit a systematic difference, governed by the coupling between stresses and internal variables in the internal energy. This effect will here be illustrated in relation to a generalized Drucker–Prager model, in which the yield surface and flow potential are represented by cones in principal stress space (Drucker and Prager, 1952). The basic model is probably the simplest plasticity model combining the mean stress and the deviatoric stress. More detailed models are discussed in Section 7.5.

In this subsection the main point is the coupling between volumetric strain effects in the internal energy and the direction of plastic flow, notably the volumetric plastic strain. The discussion is therefore limited to a two-component format consisting of the mean stress and the equivalent deviatoric stress (σ_m, σ_e) and the conjugate strain variables $(\varepsilon_v, \varepsilon_e)$, defined in (7.152)–(7.153). The yield surface, represented by the yield function

$$F(\sigma_e, \sigma_m) = \sigma_e + \alpha\,\sigma_m - \sigma_F, \tag{7.176}$$

is shown in full line in Fig. 7.16. The inclination of the line is given via the non-dimensional coefficient α. It is seen that the yield surface expands with increasing mean compression, corresponding to decreasing mean stress σ_m. This general behavior is typical of granular media, soils and rock. It is generally found that flow strains based on the normal to the yield function F will predict a too large dilation. This suggests a flow potential G with a smaller slope, as indicated by the dashed line in the figure:

$$G(\sigma_e, \sigma_m) = \sigma_e + \beta \sigma_m - \sigma_G. \tag{7.177}$$

The idea here is to illustrate how the experimentally observed behavior $\beta < \alpha$ can be generated via a physically plausible assumption about the internal energy φ.

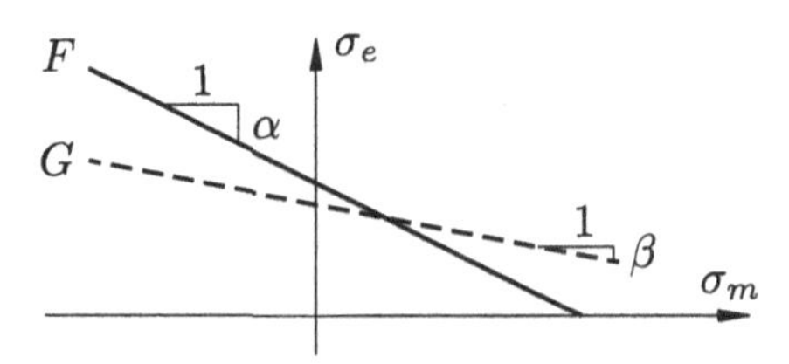

Fig. 7.16. Drucker–Prager yield function $F(\sigma_e, \sigma_m)$ and flow potential $G(\sigma_e, \sigma_m)$.

Materials represented via a Drucker–Prager yield surface of the form (7.176) often have, or develop, internal porosity. This implies that not all the externally observable volume strain ε_v reflects actual straining of the material. This can be represented, e.g. by an internal energy of the form

$$\varphi(\varepsilon_e, \varepsilon_v, \kappa) = \tfrac{1}{2}k(\varepsilon_v - \kappa)^2 + \tfrac{3}{2}\mu\varepsilon_e^2, \tag{7.178}$$

where κ represents a stress-free volumetric strain. The internal energy associated with the deviatoric strain ε_e is unaffected by the internal variable κ. The stresses follow from the internal energy by differentiation as

$$\sigma_m = \frac{\partial \varphi}{\partial \varepsilon_v} = k(\varepsilon_v - \kappa), \quad \sigma_e = \frac{\partial \varphi}{\partial \varepsilon_e} = 3\mu\,\varepsilon_e. \tag{7.179}$$

The conjugate internal variable ζ follows similarly from differentiation with respect to the 'porosity variable' κ as

$$\zeta = \frac{\partial \varphi}{\partial \kappa} = -k(\varepsilon_v - \kappa). \tag{7.180}$$

In this relation the strain ε_v can be eliminated by use of (7.179a), whereby

$$\zeta = -\sigma_m. \tag{7.181}$$

This relation permits transformation between the flow potential $G(\sigma_e, \sigma_m, \zeta)$ and the yield potential $F(\sigma_e, \sigma_m, \kappa)$ by direct substitution.

The inclination β of the flow potential is assumed to be independent of the internal variable, and the flow potential is therefore assumed of the following form in the stresses and the complementary internal variable ζ:

$$G(\sigma_e, \sigma_m, \zeta) = \sigma_e + \beta\,\sigma_m - (\sigma_0 + \gamma\zeta), \qquad (7.182)$$

where γ is a positive constant. The dissipation rate follows from (7.96) in the form

$$\mathcal{D} = \dot{\lambda}\Big(\sigma_e \frac{\partial G}{\partial \sigma_e} + \sigma_m \frac{\partial G}{\partial \sigma_m} + \zeta \frac{\partial G}{\partial \zeta}\Big) = \dot{\lambda}(G + \sigma_0) = \dot{\lambda}\,\sigma_0. \qquad (7.183)$$

The result follows from the property of G as a homogeneous function of degree one in σ_e, σ_m and ζ. This relation identifies λ as an equivalent irreversible strain.

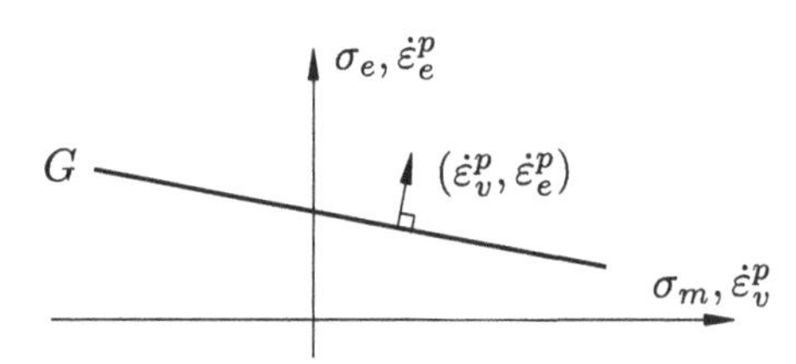

Fig. 7.17. Drucker–Prager flow potential $G(\sigma_e, \sigma_m)$ with plastic strain rates $(\dot{\varepsilon}_v^p, \dot{\varepsilon}_e^p)$.

The plastic strain rate vector $(\dot{\varepsilon}_v^p, \dot{\varepsilon}_e^p)$ is normal to the flow potential $G(\sigma_e, \sigma_m)$ as shown in Fig. 7.17. The corresponding yield function follows from the flow potential by elimination of the internal variable ζ by use of (7.181):

$$F(\sigma_e, \sigma_m) = \sigma_e + (\beta + \gamma)\sigma_m - \sigma_0 = 0. \qquad (7.184)$$

It is seen that the introduction of an internal volumetric strain variable κ leads to a change of the slope β of the flow potential G to

$$\alpha = \beta + \gamma \qquad (7.185)$$

in the yield potential F. This illustrates that a stress-free volumetric strain component κ in the internal energy leads to a smaller slope of the flow potential than the yield surface, and thereby to less plastic dilation than predicted by a corresponding associated version of the theory. Hardening can be introduced into the model by including a suitable additive function $\varphi_0(\sigma_0)$ in the internal energy.

7.4.3 General computational procedure

In the analysis of elasto-plastic behavior in solids and structures the incremental structure of the evolution equations leads to an extra level in the analysis procedure. Basically the analysis is performed by a sequence of load increments. Within the individual load step the predicted deformation may be accepted, perhaps with a check on unbalance, or equilibrium iterations may be included. In either case the load step leads to a predicted motion and thereby to a predicted strain increment $\Delta\boldsymbol{\varepsilon}$ at each material point used in the model. Due to the special structure of the evolution equations of plasticity, the finite increment of the total strain must be split into a plastic part and an elastic part, from which the stress is determined. In the implicit methods this is typically done in an extra iterative loop at each individual material point. The input to this loop is the total strain increment $\Delta\boldsymbol{\varepsilon}$, and the result is the corresponding finite increments of elastic and plastic strains as well as stress and any internal parameters. When using local material models this step does not involve coupling between the different material points, and thus this step in the analysis can be carried out sequentially at the material point level. The key point is to set up a consistent extension of the differential evolution equations to finite increments, and to obtain the necessary matrices for iterative solution of these finite increment equations.

In the local analysis the total strain increment $\Delta\boldsymbol{\varepsilon}$ has been predicted, and therefore remains fixed, while the increments of the stress $\boldsymbol{\sigma}$, the hardening parameters $\boldsymbol{\kappa}$ and the plastic multiplier λ are to be computed. The procedure consists in iterative elimination of the residuals defined by the discretized form of the strain rate equation (7.95a),

$$\mathbf{r}_{\varepsilon} = \Delta\boldsymbol{\varepsilon} - \Delta\boldsymbol{\varepsilon}_e - \Delta\lambda\, \boldsymbol{\nabla}_{\sigma} G(\boldsymbol{\sigma}, \boldsymbol{\zeta}) \tag{7.186}$$

and the discretized rate equation (7.95b) for the hardening parameters,

$$\mathbf{r}_{\kappa} = -\Delta\boldsymbol{\kappa} - \Delta\lambda\, \boldsymbol{\nabla}_{\zeta} G(\boldsymbol{\sigma}, \boldsymbol{\zeta}). \tag{7.187}$$

In these relations the symbol Δ denotes the total increment relative to the last established state of equilibrium. An iterative sequence of stress increments is illustrated in Fig. 7.18.

The iteration follows the Newton–Raphson scheme, in which the nonlinear equations are expanded in terms of current derivatives and linearized sub-increments, denoted by δ. The linearized form of the equations (7.186)

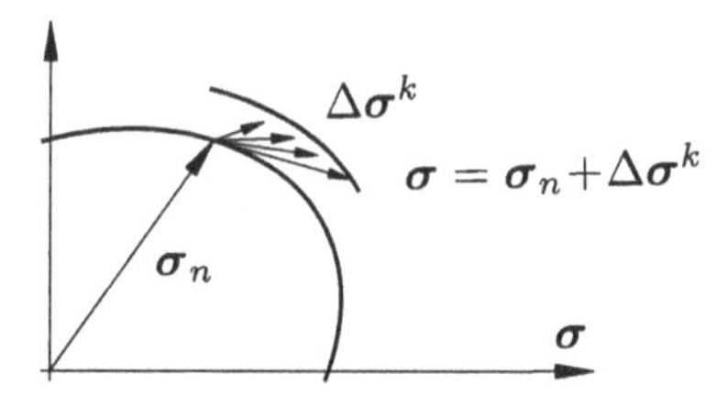

Fig. 7.18. Sequence of iterative total stress increments $\Delta\boldsymbol{\sigma}^k$.

and (7.187) is

$$\delta\boldsymbol{\varepsilon}_e + \delta\lambda\,\boldsymbol{\nabla}_\sigma G + \Delta\lambda\,\delta(\boldsymbol{\nabla}_\sigma G) = \mathbf{r}_\varepsilon, \tag{7.188}$$

$$\delta\boldsymbol{\kappa} + \delta\lambda\,\boldsymbol{\nabla}_\zeta G + \Delta\lambda\,\delta(\boldsymbol{\nabla}_\zeta G) = \mathbf{r}_\kappa. \tag{7.189}$$

The most systematic form of these equations is obtained by using the conjugate variables $\boldsymbol{\sigma}$ and $\boldsymbol{\zeta}$. Thus, the terms $\delta\boldsymbol{\varepsilon}_e$ and $\delta\boldsymbol{\kappa}$ must be recast in terms of their conjugate counterparts $\delta\boldsymbol{\sigma}$ and $\delta\boldsymbol{\zeta}$. A systematic approach is developed by use of the complementary energy density defined by

$$\psi(\boldsymbol{\sigma},\boldsymbol{\zeta}) = \sigma_{\alpha\beta}\,\varepsilon^e_{\alpha\beta} + \zeta_{\alpha\beta}\,\kappa_{\alpha\beta} - \varphi(\boldsymbol{\varepsilon}_e,\boldsymbol{\kappa}). \tag{7.190}$$

This extends the definition (7.5) to the case including internal variables. When using the complementary energy density $\psi(\boldsymbol{\sigma},\boldsymbol{\zeta})$, the roles of the primary variables $\boldsymbol{\varepsilon}_e, \boldsymbol{\kappa}$ and their conjugate variables $\boldsymbol{\sigma}, \boldsymbol{\zeta}$ are interchanged relative to the discussion in Section 7.2.1. Thus, differentiation of the complementary energy density gives

$$\boldsymbol{\varepsilon}_e = \frac{\partial\psi}{\partial\boldsymbol{\sigma}^T} = \boldsymbol{\nabla}_\sigma\psi(\boldsymbol{\sigma},\boldsymbol{\zeta}), \qquad \boldsymbol{\kappa} = \frac{\partial\psi}{\partial\boldsymbol{\zeta}^T} = \boldsymbol{\nabla}_\zeta\psi(\boldsymbol{\sigma},\boldsymbol{\zeta}), \tag{7.191}$$

while further differentiation with respect to time gives

$$\dot{\boldsymbol{\varepsilon}}_e = (\boldsymbol{\nabla}_{\sigma\sigma}\psi)\,\dot{\boldsymbol{\sigma}} + (\boldsymbol{\nabla}_{\sigma\zeta}\psi)\,\dot{\boldsymbol{\zeta}}, \tag{7.192}$$

$$\dot{\boldsymbol{\kappa}} = (\boldsymbol{\nabla}_{\zeta\sigma}\psi)\,\dot{\boldsymbol{\sigma}} + (\boldsymbol{\nabla}_{\zeta\zeta}\psi)\,\dot{\boldsymbol{\zeta}}. \tag{7.193}$$

These are the desired relations. They can also be obtained, although less elegantly, from the double derivatives of the internal energy $\varphi(\boldsymbol{\varepsilon}_e,\boldsymbol{\kappa})$ by inversion of (7.74)–(7.75).

It is now observed that the terms $\delta(\boldsymbol{\nabla}_\sigma G)$ and $\delta(\boldsymbol{\nabla}_\zeta G)$ lead to a similar format in terms of the double derivatives of the flow potential $G(\boldsymbol{\sigma},\boldsymbol{\zeta})$. The equations (7.188)–(7.189) therefore take a particularly simple and compact form, if an algorithmic complementary energy is defined as

$$\Psi(\boldsymbol{\sigma},\boldsymbol{\zeta}) = \psi(\boldsymbol{\sigma},\boldsymbol{\zeta}) + \Delta\lambda\, G(\boldsymbol{\sigma},\boldsymbol{\zeta}). \tag{7.194}$$

In terms of this potential function the equations (7.188)–(7.189) take the compact form

$$\begin{bmatrix} \boldsymbol{\nabla}_{\sigma\sigma}\Psi & \boldsymbol{\nabla}_{\sigma\zeta}\Psi \\ \boldsymbol{\nabla}_{\zeta\sigma}\Psi & \boldsymbol{\nabla}_{\zeta\zeta}\Psi \end{bmatrix} \begin{bmatrix} \delta\boldsymbol{\sigma} \\ \delta\boldsymbol{\zeta} \end{bmatrix} + \delta\lambda \begin{bmatrix} \boldsymbol{\nabla}_{\sigma} G \\ \boldsymbol{\nabla}_{\zeta} G \end{bmatrix} = \begin{bmatrix} \mathbf{r}_{\varepsilon} \\ \mathbf{r}_{\kappa} \end{bmatrix}. \tag{7.195}$$

These equations must be supplemented by the consistency condition. According to their definition in connection with the maximum dissipation formulation the yield function $F(\boldsymbol{\varepsilon}_e,\boldsymbol{\kappa})$ and the flow potential $G(\boldsymbol{\sigma},\boldsymbol{\zeta})$ are the same function, but expressed in terms of conjugate sets of variables. In the present context the iteration variables are $\boldsymbol{\sigma},\boldsymbol{\zeta}$, and it is therefore convenient to represent the yield condition in terms of the flow potential. When the yield condition is expressed via the residual

$$r_G = -G(\boldsymbol{\sigma},\boldsymbol{\zeta}) \tag{7.196}$$

the corresponding linearized equation takes the form

$$(\boldsymbol{\nabla}_{\sigma} G)^T \delta\boldsymbol{\sigma} + (\boldsymbol{\nabla}_{\zeta} G)^T \delta\boldsymbol{\zeta} = r_G. \tag{7.197}$$

When this equation is combined with (7.195) the resulting system of equations for the iterative increments is

$$\begin{bmatrix} \boldsymbol{\nabla}_{\sigma\sigma}\Psi & \boldsymbol{\nabla}_{\sigma\zeta}\Psi & \boldsymbol{\nabla}_{\sigma} G \\ \boldsymbol{\nabla}_{\zeta\sigma}\Psi & \boldsymbol{\nabla}_{\zeta\zeta}\Psi & \boldsymbol{\nabla}_{\zeta} G \\ (\boldsymbol{\nabla}_{\sigma} G)^T & (\boldsymbol{\nabla}_{\zeta} G)^T & 0 \end{bmatrix} \begin{bmatrix} \delta\boldsymbol{\sigma} \\ \delta\boldsymbol{\zeta} \\ \delta\lambda \end{bmatrix} = \begin{bmatrix} \mathbf{r}_{\varepsilon} \\ \mathbf{r}_{\kappa} \\ r_G \end{bmatrix}. \tag{7.198}$$

It is remarkable that this local system of equations is symmetric, also for non-associated plasticity.

In the case of associated plasticity the internal and external variables uncouple, and the equations can be expressed in terms of the sub-increments $\delta\boldsymbol{\sigma}, \delta\boldsymbol{\kappa}$ together with the yield potential $F(\boldsymbol{\sigma},\boldsymbol{\kappa})$. The key to the transformation is the incremental relation

$$\delta\boldsymbol{\zeta} = \Big(\frac{\partial\boldsymbol{\zeta}}{\partial\boldsymbol{\kappa}}\Big)^T \delta\boldsymbol{\kappa} = (\boldsymbol{\nabla}_{\kappa\kappa}\varphi)\delta\boldsymbol{\kappa} \tag{7.199}$$

and the corresponding differentiation rule

$$\boldsymbol{\nabla}_{\kappa} F = \Big(\frac{\partial\boldsymbol{\zeta}}{\partial\boldsymbol{\kappa}}\Big)^T \boldsymbol{\nabla}_{\zeta} G = (\boldsymbol{\nabla}_{\kappa\kappa}\varphi)\boldsymbol{\nabla}_{\zeta} G. \tag{7.200}$$

The equation system (7.198) is transformed by introducing $\delta\boldsymbol{\zeta}$ from (7.199), followed by pre-multiplication of the second row by $(\nabla_{\kappa\kappa}\varphi)$ (continues below).

$$\begin{bmatrix} \nabla_{\sigma\sigma}\Psi & \nabla_{\sigma\kappa}\Psi & \nabla_{\sigma}F \\ \nabla_{\kappa\sigma}\Psi & \nabla_{\kappa\kappa}\Psi & \nabla_{\kappa}F \\ (\nabla_{\sigma}F)^T & (\nabla_{\kappa}F)^T & 0 \end{bmatrix} \begin{bmatrix} \delta\boldsymbol{\sigma} \\ \delta\boldsymbol{\kappa} \\ \delta\lambda \end{bmatrix} = \begin{bmatrix} \mathbf{r}_{\varepsilon} \\ \mathbf{r}_{\zeta} \\ r_F \end{bmatrix}, \tag{7.201}$$

where the notation $r_F = r_G$ has been introduced for consistency. In this formulation the residual $\mathbf{r}_{\kappa}$ in the second row has been replaced by

$$\mathbf{r}_{\zeta} = (\nabla_{\kappa\kappa}\varphi)\,\mathbf{r}_{\kappa} \simeq -\Delta\boldsymbol{\zeta} - \Delta\lambda\,\nabla_{\kappa}F(\boldsymbol{\sigma},\boldsymbol{\kappa}). \tag{7.202}$$

The last approximation is given as the form, in which the first term is represented by the linear increment $\Delta\boldsymbol{\zeta}$ instead of the product $(\nabla_{\kappa\kappa}\varphi)\Delta\boldsymbol{\kappa}$. This is a slight modification of the initial formulation of the discretization, but equally valid and more consistent with the formulation in terms of the yield potential $F(\boldsymbol{\sigma},\boldsymbol{\kappa})$.

Associated plasticity is a consequence of uncoupling of internal and external variables in the internal energy, i.e. $\nabla_{\varepsilon\kappa}\varphi = \mathbf{0}$ and $\nabla_{\sigma\zeta}\psi = \mathbf{0}$. This uncoupling in terms of energy does not necessarily imply uncoupling of the yield potential $F(\boldsymbol{\sigma},\boldsymbol{\kappa})$ and the flow potential $G(\boldsymbol{\sigma},\boldsymbol{\zeta})$. However, if the flow potential function satisfies the uncoupling condition $\nabla_{\sigma\zeta}G = \mathbf{0}$, then this implies that $\nabla_{\sigma\kappa}\Psi = \mathbf{0}$ and leads to the often-used form

$$\begin{bmatrix} (\nabla_{\varepsilon\varepsilon}\varphi)^{-1} + \Delta\lambda\nabla_{\sigma\sigma}F & \mathbf{0} & \nabla_{\sigma}F \\ \mathbf{0} & (\nabla_{\kappa\kappa}\varphi) + \Delta\lambda\nabla_{\sigma\sigma}F & \nabla_{\kappa}F \\ (\nabla_{\sigma}F)^T & (\nabla_{\kappa}F)^T & 0 \end{bmatrix} \begin{bmatrix} \delta\boldsymbol{\sigma} \\ \delta\boldsymbol{\kappa} \\ \delta\lambda \end{bmatrix} = \begin{bmatrix} \mathbf{r}_{\varepsilon} \\ \mathbf{r}_{\zeta} \\ r_F \end{bmatrix}. \tag{7.203}$$

This formula is quite similar to the rate equation (7.109), when the two diagonal terms are represented by their corresponding 'algorithmic value', representing the finite increment via a contribution proportional to $\Delta\lambda$. The algorithmic tangent stiffness matrix $\mathbb{C}_a$ is defined for associated plasticity theory via the top diagonal term as

$$\mathbb{C}_a^{-1} = (\nabla_{\varepsilon\varepsilon}\varphi)^{-1} + \Delta\lambda\,\nabla_{\sigma\sigma}F = \mathbb{C}^{-1} + \Delta\lambda\,\nabla_{\sigma\sigma}F. \tag{7.204}$$

A similar algorithmic matrix for the internal variable is defined by the center term.

Consistent tangent matrices for incremental changes around a predicted state with finite multiplier increment $\Delta\lambda$ were discussed in relation to the return format by Simo and Taylor (1985). The concept arises naturally in connection with the Newton–Raphson scheme. The consistent tangent matrices play a role on two levels in the computation. They are generated on a node-by-node basis within the stress iteration loop and improve the rate of convergence here. When the stresses have been determined at all Gauss points the corresponding residual forces may require an additional global determination of the deformation, strain, etc. In that connection changes on the sub-increment level are also related via the consistent tangent stiffness. The analysis can be simplified, if the changes of the internal variables are condensed into the hardening modulus H and determined separately. However, general plasticity models with coupled internal and external variables in both energy and flow potential lead to full coupling of the local incremental equations as illustrated in (7.198).

7.5 Models for granular materials

Granular materials like soils, grain, etc. represent an interesting case, in which basic mechanics principles can be used to formulate plasticity models that reflect several special features of the stress–strain behavior with remarkable accuracy. One of the basic features is observed in a triaxial test, in which a uniform isostatic compression is overlaid by an increasing uniaxial compression. In the beginning of such a test, where the uniaxial component is small relative to the isostatic component, the material will experience compaction – i.e. negative dilation. Just before failure, characterized by rapidly increasing strains, the behavior changes and becomes dilational. It turns out that the change from compaction to dilation is a characteristic of any particular granular material, described by the ratio of the stress difference to the mean stress. A mathematical theory incorporating this transition was developed by Schofield and Wroth (1968) under the name of Cam-Clay theory and has given name to the general concept of 'critical state soil mechanics'. A simplified modern formulation using the concepts of plasticity theory can be found e.g. in Wood (1990), and later Collins and Houlsby (1997) have recast this theory into potential form, using the extremum properties of the energy dissipation rate. The original work on these theories was formulated in terms of two stress and two strain components – typically a mean value and the component difference. The following describes a fully triaxial plasticity theory for granular materials, in which the triaxial format arises as a natural extension in terms of invariants of a basic behavior in a hypothetical two-component generalized shear test (Krenk, 1998, 2000). Alternative models have been proposed e.g. by Lade and Kim (1995). The theory applies to granular materials without adhesion, and therefore is restricted to compressive stresses. In this section it has therefore been found convenient to use a notation in which stresses are positive in compression, and strains are positive in contraction. The final equations are easily converted by changing the sign on all stress and strain components.

7.5.1 Flow potential and yield surface

A typical yield surface of a granular material is shown in Fig. 7.19(a) in principal stress space. A suitable mathematical format should account for the smooth triangular shape in the deviatoric sections and the closed rounded shape with increasing mean stress. This is illustrated by the two generating contours in Fig. 7.19(b). It is convenient first to identify a suitable format for the deviatoric contour and subsequently determine the dependence of the size of the contour on the mean stress, see e.g. Collins (2003).

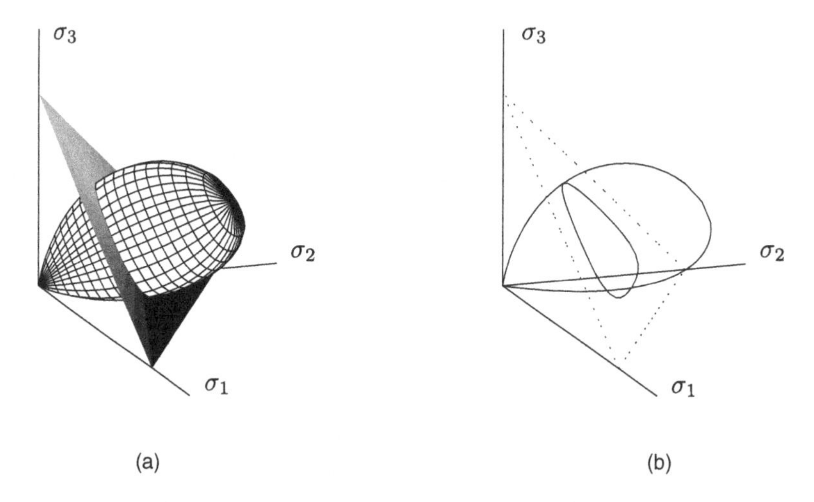

Fig. 7.19. (a) Yield surface in principal stress space, (b) generating curves.

In the formulation of the model it is convenient to use the mean stress σ_m and the deviatoric stress components $\sigma'_{\alpha\beta}$, defined by (7.26) and (7.28). The simplest mathematical form of a smooth triangular contour that satisfies the symmetry requirements with respect to the three principal deviatoric components $(\sigma'_1, \sigma'_2, \sigma'_3)$ is the cubic polynomial equation

$$(\sigma'_1 + d)(\sigma'_2 + d)(\sigma'_3 + d) \;=\; \eta\, d^3, \qquad (7.205)$$

where d determines the size of the circumscribing triangle, and η is a non-dimensional shape parameter (Krenk, 1996). It may be argued theoretically that the triangle should be the intersection of the deviatoric plane with the coordinate planes in principal stress space, whereby

$$d \;=\; \sigma_m. \qquad (7.206)$$

This is supported by experimental evidence, e.g. by Lade and Duncan (1975). The implication is that in any particular deviatoric plane the smooth triangular contour is defined in terms of a single parameter η. Thus, the full surface is determined if this parameter is prescribed as a function of the mean stress, $\eta = \eta(\sigma_m)$. A set of deviatoric stress contours is shown in Fig. 7.20.

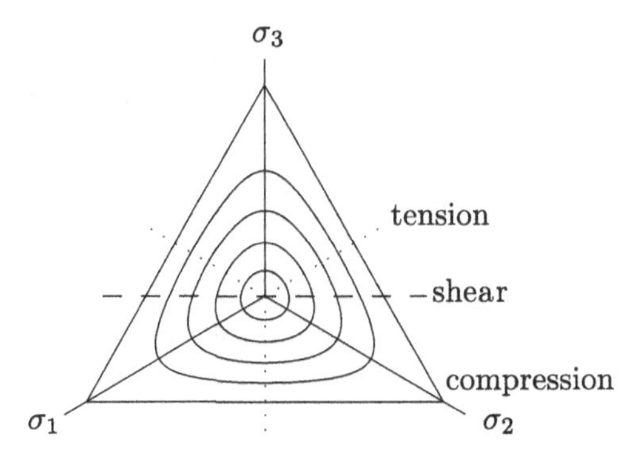

Fig. 7.20. Smooth triangular contours in deviatoric plane.

The surface is easily reformulated in full component format by using the stress invariants discussed in Section 7.1.1. The particular choice $d = \sigma_m$ implies that the left side of (7.205) is simply the product of the three principal stress components, and it then follows from the definition of the third stress invariant I_3 by (7.19) that the deviatoric contour is given by

$$I_3 \; = \; \eta(\sigma_m)\,\sigma_m^3. \qquad (7.207)$$

Several other formats for the deviatoric contour have been used in the literature, e.g. the Matsuoka and Nakai (1985) format with σ_m^3 replaced by $\sigma_m J_2$. However, not all of these formats are convex in their full parameter range. In the present formulation the deviatoric contour can be expressed in terms of the deviatoric stress invariants J_2 and J_3 from (7.30) and (7.31) as

$$J_3 - \sigma_m J_2 \, + \, (1 - \eta(\sigma_m))\sigma_m^3 \; = \; 0. \qquad (7.208)$$

This latter format in terms of deviatoric invariants is particularly useful for identification of suitable functions $\eta(\sigma_m)$ for the yield surface and the flow potential.

Most triaxial experiments on granular materials have been made with two of the principal stresses equal, either as 'triaxial compression' with e.g. $\sigma_1 > \sigma_2 = \sigma_3$, or as 'triaxial tension' with e.g. $\sigma_1 < \sigma_2 = \sigma_3$. These particular conditions lie on the lines connecting the center of the triangle in Fig. 7.20 with either a corner (full line) or the mid-point of a side (dotted line) respectively. Theoretically it is more convenient to refer to a condition of 'generalized shear', in which an isostatic state of compression is superposed by a state of pure shear. These conditions correspond to

$$\sigma_1 > \sigma_3, \qquad \sigma_2 = \tfrac{1}{2}(\sigma_1 + \sigma_3) \tag{7.209}$$

or similar states obtained by index permutation. This state of stress is illustrated by the dashed line in Fig. 7.20. It has the interesting property that $\sigma_2' = 0$, and

$$\sqrt{J_2} = \tfrac{1}{2}(\sigma_1 - \sigma_3) = \tau, \qquad J_3 = 0 \tag{7.210}$$

where τ is the maximum value of the shear stress. When these expressions are substituted into equation (7.207), the following formula is obtained:

$$\tau/\sigma_m = \sqrt{1 - \eta(\sigma_m)} = \gamma(\sigma_m). \tag{7.211}$$

According to the friction hypothesis of Coulomb the ratio τ/σ_m is equal to $\sin\varphi$, where φ is the friction angle. In the Coulomb theory the friction angle φ is a constant. However, it has been argued by Krenk (1998) that when plastic flow takes place the yield condition must be satisfied, and thus an effective angle of friction φ_* can be determined from the Coulomb condition. When combined with a physically argued dilation condition this leads to an analytical determination of the flow potential $G(\sigma_m, \tau)$ for the state of generalized shear. However, the result involves the combination of the sine and the ln function, and it is therefore of interest to develop a similar but simpler form of the flow potential from the expression for the rate of dissipation.

The following argument is based on a state of generalized shear stress as defined in (7.209). The maximum and minimum stress are combined into mean stress and maximum shear stress:

$$\sigma_m = \tfrac{1}{2}(\sigma_1 + \sigma_3), \qquad \tau = \tfrac{1}{2}(\sigma_1 - \sigma_3). \tag{7.212}$$

Corresponding conjugate plastic strain rates are defined by

$$\dot{\varepsilon}_p = \dot{\varepsilon}_1^p + \dot{\varepsilon}_3^p, \qquad \dot{\gamma}_p = \dot{\varepsilon}_1^p - \dot{\varepsilon}_3^p. \tag{7.213}$$

It is assumed that the dissipation from the intermediate stress and strain components can be neglected. The dissipation rate from (7.96) then takes the form

$$\mathcal{D} = \sigma_1\dot{\varepsilon}_1^p + \sigma_3\dot{\varepsilon}_3^p = \sigma_m\dot{\varepsilon}_p + \tau\dot{\gamma}_p. \tag{7.214}$$

After division by $\sigma_m \dot{\gamma}_p$ this equation takes the non-dimensional form

$$\frac{\tau}{\sigma_m} + \frac{\dot{\varepsilon}_p}{\dot{\gamma}_p} = \frac{\mathcal{D}}{\sigma_m \dot{\gamma}_p}. \tag{7.215}$$

The plastic strain rate vector $(\dot{\varepsilon}_p, \dot{\gamma}_p)$ is along the gradient of the flow potential $G(\sigma_m, \tau)$, and thus satisfies the equation

$$d\sigma_m \, \dot{\varepsilon}_p + d\tau \, \dot{\gamma}_p = 0, \tag{7.216}$$

where $(d\sigma_m, d\tau)$ is an increment along an equipotential curve of $G(\sigma_m, \tau)$. Substitution of this result into (7.215) leads to the following differential equation for the function $\gamma(\sigma_m) = \tau/\sigma_m$:

$$\sigma_m \frac{d\gamma}{d\sigma_m} = \frac{\mathcal{D}}{\sigma_m \dot{\gamma}_p}. \tag{7.217}$$

This function determines the shape of the surface as a function of the mean stress σ_m, and the physics enters the theory via an assumed expression for the dissipation rate $\mathcal{D}$ in terms of the stresses and plastic strain rates.

In the original Cam-Clay theory it was assumed that the rate of dissipation was due to a friction-like mechanism and expressed via the product of the mean stress σ_m and the shear strain rate $\dot{\gamma}_p$ and a friction coefficient n. Hereby the right side of the equation (7.217) takes the constant value n and is easily integrated to give

$$\frac{\tau}{\sigma_m} = \gamma(\sigma_m) = n \ln\left(\frac{\sigma_0}{\sigma_m}\right). \tag{7.218}$$

In this formula σ_0 represents the value of σ_0 where the curve intersects the isostatic axis, i.e. the intersection of the surface as shown in Fig. 7.19 with the principal diagonal, and thereby constitutes a direct measure of the size of the current surface. The surface described by this function has an apex at the intersection with the principal diagonal. It has later been modified to a smooth shape in an associated plasticity theory for granular materials, see e.g. Wood (1990). However, the real problem with this form is not the apex, but the fact that the surface encloses regions with tension due to the rather simple assumption regarding the friction mechanism. This can be remedied in a fairly simple way by changing the assumption regarding the dissipation $\mathcal{D}$. If the representative stress in the friction assumption is changed from the mean stress σ_m to the minimum stress $\sigma_{\min} = \sigma_m - \tau$, the curve is found to be

$$\frac{\tau}{\sigma_m} = \gamma(\sigma_m) = 1 - \left(\frac{\sigma_m}{\sigma_0}\right)^n. \tag{7.219}$$

This generalized shear contour is used to define the plastic flow potential $G(\boldsymbol{\sigma})$ according to the surface format (7.207). Direct substitution gives

$$G(\boldsymbol{\sigma}) = -I_3 + \eta_G(\sigma_m)\,\sigma_m^3 \tag{7.220}$$

with the contour described by the function

$$\eta_G(\sigma_m) = 1 - \gamma(\sigma_m)^2 = (\sigma_m/\sigma_G)^n\big[1 - (\sigma_m/\sigma_G)^n\big]. \tag{7.221}$$

The corresponding surface is illustrated in Fig. 7.21, clearly showing the apex at $\sigma_m = \sigma_G$. This format was used in the implementation of Ahadi and Krenk (2000, 2003). However, from the point of numerical analysis it would be desirable to have the mathematical format of the surface in a form in which the stresses in the J_2 and J_3 terms are of degree one.

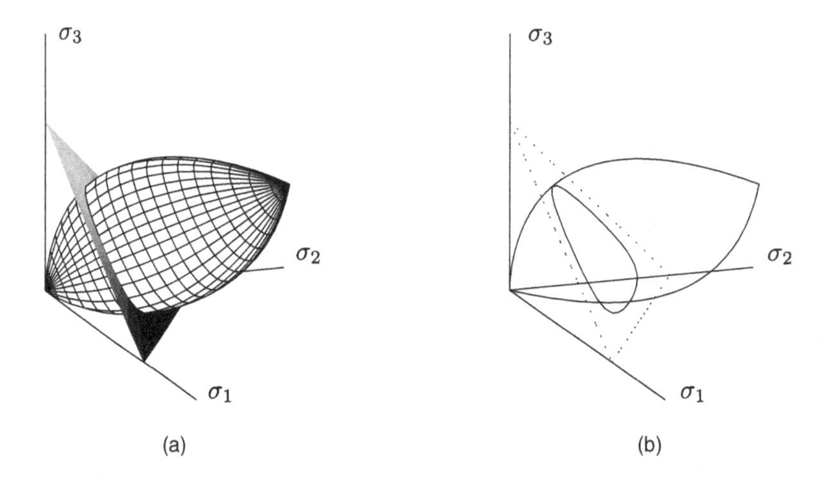

Fig. 7.21. (a) Flow potential in principal stress space, (b) generating curves.

It would now be desirable to introduce the internal elastic energy as described in Section 7.2 with suitable internal kinematic parameters, and then to derive the resulting yield surface F. However, no complete theory in terms of conjugate internal and external variables appears to be available at the present time. The yield surface is therefore introduced by retaining the function format

$$F(\boldsymbol{\sigma}) = -I_3 + \eta_F(\sigma_m)\,\sigma_m^3, \tag{7.222}$$

whereby the rounded triangular contour family is retained. The dependence on the mean stress is represented by the function

$$\eta_F(\sigma_m) = (\sigma_m/\sigma_F)^m. \tag{7.223}$$

This surface is rounded at the intersection $\sigma_m = \sigma_F$ with the principal diagonal. It was used for the illustration in Fig. 7.19.

7.5.2 Elasticity and hardening

The elastic stress–strain relations are chosen in accordance with the original Cam-Clay theory, in which the bulk modulus is proportional to the mean stress σ_m and the shear modulus is constant:

$$\dot{\varepsilon}_v^e = \frac{\kappa}{\sigma_m}\,\dot{\sigma}_m, \qquad \dot{\varepsilon}_{\alpha\beta}^{e\prime} = \frac{1}{2\mu}\,\dot{\sigma}_{\alpha\beta}', \tag{7.224}$$

where the volumetric flexibility is described by the non-dimensional parameter κ. In the elasto-plastic regime the total volumetric strain rate is given by the relation

$$\dot{\varepsilon}_v = \dot{\varepsilon}_v^e + \dot{\varepsilon}_v^p = \frac{\chi}{\sigma_m}\,\dot{\sigma}_m, \tag{7.225}$$

where $\chi > \kappa$ represents the increased volumetric flexibility in the elasto-plastic range. When subtracting the elastic part, the plastic strain rate is found as

$$\dot{\sigma}_m = \frac{\sigma_m}{\chi - \kappa}\,\dot{\varepsilon}_v^p. \tag{7.226}$$

In the case of isostatic loading $\dot{\sigma}_m = \dot{\sigma}_F$, and thus the relation may be considered as the evolution equation for the stress σ_F determining the current size of the yield surface. This is the hardening evolution rule adopted in the original Cam-Clay theory (Schofield and Wroth, 1968). In that theory the yield surface expansion was controlled by the rate of work of the mean stress $\sigma_m\dot{\varepsilon}_v^p$. Thus hardening stops, and unlimited plastic deformation can take place, once the critical state of zero plastic dilation is reached. This does not reflect the experimentally observed behavior in typical tests, where the material passes into a dilative state accompanied by large plastic strains. This can be incorporated into the model by adding a weighted contribution from the deviatoric components to the plastic work, whereby the hardening equation becomes

$$\dot{\sigma}_F = \frac{1}{\chi - \kappa}\left(\sigma_m\dot{\varepsilon}_v^p + w\,\sigma_{\alpha\beta}'\dot{\varepsilon}_{\alpha\beta}^{p\prime}\right). \tag{7.227}$$

In this equation w is a weight factor leading to a modest contribution to the hardening from the plastic work from the deviatoric components.

The plastic hardening is characterized via the hardening modulus H as defined via $\dot{F}$ in (7.110),

$$H = -\frac{\partial F}{\partial \sigma_F}\frac{\partial \sigma_F}{\partial \lambda} = H_1 H_2. \tag{7.228}$$

The factor H_1 gives the dependence of the yield function on the size parameter σ_F. It follows from (7.222) by differentiation,

$$H_1 = -\frac{\partial F}{\partial \sigma_F} = m\,\sigma_m^2 \big(\sigma_m/\sigma_F\big)^{m+1}. \tag{7.229}$$

The factor H_2 describes the plastic hardening via the hardening rule (7.227). After substitution of the plastic strain rates in terms of the flow potential G, the following result is obtained:

$$H_2 = \frac{\partial \sigma_F}{\partial \lambda} = \frac{1}{\chi - \kappa}\Big(\sigma_m \frac{\partial G}{\partial \sigma_m} + w\,\sigma'_{\alpha\beta}\frac{\partial G}{\partial \sigma'_{\alpha\beta}}\Big). \tag{7.230}$$

The differentiations are most easily carried out by using the format (7.208) for G in terms of mean and deviator stress components,

$$H_2 = \frac{1}{\chi - \kappa}\Big[(1-w)\big(3J_3 - 2\sigma_m J_2\big) + \sigma_m^4 \frac{d\eta_G}{d\sigma_m}\Big]. \tag{7.231}$$

This completes the formulation of the model. It contains a total of six parameters – three parameters μ, κ and χ relating to stiffness, and three non-dimensional parameters n, m and w specifying the potential functions in the theory. Parameter calibration and numerical implementation are discussed by Ahadi and Krenk (2000, 2003).

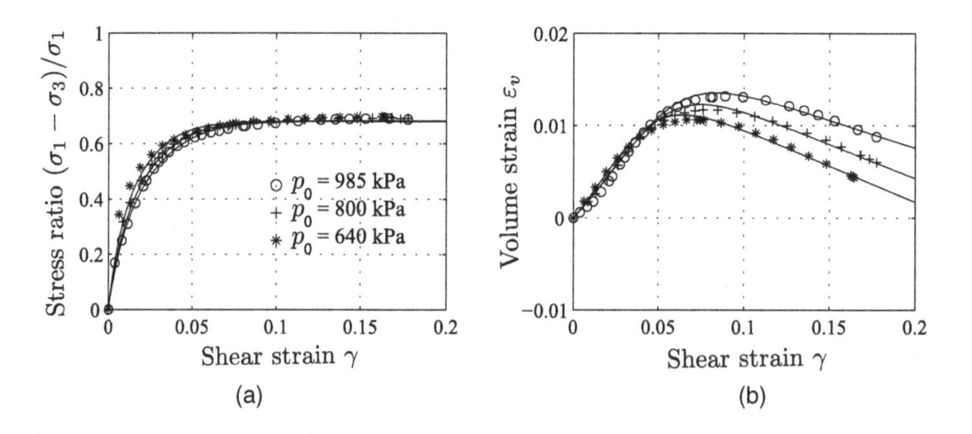

Fig. 7.22. Stress ratio and volumetric strain for loose Baskarp sand.

Figure 7.22 shows the ability of the model to represent the development of stress and volumetric strain in a typical triaxial cylinder test on sand by Borup and Hedegaard (1995). The development of volumetric strain in Fig. 7.22(b) clearly shows the initial contraction phase, followed by dilation. The parameter p_0 refers to the uniform compressive stress at the beginning of the test.

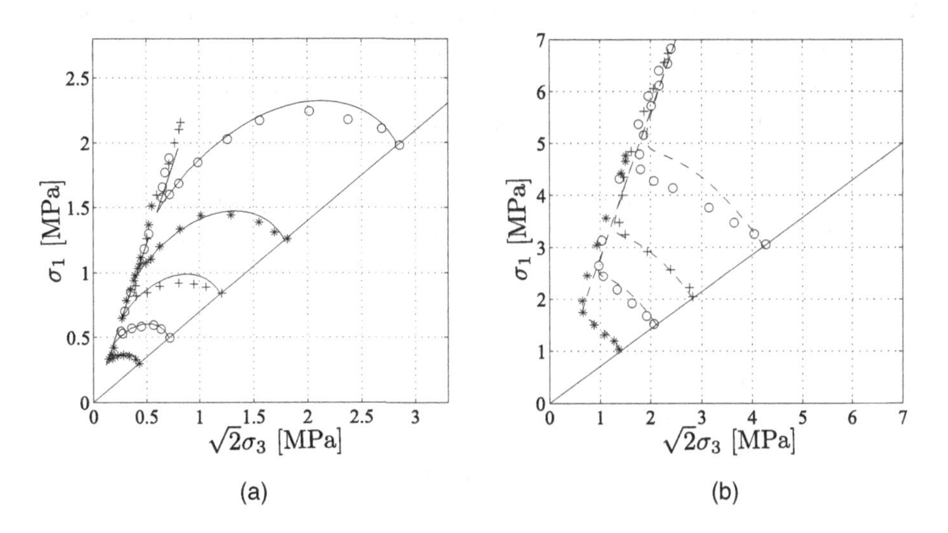

Fig. 7.23. Constant volume tests on Sacramento sand: (a) loose, (b) dense.

Figure 7.23 illustrates the ability of the model to represent the results from so-called undrained tests on sand by Lee and Seed (1967). In the undrained test the sample is saturated with water and sealed, whereby the total volume is kept approximately constant. The figure shows a vertical plane containing the principal diagonal in stress space. The tests are started by loading the specimen to an isostatic state of compression, here appearing as the starting point of the individual curve on the principal diagonal. The first curved part of the curve depends on the shape of the yield potential G, and the excellent agreement supports the present shape derived from a simplified friction hypothesis. A shortcoming of the present model is the lack of explicit internal variables, representing e.g. the development of the void volume fraction during plastic deformation. An extension to include the evolution of the void volume could be made along the lines indicated for the simple Drucker–Prager model in Section 7.4.2.

7.6 Finite strain plasticity

In the previous sections plasticity theory has been presented within the framework of infinitesimal strains. In the generalization of the theory to finite strains a central role is played by the structure of the evolution equations (7.95) in which the total strain rate is the sum of the rate of elastic recoverable strains and the rate of irreversible plastic strains,

$$\dot{\varepsilon} = \dot{\varepsilon}_e + \dot{\varepsilon}_p. \tag{7.232}$$

When extending the theory to finite strains one option is to consider this as a spatial rate equation and to replace the time derivatives with a corresponding objective strain rate along the lines discussed in Section 6.2.3. However, in the discretized form problems may arise with full reversibility of the elastic strains.

An alternative formulation in terms of material components was proposed by Lee (1969). An extensive discussion of this format in relation to numerical computations has been presented by Simo and Ortiz (1985) and Simo (1988a,b). The basic idea is to consider the deformation gradient at a point. If a small volume of material around the point were separated and the stresses were released, then the deformation gradient due to plastic straining would remain. This property is expressed by the factored format

$$\mathbf{F} = \mathbf{F}_e \mathbf{F}_p. \tag{7.233}$$

Here $\mathbf{F}_p$ is the deformation gradient associated with plastic straining, while $\mathbf{F}_e$ is an extra factor due to the elastic deformation. Release of the stresses would lead to elastic unloading and introduce the factor $\mathbf{F}_e^{-1}$, leaving the residual deformation gradient $\mathbf{F}_p$. The relation can be illustrated by use of component format, where a material particle initially at $\mathbf{x}_0$ moves to $\mathbf{x}$ in elasto-plastic loading, and then moves to $\mathbf{x}_p$ upon elastic unloading. The factored format (7.233) then takes the form

$$\frac{\partial x_\alpha}{\partial x_\beta^0} = \frac{\partial x_\alpha}{\partial x_\gamma^p}\frac{\partial x_\gamma^p}{\partial x_\beta^0}. \tag{7.234}$$

It is seen that $\mathbf{F}$ on the left side and the last factor $\mathbf{F}_p$ on the right side both represent well-defined deformation gradients with respect to the initial state represented by $\mathbf{x}_0$. In contrast, the factor $F_{\alpha\gamma}^e = \partial x_\alpha / \partial x_\gamma^p$ is a deformation gradient relative to an initial state described by the residual state $\mathbf{x}_p$.

In the combined elasto-plastic state the Green strain tensor $\mathbf{E}$ is expressed in terms of the Green deformation tensor $\mathbf{C}$ as

$$\mathbf{C} = \mathbf{F}^T\mathbf{F}, \qquad \mathbf{E} = \tfrac{1}{2}(\mathbf{C} - \mathbf{I}). \tag{7.235}$$

In the hypothetical stress released state the Green strain $\mathbf{E}_p$ is given similarly as

$$\mathbf{C}_p = \mathbf{F}_p^T\mathbf{F}_p, \qquad \mathbf{E}_p = \tfrac{1}{2}(\mathbf{C}_p - \mathbf{I}). \tag{7.236}$$

While the total and the stress released states are given unambiguously, the interpretation of an elastic state and its associated straining is slightly more open. The simplest formulation is obtained by defining the elastic strain $\mathbf{E}_e$ as the additional Green strain introduced by the deformation from the residual plastic state to the total state of deformation. In terms of deformation gradients this definition amounts to

$$\mathbf{E}_e = \tfrac{1}{2}(\mathbf{F}^T\mathbf{F} - \mathbf{F}_p^T\mathbf{F}_p) = \tfrac{1}{2}(\mathbf{C} - \mathbf{C}_p). \qquad (7.237)$$

This definition leads to additive elastic and plastic strains,

$$\mathbf{E} = \mathbf{E}_e + \mathbf{E}_p. \qquad (7.238)$$

The implication of this property is that the complete structure of the plasticity equations developed above is retained, when using the appropriate material definitions of stresses and strains. It is noted that the factored format (7.233) of the deformation gradient only defines the plastic deformation gradient in the form $\mathbf{R}\mathbf{F}_p$, where $\mathbf{R}$ is an undetermined rotation. However, it follows from the definition (7.237) that the elastic strain $\mathbf{E}_e$ is independent of this rotation.

The definition (7.237) implies that the elastic strains are defined relative to an initial state including the total accumulated plastic straining. This is reflected in the appearance of the tensor $\mathbf{C}_p$ instead of the unit tensor $\mathbf{I}$ in the definition of the strain $\mathbf{E}_e$. Alternatively, an elastic strain can be defined from the deformation gradient $\mathbf{F}_e$ by relations similar to (7.235) and (7.236) for the total and the plastic strain:

$$\tilde{\mathbf{C}}_e = \mathbf{F}_e^T\mathbf{F}_e, \qquad \tilde{\mathbf{E}}_e = \tfrac{1}{2}(\tilde{\mathbf{C}}_e - \mathbf{I}), \qquad (7.239)$$

where a tilde has been used to distinguish this definition from the previous. It follows directly from the previous elastic strain definition (7.237) and the factored form of the deformation gradient that

$$\mathbf{E}_e = \mathbf{F}_p^T\tilde{\mathbf{E}}_e\mathbf{F}_p. \qquad (7.240)$$

This definition complicates the previous additive strain relation (7.238) and also introduces a need for determining the rotation component of $\mathbf{F}_p$ in the case of anisotropic materials. The tensor $\tilde{\mathbf{E}}_e$ was adopted as the elastic strain by Lee (1969) and Halphen and Son (1975). This elastic strain definition has since been used fairly extensively in computational models in spite of the complications arising from the non-additive strain format.

The extension of the small displacement plasticity theory to finite displacements is straightforward, when using the elastic strain definition (7.237).

The first step is the introduction of an internal energy density in terms of finite strains in the form $\varphi(\mathbf{E}_e, \boldsymbol{\kappa})$, where the plastic strain $\mathbf{E}_p$ may be included among the internal variables $\boldsymbol{\kappa}$. The complementary variables are defined via the partial derivatives of the internal energy, giving the Piola–Kirchhoff stress of the second kind as

$$\mathbf{S} = \frac{\partial \varphi(\mathbf{E}_e, \boldsymbol{\kappa})}{\partial \mathbf{E}_e^T}. \tag{7.241}$$

The definition of the complementary internal variables $\boldsymbol{\zeta}$ is unchanged from (7.73b). It is common within finite strain continuum mechanics to express the potentials in terms of the Green deformation tensor $\mathbf{C}$ instead of the strain $\mathbf{E}$. If that tradition is followed, the internal energy function is expressed in the form $\varphi(\mathbf{C}, \mathbf{C}_p, \boldsymbol{\kappa})$, whereby the stress definition (7.241) becomes

$$\mathbf{S} = 2\frac{\partial \varphi(\mathbf{C}, \mathbf{C}_p, \boldsymbol{\kappa})}{\partial \mathbf{C}^T}. \tag{7.242}$$

Clearly, the formulation could also be made in terms of an equivalent $\mathbf{C}_e$ defined from the elastic strain in (7.237), but it should be noted that this definition is different from that of $\tilde{\mathbf{C}}_e$.

With the elastic strain $\mathbf{E}_e$ defined by (7.237) and the Piola–Kirchhoff stress $\mathbf{S}$ defined from the internal energy by (7.241), the theory of elasto-plastic materials developed in this chapter for small displacement theory can be extended directly to finite displacements, and the corresponding finite element formulation follows the principles outlined in Chapter 6. Alternatively, the elastic strain may be constructed from the current state as described in detail by Simo and Miehe (1992) using the elastic model of Section 7.1.3.

7.7 Summary

Many types of material response can be modeled adequately by constitutive relations that are independent of the rate of deformation. The two main types of models within this class are the elastic materials, exhibiting reversible response, and the elasto-plastic models, including the effect of irreversible plastic deformation. These models can be combined into a common framework based on the concept of an internal energy $\varphi(\boldsymbol{\varepsilon}_e, \boldsymbol{\kappa})$ and a yield potential $F(\boldsymbol{\varepsilon}_e, \boldsymbol{\kappa})$ expressed in terms of the reversible elastic strain $\boldsymbol{\varepsilon}_e$ and a set of internal variables $\boldsymbol{\kappa}$. The elastic response, taking place for states inside the yield surface, is governed solely by the internal energy, while combined elasto-plastic response can occur for states on the yield surface. Both types of response are governed by evolution equations expressed in terms of conjugate variables $\boldsymbol{\sigma}$ and $\boldsymbol{\zeta}$, defined by the derivatives of the internal energy. The plastic flow is determined by a flow potential $G(\boldsymbol{\sigma}, \boldsymbol{\zeta})$, obtained by exchanging the original variables in the yield potential $F(\boldsymbol{\varepsilon}_e, \boldsymbol{\kappa})$ with their conjugate counterparts. An important consequence of this result is that materials with uncoupled internal and external variables will exhibit plastic flow along the normal to the yield function F, while coupling between internal and external variables leads to plastic flow in a different direction, determined by the normal to G. A simple illustration of this effect is that irreversible volume strain, such as change of pore volume, typically will lead to non-associated plastic dilation, an effect observed for granular materials.

Numerical integration of the elasto-plastic evolution equations requires the introduction of a finite increment representation. This is developed on two levels: an elementary 'return algorithm' formulation for von Mises plasticity, and a general formulation for iterative determination of stress and internal variables corresponding to an imposed total strain increment. It turns out that even in the case of non-associated plasticity this problem can be formulated via a symmetric equation system, when using conjugate variables.

Finally, the question of finite strains is addressed, and two alternative formulations are indicated. In the first formulation the finite elastic strain appears as the difference between the total strain and a finite plastic strain. This definition permits direct interpretation of the small strain formulation in terms of finite Green strain in connection with the second Piola–Kirchhoff stress. In the second formulation the elastic strain is defined as the additional straining after imposing the plastic strain. Thus, the second definition uses the residual plastic state as reference configuration in the Green strain definition.

7.8 Exercises

Exercise 7.1 Show that the definition (7.17)–(7.19) of the principal stress invariants leads to the explicit component representations

$$
\begin{aligned}
I_1 &= \sigma_{11} + \sigma_{22} + \sigma_{33}, \\
I_2 &= \sigma_{22}\sigma_{33} + \sigma_{33}\sigma_{11} + \sigma_{11}\sigma_{22} - \sigma_{23}\sigma_{32} - \sigma_{13}\sigma_{31} - \sigma_{12}\sigma_{21}, \\
I_3 &= \sigma_{11}\sigma_{22}\sigma_{33} + 2\sigma_{23}\sigma_{13}\sigma_{12} - \sigma_{11}\sigma_{23}\sigma_{32} - \sigma_{22}\sigma_{13}\sigma_{31} - \sigma_{33}\sigma_{12}\sigma_{21}.
\end{aligned}
$$

Exercise 7.2 Find the tensor component derivatives $\partial I_1/\partial\sigma_{\alpha\beta}$, $\partial I_2/\partial\sigma_{\alpha\beta}$, $\partial I_3/\partial\sigma_{\alpha\beta}$ of the principal stress invariants from the component form in

Exercise 7.1, and arrange the result in matrix format. Compare with the components obtained from the differentiation formulae (7.23)–(7.25).

Exercise 7.3 The stress invariants and their derivatives can be handled in the compact array format as illustrated by the following calculations.

(a) Express the stress principal invariants I_1, I_2 and I_3 in terms of the six independent stress components used in the compact stress vector format (7.8).

(b) Find the derivatives $\partial I_1/\partial\boldsymbol{\sigma}$, $\partial I_2/\partial\boldsymbol{\sigma}$, $\partial I_3/\partial\boldsymbol{\sigma}$ corresponding to the independent stress components and arrange the results in one-dimensional array format.

(c) Compare the array format components with the tensor components obtained in Exercise 7.2, and explain the result using the scalar increments $dI_j = (\partial I_j/\partial\boldsymbol{\sigma})d\boldsymbol{\sigma}$.

Exercise 7.4 Prove the results (7.32) and (7.33) for the principal stress invariants J_2 and J_3 by using the zero mean deviatoric stress property.

Exercise 7.5 Prove the following relations, used in the derivation of the von Mises elasto-plastic tangent stiffness, by use of the tensor component relations (7.49) and (7.128):

$$C_{\alpha\beta\gamma\delta}\,\frac{\partial\sigma_e}{\partial\sigma_{\gamma\delta}} = 3\mu\,\frac{\sigma'_{\alpha\beta}}{\sigma_e}, \qquad \frac{\partial\sigma_e}{\partial\sigma_{\alpha\beta}}\,C_{\alpha\beta\gamma\delta}\,\frac{\partial\sigma_e}{\partial\sigma_{\gamma\delta}} = 3\mu.$$

Exercise 7.6 Use the expression (7.96) for the rate of plastic dissipation to prove the following result for perfect von Mises plasticity: $\boldsymbol{\sigma}^T\dot{\boldsymbol{\varepsilon}}_p = \dot{\lambda}\sigma_e = \dot{\lambda}\sigma_0$. This in turn implies the interpretation of the plastic multiplier as $\dot{\lambda} = \dot{\varepsilon}_p$, demonstrated in (7.133) by other means.

Exercise 7.7 Let the elasto-plastic stiffness $\mathbb{C}_{ep}$ be defined by the rate equation (7.118). Consider a total strain rate proportional to the gradient of the plastic potential, i.e. $\dot{\boldsymbol{\varepsilon}} \propto \boldsymbol{\nabla}_\sigma G$. For this particular strain direction the elastic and plastic strain increments are proportional. Find the reduction in stiffness due to plasticity, and give an interpretation of the non-dimensional ratio $H/(\boldsymbol{\nabla}_\sigma G)^T\mathbb{C}(\boldsymbol{\nabla}_\sigma F)$. Compare with the uniaxial von Mises plasticity formula (7.135) and Fig. 7.11.

Exercise 7.8 Repeat the calculations of Example 7.3 illustrating explicit stress updating methods, but this time with half the strain increment. Comment on the effect of using half the increment length.

Exercise 7.9 The relations of von Mises plasticity theory can be derived directly in matrix–vector format. The following questions illustrate the procedure.

(a) Use the six-dimensional stress vector $\boldsymbol{\sigma}$ introduced in (7.8) and find the symmetric matrix $\mathbf{M}$ for which $\sigma_e^2 = \frac{1}{2}\boldsymbol{\sigma}^T\mathbf{M}\,\boldsymbol{\sigma}$.
(b) Use the matrix relation (a) to find the gradient vector $\boldsymbol{\nabla}_\sigma\sigma_e$ and compare with (7.128).
(c) Use the gradient $\boldsymbol{\nabla}_\sigma\sigma_e$ from (b) and the elastic stiffness matrix $\mathbb{C}$ from (7.53) to obtain $\mathbb{C}(\boldsymbol{\nabla}_\sigma\sigma_e)$.
(d) Combine (b) and (c) to obtain $(\boldsymbol{\nabla}_\sigma\sigma_e)^T\mathbb{C}(\boldsymbol{\nabla}_\sigma\sigma_e)$.

Compare the results in (c) and (d) with the tensor component results in Exercise 7.5. Note that while the gradient (b) is in strain vector format, the product (c) with the stiffness matrix brings the vector into stress vector format.

Exercise 7.10 Consider an associated Drucker–Prager material with the yield condition (7.176). The yield stress in uniaxial tension is σ_t and the yield stress in uniaxial compression is σ_c ($> \sigma_t$). Determine the parameter α in terms of σ_c and σ_t, and indicate the stress path of the uniaxial tension and compression tests on a figure like Fig. 7.16.

Exercise 7.11 Show that the elasto-plastic evolution equations (7.170) can be recast in terms of the flow potential $G(\boldsymbol{\sigma}, \boldsymbol{\zeta}, \zeta_0)$ and the multiplier $\dot{\lambda}$ in the form

$$\begin{bmatrix} \dot{\boldsymbol{\varepsilon}} \\ \mathbf{0} \\ \dot{F} \end{bmatrix} = \begin{bmatrix} \mathbb{C}^{-1} & \mathbf{0} & (\boldsymbol{\nabla}_\sigma G) \\ \mathbf{0} & (\nabla_{\kappa\kappa}\varphi_1)^{-1} & (\boldsymbol{\nabla}_\zeta G) \\ (\boldsymbol{\nabla}_\sigma G)^T & (\boldsymbol{\nabla}_\zeta G)^T & -\varphi_0'' \end{bmatrix} \begin{bmatrix} \dot{\boldsymbol{\sigma}} \\ \dot{\boldsymbol{\zeta}} \\ \dot{\lambda} \end{bmatrix}.$$

Exercise 7.12 The dependence on the internal tensor variable $\kappa_{\alpha\beta}$ is expressed by the potential $\varphi_1(\kappa_e)$. The resulting formulation simplifies if the potential is expressed in the form $\varphi_1(\kappa_e) = \Phi_1(\frac{1}{2}\kappa_e^2)$, where the argument κ_e has been replaced by $\frac{1}{2}\kappa_e^2$. Show that in this formulation the formulae (7.158) and (7.161) simplify to

$$\zeta_{\alpha\beta} = \kappa_{\alpha\beta}'\,\Phi_1', \qquad \frac{\partial\zeta_{\alpha\beta}}{\partial\kappa_{\alpha\beta}} = \delta_{\alpha\gamma}\delta_{\beta\delta}\,\Phi_1' + \kappa_{\alpha\beta}'\,\kappa_{\gamma\delta}'\,\Phi_1''.$$

This result leads directly to the conclusion of isotropy for $\Phi_1'' = 0$.

Exercise 7.14 Use the differential equation (7.217) for the $\tau/\sigma_m = \gamma(\sigma_m)$ contour to derive the result (7.218) for the original Cam-Clay hypothesis $\mathcal{D} = n\sigma_m\dot{\gamma}_p$, and the result (7.219) for the modified friction hypothesis $\mathcal{D} = n\sigma_{\min}\dot{\gamma}_p$.

8

Numerical solution techniques

The solution of non-linear problems relies heavily on numerical methods. Only few non-linear problems allow direct solution, and most often an iterative strategy must be used. A simple example of such an iterative strategy is the Newton–Raphson method described in Section 1.2. In its standard form it consists of a series of prescribed load increments combined with an iterative solution of the equilibrium equations for the corresponding displacement increments. This strategy has a number of disadvantages. In the full Newton–Raphson method each step in the iterative solution requires solution of a linearized set of equations. This may involve a very high computational effort, and therefore modified versions, in which the equations are not reformulated in each step, can be an alternative. The modified Newton–Raphson method has slower convergence, and this may offset some of the gain from the simplified solution of the equations. In addition to concerns about the numerical efficiency, the Newton–Raphson method also encounters problems in passing limit and bifurcation points.

In the Newton–Raphson method the load increment is specified at the beginning of the load step and kept constant during equilibrium iterations. This leads to lack of efficiency and possibly complications when the stiffness changes rapidly, and in particular around load limit points where the sign of the load increment changes. Several techniques have been developed to deal with this problem. Here two types of techniques will be described. In the first the magnitude of the load increment is adjusted in such a way that the residual force vector attains an optimal property. This approach was proposed by Bergan (1980, 1981), who used minimum length of the residual force vector as optimality criterion. Alternatively, the residual force may be required to be orthogonal to the current displacement increment (Krenk, 1995a). The other approach is based on the concept of an arc-length of the equilibrium curve in combined load–displacement space (Riks, 1979;

Crisfield, 1981; Ramm, 1981; Forde and Stiemer, 1987; and others). The key point in both of these techniques is that the prescribed load increment is considered as a first estimate subject to adjustment by the algorithm. The fact that both loads and displacements are adjusted during the iteration process makes these techniques robust and suitable for automatic selection of increments. The basic problem of variable load step and passage of local extremes is discussed in Section 8.1, while residual force and arc-length methods are presented in Sections 8.2 and 8.3, respectively.

The non-linear solution methods discussed here all use a predictor for the displacement increment, obtained from the residual force and some representation of the current stiffness. This format may be modified by use of a modified stiffness obtained by incorporating information from the force residual, briefly described in Section 8.4. Methods of this type, often called quasi-Newton methods, have their origin in the theory of optimization, see e.g. Luenberger (1984). The quasi-Newton methods were introduced into finite element computations by Matthies and Strang (1979). When the equilibrium equations can be derived from an energy functional, e.g. the potential energy, the problem can be considered as one of finding the minimum of the energy functional, i.e. a standard problem of optimization. However, also problems without a potential formulation, e.g. incremental plasticity problems, may be solved by using these techniques.

8.1 Iterative solution of equilibrium equations

The non-linear problems considered here consist in finding a sequence of equilibrium states of a non-linear system with internal forces $\mathbf{g}(\mathbf{u})$, when considering a series of load states $\mathbf{f}$, where $\mathbf{f}$ is a load pattern, typically incremented in a proportional way. This corresponds to a sequence of non-linear equations of the form

$$\mathbf{g}(\mathbf{u}_n) = \mathbf{f}_n, \quad n = 0, 1, \ldots, \tag{8.1}$$

where the subscript n indicates the solution after load step n. The solution strategies make use of the residual force, defined as

$$\mathbf{r} = \mathbf{f} - \mathbf{g}(\mathbf{u}). \tag{8.2}$$

The residual force describes the part of the load $\mathbf{f}$ that is not balanced by the internal forces in the body $\mathbf{g}(\mathbf{u})$, and the solution typically consists of a sequence of steps by which the residual force is reduced to negligible magnitude.

A typical load step starts from the last established state of equilibrium

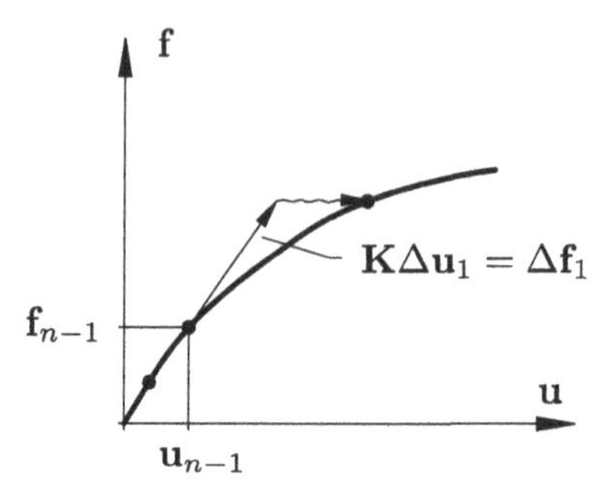

Fig. 8.1. Initial linear step followed by sub-increments.

$(\mathbf{u}_{n-1}, \mathbf{f}_{n-1})$ as illustrated in Fig. 8.1. The load step starts with imposing the additional load $\Delta\mathbf{f}_1$, where the subscript 1 refers to the iteration number. The corresponding displacement increment $\Delta\mathbf{u}_1$ is found from a linearized form of the equilibrium equation (8.1),

$$\mathbf{K}\,\Delta\mathbf{u}_1 = \Delta\mathbf{f}_1. \tag{8.3}$$

The matrix $\mathbf{K}$ represents the stiffness of the body. It will often be the tangent stiffness, defined by $\mathbf{K} = \partial\mathbf{g}/\partial\mathbf{u}$, but may also be a modified form as discussed in Section 8.4.

In a non-linear problem the increments $\Delta\mathbf{u}_1, \Delta\mathbf{f}_1$ determined by the linearized equation (8.3) will not produce an equilibrium solution, and thus iterative corrections are needed to eliminate the residual force vector $\mathbf{r}$ corresponding to the predicted state. After iteration i the displacement increment is $\Delta\mathbf{u}_i$, and the load increment residual is

$$\mathbf{r}_i = \mathbf{f}_{n-1} + \Delta\mathbf{f}_i - \mathbf{g}(\mathbf{u}_{n-1} + \Delta\mathbf{u}_i). \tag{8.4}$$

It is an important feature of general procedures for tracing equilibrium paths that the load increment $\Delta\mathbf{f}_i$ can be modified by the algorithm as part of the iteration cycle. Thus, a general iteration step may involve a load sub-increment $\delta\mathbf{f}_i$ as well as a displacement sub-increment $\delta\mathbf{u}_i$. The total increments after iteration i are obtained by adding the sub-increments

$$\Delta\mathbf{u}_i = \Delta\mathbf{u}_{i-1} + \delta\mathbf{u}_i, \qquad \Delta\mathbf{f}_i = \Delta\mathbf{f}_{i-1} + \delta\mathbf{f}_i. \tag{8.5}$$

The introduction of sub-increments of the load $\delta\mathbf{f}$ in addition to the displacements $\delta\mathbf{u}$ generates a considerable freedom in the development of non-linear algorithms for tracing an equilibrium path.

8.1.1 Non-linear iteration strategies

Three different non-linear solution strategies are illustrated in Fig. 8.2. Each sub-figure shows a single load step, which is initiated by imposing the load increment $\Delta\mathbf{f}_1$, leading to the corresponding displacement increment $\Delta\mathbf{u}_1$ by (8.3). Figure 8.2(a) illustrates the Newton–Raphson method, described in Section 1.2. In this method the load increment is prescribed as constant, and the displacement is updated via sub-increments $\delta\mathbf{u}_i$ found from the residual force. Thus, the Newton–Raphson method can be characterized by the sub-increment relations

$$\delta\mathbf{f}_i = \mathbf{0}, \qquad \mathbf{K}\,\delta\mathbf{u}_i = \mathbf{r}_i. \tag{8.6}$$

In the full Newton–Raphson method the stiffness matrix $\mathbf{K}$ represents the current tangent stiffness, while modified forms can be developed using e.g. the stiffness matrix at the latest determined equilibrium state. Clearly, these methods are limited by the need for selecting a load increment $\Delta\mathbf{f}$ that is suitable with respect to magnitude and sign.

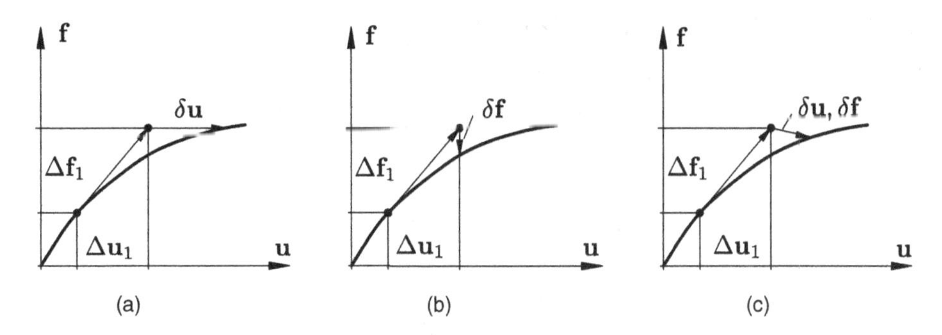

Fig. 8.2. (a) Newton–Raphson methods, (b) residual force methods, (c) arc-length methods.

An alternative method that includes automatic load increment adjustment is illustrated in Fig. 8.2(b). In this method a displacement increment is calculated corresponding to a prescribed load increment $\Delta\mathbf{f}_i$. Subsequently the load increment $\Delta\mathbf{f}_i$ is adjusted to optimize the properties of the residual force $\mathbf{r}_i$. Consequently, this method is called the residual force method. The strategy can be briefly indicated in the form

$$\delta\mathbf{f}_i \neq \mathbf{0}, \qquad \mathbf{K}\,\delta\mathbf{u}_i = \mathbf{r}_i + \delta\mathbf{f}_i. \tag{8.7}$$

The load adjustment is indicated in the figure, where it should be noted that for multi-dimensional problems the load adjustment does not by itself lead to a point on the equilibrium curve. This procedure is described in Section 8.2.

The third class of solution strategies makes use of simultaneous load and displacement sub-increment $(\delta\mathbf{u}_i, \delta\mathbf{f}_i)$, indicated in Fig. 8.2(c) as return to the equilibrium path by an inclined curve. The simultaneous change of load and displacement increment is governed by a relation of the form

$$c(\Delta\mathbf{u}_i, \Delta\mathbf{f}_i) = 0. \tag{8.8}$$

The load is typically given in terms of a load pattern multiplied by a scalar load intensity factor, and the constraint equation (8.8) is then a scalar equation. This equation is used to impose a constraint on the magnitude of the total increment $(\Delta\mathbf{u}_i, \Delta\mathbf{f}_i)$ considered as a vector, and these methods therefore go by the name of arc-length methods. The arc-length methods are currently the most used methods for equilibrium analysis of non-linear structures and solids. They are described in Section 8.3, and issues relating to appropriate definition of sub-increments and the constraint equation are discussed.

8.1.2 Direction and step-size control

A typical load–displacement curve with load and displacement limit points is shown in Fig. 8.3(a). First a load limit point is passed at A. Then the

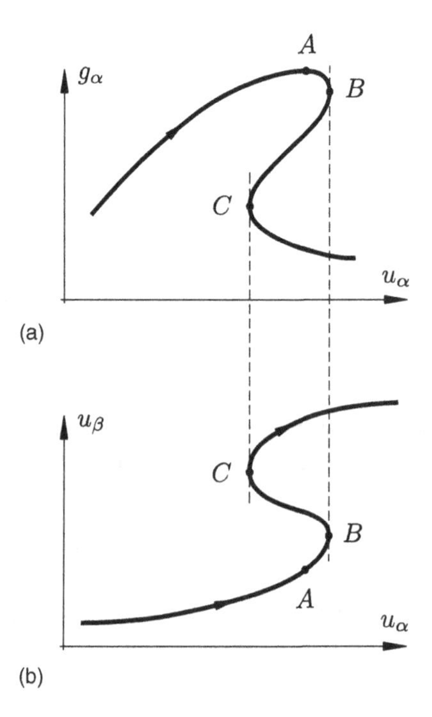

Fig. 8.3. Load and displacement limit points.

displacement component u_α shown in the figure reaches a local maximum at B and a local minimum at C. Figure 8.3(b) shows a corresponding two-dimensional displacement path. In non-linear displacement analysis without bifurcations, the load–displacement path does not cross itself, and therefore these displacement limit points are generated by the particular projections of load–displacement space used in Fig. 8.3. The equilibrium path-following techniques discussed here move from one equilibrium state to the following by an initial prediction $(\Delta\mathbf{u}, \Delta\mathbf{f})$ based on the tangent stiffness, or an approximation to the tangent stiffness. For this predictor to be an efficient origin for the following iteration process, it is necessary that it has the right orientation along the equilibrium path. Furthermore, the curvature of the equilibrium path typically imposes limitations on the size of the initial incremental predictors. These issues are briefly dealt with here, and implemented in the orthogonal residual algorithm and the arc-length algorithm in the following sections.

A load step is initiated by incrementing the load by $\Delta\mathbf{f}_1$ and finding the corresponding displacement increment $\Delta\mathbf{u}_1$ as a solution of the stiffness relation (8.3). If the previous increment has passed a load limit point, there may be a need for changing the direction of load incrementation. The problem is illustrated in Fig. 8.4, showing the displacement increment $\Delta\mathbf{u}_0$ of the previous load step, connecting two points on the equilibrium curve. In the current step, use of a tangent stiffness relation will produce a displacement increment $\Delta\mathbf{u}_1$ along the tangent of the equilibrium curve. It is important that the predicted displacement increment $\Delta\mathbf{u}_1$ points in the direction that continues the equilibrium curve, as shown in Fig. 8.4(b), and not in the reverse direction as indicated in Fig. 8.4(a). A simple implementation of direction control consists in checking the condition

$$\Delta\mathbf{u}_0^T \Delta\mathbf{u}_1 < 0 \tag{8.9}$$

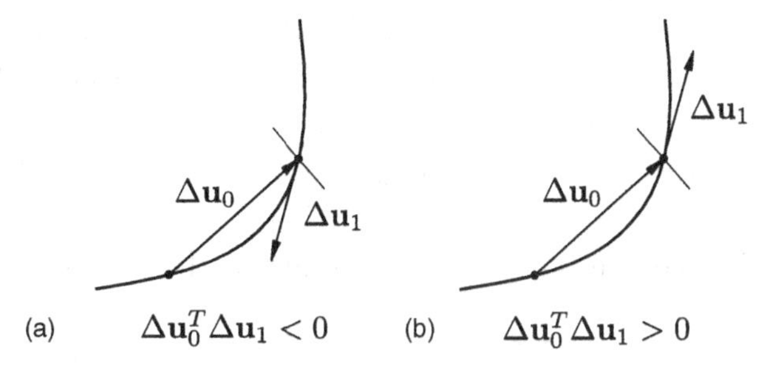

Fig. 8.4. (a) Doubling back, (b) continuation of equilibrium path.

on the projection of the previous on the current displacement increment. If this condition is satisfied, the sign is changed on both the current load and displacement increments,

$$\Delta \mathbf{u}_1 = -\Delta \mathbf{u}_1, \quad \Delta \mathbf{f}_1 = -\Delta \mathbf{f}_1. \tag{8.10}$$

Direct sign reversal as indicated here is only relevant in the first prediction step, as the following process is governed by the sub-increments.

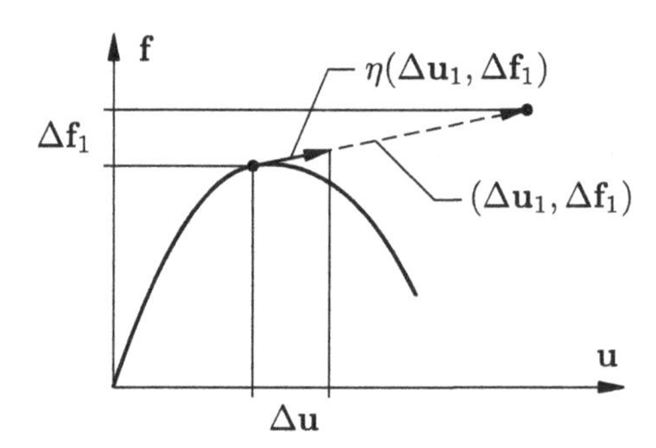

Fig. 8.5. Scaling the first displacement increment.

The efficiency of a solution method typically depends on the use of suitable initial load (or displacement) increments in each load step. The problem is illustrated in Fig. 8.5, where the initial load and displacement increments $(\Delta \mathbf{u}_1, \Delta \mathbf{f}_1)$ are scaled by the factor η. There is no general optimal predictor for the initial increment of a load step based on information from the previous steps as e.g. a bifurcation point may change the behavior dramatically within a single load step. However, for smooth equilibrium paths without bifurcation points several methods have been developed for selection of a suitable initial load or displacement increment, e.g. the 'effective stiffness' concept of Bergan (1980). A simple method that often works satisfactorily is based on the observation that more iterations are typically needed when encountering large changes of the equilibrium path. Thus, the number of iterations needed for convergence of the previous load step can serve as an indicator of the suitability of the step size (Crisfield, 1991). Typically the problem consists in decreasing stiffness leading to excessive displacement increments, suggesting a load step control based on the initial displacement increment. The most common form is a power function of the number of iterations in the previous step, corresponding to the relation

$$||\Delta \mathbf{u}_1|| = (i_d/i_0)^\alpha \, ||\Delta \mathbf{u}_0|| \tag{8.11}$$

where $||\Delta \mathbf{u}_0||$ is a suitable norm of the previous step, i_0 is the number of iterations in this previous load step, and i_d is the 'desired number of iterations'. A typical value of the exponent is $\alpha \simeq 0.5$. The number i_d controls the step size. It depends on the requested accuracy on the iteration process and may depend on the problem size. In the present simple form (8.11) controls the displacement part of the increments $(\Delta\mathbf{u}_1, \Delta\mathbf{f}_1)$. More elaborate formats involving displacement and load components are discussed in connection with arc-length algorithms in Section 8.3.

8.2 Orthogonal residual method

The basic idea in the residual force methods is that after having obtained a new displacement increment $\Delta\mathbf{u}_i$, the load increment is modified such as to obtain some desirable property of the residual force. This approach was originally proposed by Bergan (1980, 1981) in connection with a minimum criterion on the residual force. It turns out that it is difficult to define a suitable norm of the residual force to be minimized, independent of the problem under consideration. Also, the use of a minimum criterion may lead to locking of the iteration procedure. In the following an alternative formulation of the residual force procedure in terms of an orthogonality condition is presented. The original form of this procedure was proposed by Krenk (1993), but here the method is supplemented with displacement increment control, leading to improved robustness of the algorithm. An alternative modification has recently been presented by Kouhia (2008).

Let the equilibrium equation to be solved be of the form

$$\mathbf{g}(\mathbf{u}) = \mathbf{f}, \tag{8.12}$$

where $\mathbf{f}$ is the load vector and $\mathbf{u}$ the corresponding generalized displacement vector. The objective is to generate a sequence of equilibrium states $(\mathbf{u}_n, \mathbf{f}_n)$, $n = 1, 2, \ldots$ To simplify the notation only a single load step is considered. The load step starts at the last established equilibrium state which, for ease of notation, is here simply denoted $(\mathbf{u}_0, \mathbf{f}_0)$. A load increment $\Delta\mathbf{f}$ is then applied, and the corresponding displacement increment is determined from the stiffness relation

$$\mathbf{K}\,\Delta\mathbf{u} = \Delta\mathbf{f}, \tag{8.13}$$

where $\mathbf{K}$ is a suitable stiffness matrix. This leads to the predicted displacement $\mathbf{u}_0 + \Delta\mathbf{u}$ with the internal force vector $\mathbf{g}(\mathbf{u}_0 + \Delta\mathbf{u})$. Generally this internal force will not equal the external load $\mathbf{f}_0 + \Delta\mathbf{f}$, but generates a residual force $\mathbf{r}$. The load level is now adjusted by considering the scaled load increment $\xi\Delta\mathbf{f}$ instead of the original $\Delta\mathbf{f}$ as illustrated in Fig. 8.6(a). The residual force corresponding to the scaled load increment $\xi\Delta\mathbf{f}$ is

$$\mathbf{r} = \mathbf{f}_0 + \xi\Delta\mathbf{f} - \mathbf{g}(\mathbf{u}_0 + \Delta\mathbf{u}) = \xi\Delta\mathbf{f} - \Delta\mathbf{g}, \tag{8.14}$$

where $\Delta\mathbf{g}$ is the increment of the internal force from the last equilibrium state $\mathbf{u}_0$:

$$\Delta\mathbf{g} = \mathbf{g}(\mathbf{u}_0 + \Delta\mathbf{u}) - \mathbf{g}(\mathbf{u}_0). \tag{8.15}$$

In this formulation the parameter ξ acts as a load factor that can be adjusted to provide optimal algorithmic properties.

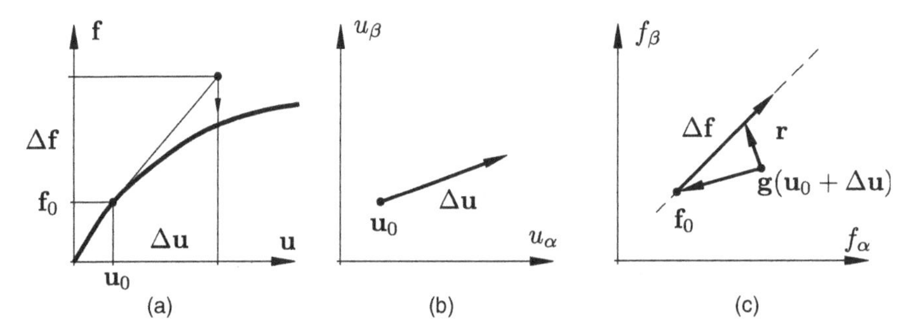

Fig. 8.6. Orthogonality between residual force $\mathbf{r}$ and displacement increment $\Delta\mathbf{u}$.

An optimal value of the scaling factor ξ is determined from the following argument. The residual force represents an unbalance that will lead to additional displacements. The residual force $\mathbf{r}$ will increase or decrease the magnitude of the current displacement increment $\Delta\mathbf{u}$ according to the sign of $\mathbf{r}^T\Delta\mathbf{u}$, i.e. the sign of the projection of the residual force on the current displacement increment. The magnitude of the current displacement increment is therefore optimal under the condition

$$\mathbf{r}^T\Delta\mathbf{u} = 0. \tag{8.16}$$

This condition is used to determine the scaling factor ξ. Substitution of the residual force (8.14) into the orthogonality condition (8.16) determines the optimal load scaling factor as

$$\xi = \frac{\Delta\mathbf{g}^T\Delta\mathbf{u}}{\Delta\mathbf{f}^T\Delta\mathbf{u}}. \tag{8.17}$$

The displacement increment $\Delta\mathbf{u}$ and the residual $\mathbf{r}$ are illustrated in Fig. 8.6(b,c), showing orthogonality between the displacement increment $\Delta\mathbf{u}$ and the residual $\mathbf{r}$ for the optimal value of ξ.

ALGORITHM 8.1. Orthogonal residual algorithm.

Initial state: $\mathbf{u}_0,\ \mathbf{f}_0,\ \Delta\mathbf{u} = \mathbf{0}$

Load increments $n = 1, 2, \ldots, n_{\max}$

$\Delta\mathbf{u}_1 = \mathbf{K}^{-1}\Delta\mathbf{f}$

if $\Delta\mathbf{u}_1^T\Delta\mathbf{u} < 0$ *then* $\Delta\mathbf{u}_1 = -\Delta\mathbf{u}_1,\quad \Delta\mathbf{f} = -\Delta\mathbf{f}$

$\Delta\mathbf{u} = \min(1, u_{\max}/\|\Delta\mathbf{u}_1\|)\,\Delta\mathbf{u}_1$

Iterations $i = 1, 2, \ldots, i_{\max}$

$\Delta\mathbf{g} = \mathbf{g}(\mathbf{u}_{n-1} + \Delta\mathbf{u}) - \mathbf{f}_{n-1}$

$\xi = \Delta\mathbf{g}^T\Delta\mathbf{u}/\Delta\mathbf{f}^T\Delta\mathbf{u}$

$\mathbf{r} = \xi\Delta\mathbf{f} - \Delta\mathbf{g}$

$\delta\mathbf{u} = \mathbf{K}^{-1}\mathbf{r}$

$\Delta\mathbf{u} = \Delta\mathbf{u} + \delta\mathbf{u}$

Stop iteration when $\|\mathbf{r}\| < \varepsilon_f\|\Delta\mathbf{f}\|$

$\mathbf{u}_n = \mathbf{u}_{n-1} + \Delta\mathbf{u}$

$\mathbf{f}_n = \mathbf{f}_{n-1} + \xi\Delta\mathbf{f}$

Stop load incrementation when $\|\mathbf{f}_n\| > f_{\max}$

When the optimal residual $\mathbf{r}$ has been obtained, a displacement sub-increment $\delta\mathbf{u}$ is determined from a stiffness relation

$$\mathbf{K}\,\delta\mathbf{u} = \mathbf{r} \tag{8.18}$$

as usual. Here the stiffness matrix $\mathbf{K}$ may correspond to the last equilibrium state $\mathbf{u}_0$, the current displacement state $\mathbf{u}_0 + \Delta\mathbf{u}$, or be an updated stiffness matrix of quasi-Newton type as discussed in Section 8.4.

The orthogonal residual algorithm is summarized in Algorithm 8.1. It starts at an equilibrium state $\mathbf{u}_0, \mathbf{f}_0$. From this state a sequence of load increments $\Delta\mathbf{f}_n$, $n = 1, 2, \ldots$ are applied. These load increments are modified by the procedure, and thus do not add up to the total increase of load. After a load increment is applied the corresponding displacement increment $\Delta\mathbf{u}_1$ is calculated from the linearized stiffness relation (8.13). The direction of the increment is checked against the result from the previous load increment. It is often desirable to prescribe the magnitude of the load increment, and the algorithm is shown with a simple scaling to a constant size of the first increment of all load steps $u_{\max} = ||\Delta\mathbf{u}||$. Alternatively, a suitable length of the first increment of a load step can be estimated from the previous load step, e.g. as indicated in (8.11).

The iteration loop starts with calculation of the internal force increment $\Delta\mathbf{g}$ since the last equilibrium. The internal force increment is then used to calculate the scaling of the load increment by the factor ξ, and the scaled load increment $\xi\Delta\mathbf{f}$ is used to determine the residual load. The residual load $\mathbf{r}$ determines the displacement sub-increment $\delta\mathbf{u}$ by the stiffness relation (8.18). In principle, the stiffness matrix $\mathbf{K}$ in this relation can be selected in several ways. However, if the load step includes a change of loading direction, use of the current stiffness may lead to very large displacement sub-increments, and experience suggests that a more robust formulation is obtained around load limit points by using the stiffness matrix from the start of the load increment. This corresponds closely to the modified Newton–Raphson procedure, but in the orthogonal residual algorithm the efficiency is increased by the sequential adjustment of the load increment during the iteration process.

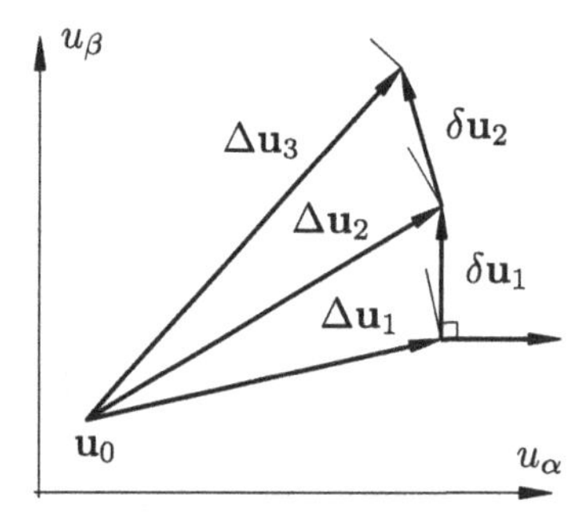

Fig. 8.7. Displacement increments in the orthogonal residual method.

Figure 8.7 shows a characteristic sequence of equilibrium iteration steps. In each step the load increment $\xi_i\Delta\mathbf{f}$ is adjusted such that the residual $\mathbf{r}$ is orthogonal to the *current* displacement increment $\Delta\mathbf{u}$. The orthogonal residuals are indicated by thin lines in Fig. 8.7. The residuals produce displacement sub-increments $\delta\mathbf{u}$ as shown in the figure. For a positive definite stiffness matrix the variables could be normalized corresponding to a unit stiffness matrix. In these normalized variables the displacement sub-increments would be identical to the residuals, and the iteration would trace a broken spiral in which orthogonality between increments $\Delta\mathbf{u}$ and sub-increments $\delta\mathbf{u}$ limits the size of the final displacement increment.

EXAMPLE 8.1. TWO-ELEMENT TRUSS. This example is concerned with the two-bar truss shown in Fig. 8.8 and analyzed analytically in Example 2.1.

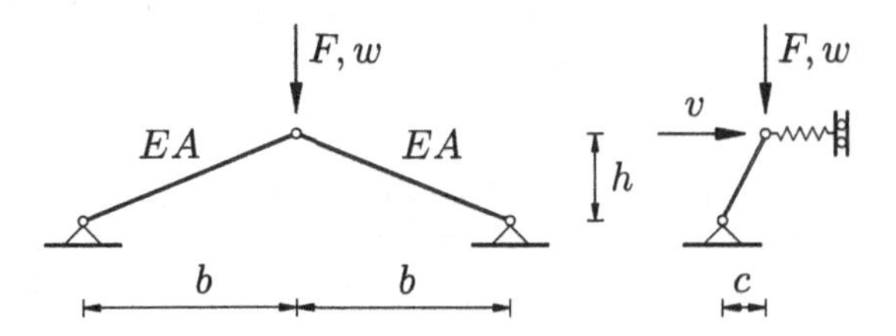

Fig. 8.8. Inclined two-bar truss.

It demonstrates the ability of the orthogonal residual algorithm to deal with limit points and sudden changes of direction of the equilibrium path in displacement space. The relative height of the truss is $h/b = 0.2$, the inclination is $c/h = 0.01$, and the relative stiffness of the lateral spring is $kb/EA = 0.02$. Figure 8.9 shows the equilibrium path for $\Delta F/EA = 0.001$. When approaching the maximum of the w–F curve the equilibrium path branches off in the transverse direction, with the middle node approximately following a circular path. It is seen on the w–v graph that the bifurcation point of the ideal structure is replaced by a smooth transition that depends on the inclination c/h. The step control, in which the magnitude of the first displacement

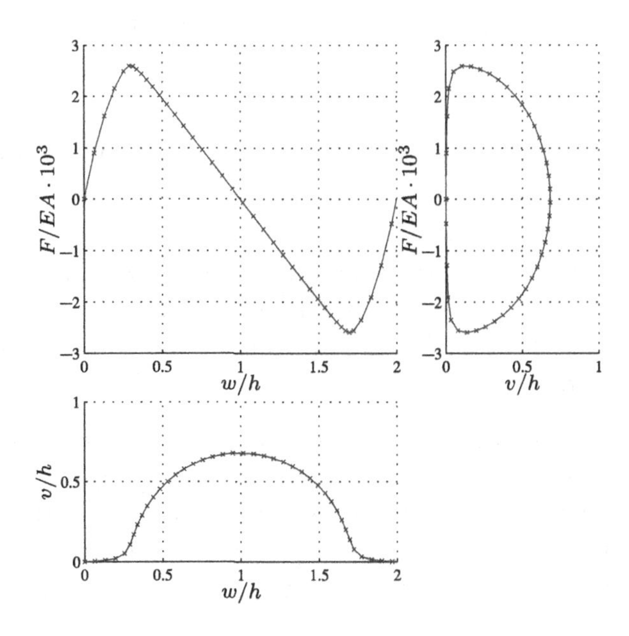

Fig. 8.9. Equilibrium path of inclined two-bar truss, $\Delta F/EA = 10^{-3}$.

increment is adjusted to the constant length u_{max}, leads to nearly equally spaced points on the displacement contour.

Table 8.1. *Two-element truss.*

$\Delta F/EA$	n_{total}	$\varepsilon_f = 10^{-4}$ i_{total}	$\varepsilon_f = 10^{-4}$ i_{max}	$\varepsilon_f = 10^{-6}$ i_{total}	$\varepsilon_f = 10^{-6}$ i_{max}
0.5×10^{-3}	80	260	6	414	10
1.0×10^{-3}	40	190	17	298	20
1.5×10^{-3}	27	164	20	258	20

Table 8.1 gives some solution statistics for three different load increments. n_{total} is the total number of load steps used to pass through the part of the equilibrium path shown in the figure. i_{total} is the total number of iterations, including the first displacement update in each load step, and i_{max} is the maximum number of equilibrium iterations needed for convergence. It is seen that while larger load steps lead to a larger number of iterations per load step, the total number of iterations decreases. However, further increase of the load step size will lead to a need for step adjustment around the load extremum points. The table contains results for relative error tolerance $\varepsilon_f = 10^{-4}$ and $\varepsilon_f = 10^{-6}$. It is seen that increasing the number of digits by 1.5 leads to an approximate increase of the total number of iterations by a factor of 1.5.

The present formulation of the orthogonal residual algorithm does not explicitly address the problem of bifurcation. Nonetheless, the present formulation may be used to analyze bifurcations, either by introducing small perturbations in the loads or geometry, or directly as illustrated in the following example. Special methods for identification and passage of limit points have been developed, e.g. by Kouhia and Mikkola (1999a,b) and Magnusson and Svensson (1998).

EXAMPLE 8.2. BIFURCATIONS OF SHALLOW DOME. In this example the shallow truss dome shown in Fig. 8.10 is analyzed using the orthogonal residual algorithm. The dimensions of the dome used in this example are given by $h_2/h_1 = 1.3$, $b_1/h_1 = 4.0$ and $b_2 = 8.0$. All the bars are assumed to be linear elastic with stiffness EA. A symmetric downward load pattern is defined by $f_1 = F$ and $f_2 = f_3 = f_4 = f_5 = f_6 = f_7 = 2F$. The deformation pattern is dominated by various types of bifurcations. The

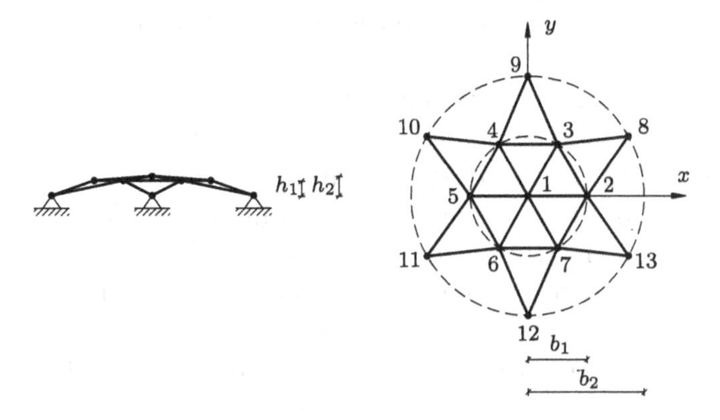

Fig. 8.10. Shallow truss dome.

equilibrium paths are shown in Fig. 8.11. The ideal structure, denoted by I, reaches a maximum load at $F/EA = 0.00093$, and then snaps through in a fully symmetric mode. The vertical and radial displacements of nodes 2–7 are indicated by a dashed curve in Fig. 8.11. Before reaching this load two bifurcation points are passed: a bifurcation with a plane of symmetry denoted by S, and a bifurcation with 3-fold symmetry, denoted by T. The

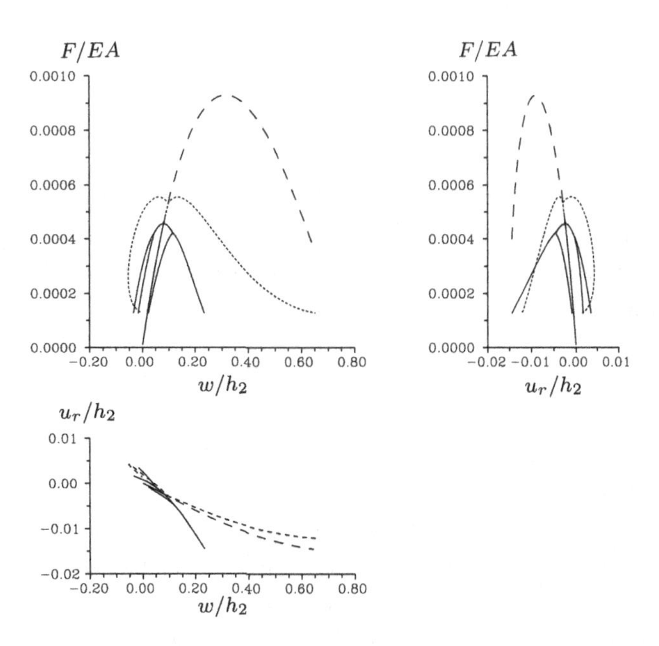

Fig. 8.11. Equilibrium paths of shallow truss dome.

3-fold symmetric buckling mode becomes unstable and bifurcates further as indicated in the figure.

The algorithm is not specifically developed for treating bifurcation problems, and must therefore be guided through the various equilibrium paths shown in Fig. 8.11. In the ideal structure with identical stiffness of all bars the coordinates of the nodes are given with finite precision – in the present example 5 digits. This gives symmetry with respect to the coordinate planes, and thus the model of the ideal truss buckles in the symmetric mode denoted S. The ideal curve I was produced by introducing an extra direction check of the form $\Delta\mathbf{u}_0^T\Delta\mathbf{u}_{i+1} > 0$ at the end of each equilibrium iteration. In practice the main concern is that no bifurcation point is passed over by the algorithm. The problem arises because the idealized structure possesses a perfect symmetry not shared by the buckling mode. In the present example this problem is overcome by introducing a random perturbation in the stiffnesses of the bars of relative magnitude 10^{-3}. This produces the 3-fold symmetric bifurcation and the following further bifurcations shown as T in Fig. 8.11.

Table 8.2. *Shallow truss dome under symmetric load.*

	$\Delta p/EA$	n_{total}	i_{total}	restarts
I	4×10^{-5}	61	–	0
S	2×10^{-5}	122	221	0
T	1×10^{-5}	104	269	9

A summary of the numerical details of the computation is given in Table 8.2. With the selected load increments only few iterations are needed in each load step, and only the double bifurcations of T require restart with locally smaller load increment.

8.3 Arc-length methods

Arc-length methods are developed from the idea that the 'length' of the combined displacement–load increment $(\Delta\mathbf{u}, \Delta\mathbf{f})$ should be controlled during equilibrium iterations (Riks, 1979). There are several variations on this basic theme. A presentation of arc-length methods including many references has been given by Crisfield (1991). The basic idea is illustrated in Fig. 8.12. The most recent equilibrium state is denoted $(\mathbf{u}_0, \mathbf{f}_0)$. After the initial load and displacement increment the equilibrium iterations are restricted to a hypersphere in the combined displacement–load space $(\mathbf{u}, \mathbf{f})$. The next equilibrium state is marked at the intersection of the equilibrium path with the hypersphere.

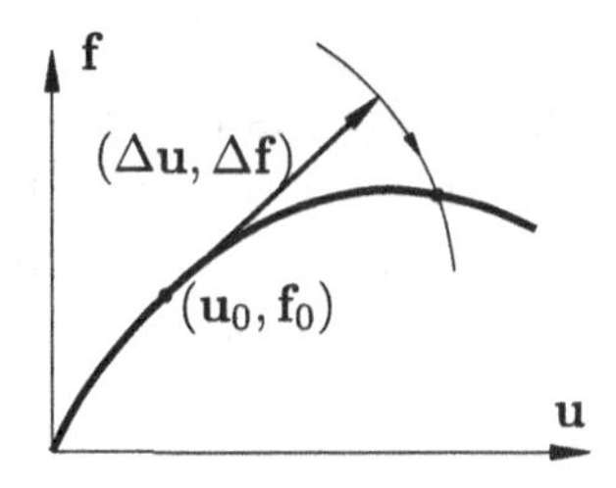

Fig. 8.12. Arc constraint on total increment.

In the first step the displacement increment $\Delta \mathbf{u}_1$ is determined from a load increment $\Delta \mathbf{f}_1$ by the linearized stiffness relation

$$\mathbf{K}\, \Delta \mathbf{u}_1 = \Delta \mathbf{f}_1. \tag{8.19}$$

This is followed by simultaneous combined displacement and load sub-increments $(\delta \mathbf{u}_i, \delta \mathbf{f}_i)$, after which the increments are updated as

$$\Delta \mathbf{u}_i = \Delta \mathbf{u}_{i-1} + \delta \mathbf{u}_i, \qquad \Delta \mathbf{f}_i = \Delta \mathbf{f}_{i-1} + \delta \mathbf{f}_i. \tag{8.20}$$

The key point is the strategy for selecting the sub-increments. Crisfield (1981) introduced a constraint in the form of a hypersphere, while Forde and Stiemer (1987) demonstrated how several constraint conditions could be constructed by starting from an orthogonality condition in the combined displacement–load space. Schweizerhof and Wriggers (1986) presented a Newton procedure for a general constraint surface. In the following a general iterative procedure, sometimes called the bordering algorithm, is described. The idea is to carry out simultaneous iterations for equilibrium conditions and the constraint condition. Alternatively, the so-called arc-length methods use iterations for equilibrium, while satisfying the constraint condition explicitly during each step of the iteration process. In practice this requires constraints that permit explicit solution for the load increment parameter, and therefore arc-length methods mainly use a constraint condition consisting of one or more hyperplanes, or a constraint condition in the form of a hypersphere. Arc-length methods are currently considered to be the most robust general purpose methods for solution of non-linear finite element equations in solid and structural mechanics, see e.g. the comparative studies by Forde and Stiemer (1987) and Clarke and Hancock (1990).

8.3.1 General constraint formulation

Let the equilibrium of the discrete system be described by a non-linear vector equation of the form

$$\mathbf{g}(\mathbf{u}) = \mathbf{f}, \tag{8.21}$$

where $\mathbf{f}$ is the external load and $\mathbf{g}(\mathbf{u})$ is the internal force, given as a function of the displacement vector $\mathbf{u}$. Let $\mathbf{u}_0, \mathbf{f}_0$ denote the last established equilibrium state. A load increment $\Delta\mathbf{f}$ is imposed, and a corresponding displacement increment $\Delta\mathbf{u}$ is determined from a stiffness relation of the form

$$\mathbf{K}\,\Delta\mathbf{u} = \Delta\mathbf{f}, \tag{8.22}$$

where $\mathbf{K}$ typically is the current tangent stiffness matrix. In the subsequent iterative procedure the load magnitude is adjusted via a scalar parameter ξ while the relative magnitude of the individual load increment components is retained. Thus, the general form of the displacement and load vectors is

$$\mathbf{u} = \mathbf{u}_0 + \Delta\mathbf{u}, \qquad \mathbf{f} = \mathbf{f}_0 + \xi\Delta\mathbf{f}. \tag{8.23}$$

In a non-linear problem the initial prediction of $\Delta\mathbf{u}$ by the linear equation (8.22) will not lead to a state of equilibrium, and the solution is therefore adjusted by displacement sub-increments $\delta\mathbf{u}$ and corresponding increments $\delta\xi$ of the load factor ξ.

An iterative procedure is then constructed from the residual force

$$\mathbf{r} = \mathbf{f} - \mathbf{g}(\mathbf{u}) = \xi\Delta\mathbf{f} - \Delta\mathbf{g} \tag{8.24}$$

and a constraint equation

$$c(\Delta\mathbf{u}, \xi\Delta\mathbf{f}) = 0. \tag{8.25}$$

The constraint equation connects the current displacement $\Delta\mathbf{u}$ to the current load increment $\xi\Delta\mathbf{f}$ as illustrated in Fig. 8.12. A Newton–Raphson iteration procedure for the solution to combined equilibrium and constraint equations proceeds via the corresponding linearized equations

$$\mathbf{r} + \delta\mathbf{r} = \mathbf{0}, \qquad c + \delta c = 0. \tag{8.26}$$

where $\mathbf{r}$ and c are the current values, while $\delta\mathbf{r}$ and δc are the first-order increments. The independent variables of the problem are the displacement vector $\mathbf{u}$ and the load factor ξ, and thus the linearized equations take the form

$$\begin{aligned} -\frac{\partial \mathbf{r}}{\partial \mathbf{u}}\delta\mathbf{u} - \frac{\partial \mathbf{r}}{\partial \xi}\delta\xi &= \mathbf{r}, \\ -\frac{\partial c}{\partial \mathbf{u}}\delta\mathbf{u} - \frac{\partial c}{\partial \xi}\delta\xi &= c. \end{aligned} \tag{8.27}$$

The partial derivatives of the residual $\mathbf{r}$ follow from (8.24) as

$$\partial\mathbf{r}/\partial\mathbf{u} = -\mathbf{K}, \qquad \partial\mathbf{r}/\partial\xi = \Delta\mathbf{f}, \tag{8.28}$$

where $\mathbf{K}$ is the current stiffness matrix and $\Delta\mathbf{f}$ the imposed load increment. The following notation is introduced for the partial derivatives of the constraint condition:

$$\partial c/\partial\mathbf{u} = \mathbf{c}_u^T, \qquad \partial c/\partial\xi = c_\xi. \tag{8.29}$$

With this notation the linearized equations (8.27) for the iterative increments take the form

$$\begin{bmatrix} \mathbf{K} & -\Delta\mathbf{f} \\ -\mathbf{c}_u^T & -c_\xi \end{bmatrix} \begin{bmatrix} \delta\mathbf{u} \\ \delta\xi \end{bmatrix} = \begin{bmatrix} \mathbf{r} \\ c \end{bmatrix}. \tag{8.30}$$

In the present context it is convenient to solve these equations using the block format. A formal solution of the first equation is

$$\delta\mathbf{u} = \mathbf{K}^{-1}\mathbf{r} + \delta\xi\,\mathbf{K}^{-1}\Delta\mathbf{f}. \tag{8.31}$$

This solution illustrates that the displacement sub-increment consists of two vector contributions, conveniently denoted by

$$\delta\mathbf{u}_r = \mathbf{K}^{-1}\mathbf{r}, \qquad \Delta\mathbf{u}_f = \mathbf{K}^{-1}\Delta\mathbf{f}, \tag{8.32}$$

whereby

$$\delta\mathbf{u} = \delta\mathbf{u}_r + \delta\xi\,\Delta\mathbf{u}_f. \tag{8.33}$$

The first term is the displacement sub-increment $\delta\mathbf{u}_r$, generated by the residual force $\mathbf{r}$ and corresponding to that used in the Newton–Raphson method. The second term represents the displacement increment following from the adjustment of the load increment. Substitution of this expression for the displacement sub-increment into the second of the equations (8.30) gives the load factor increment

$$\delta\xi = -\frac{\mathbf{c}_u^T\delta\mathbf{u}_r + c}{\mathbf{c}_u^T\Delta\mathbf{u}_f + c_\xi}. \tag{8.34}$$

The displacement sub-increment $\delta\mathbf{u}$ finally follows from substitution of this expression into (8.33).

For a non-linear constraint condition the process is a simultaneous iteration towards satisfaction of equilibrium and the constraint. There are three important exceptions: linear constraint, piecewise linear constraint, and hypersphere constraint. The linear constraints are special cases of the general procedure, while the hypersphere constraint makes use of a quadratic equation for the load factor increment $\delta\xi$ to bring the combined displacement load vector back on the constraining hypersphere. These special forms are the most used in connection for solid and structural mechanics problems and are therefore discussed in some detail in the following.

8.3.2 Hyperplane constraints

The linear constraint is illustrated in Fig. 8.13(a). In the figure the last established state of equilibrium is shown as $(\mathbf{u}_0, \mathbf{f}_0)$, and the current combined displacement and load increment is shown as the 'vector' $(\Delta\mathbf{u}, \Delta\mathbf{f})$. The sub-increment $(\delta\mathbf{u}, \delta\mathbf{f})$ is constructed from the displacement $(\delta\mathbf{u}_r, \mathbf{0})$ from the residual force and the contribution $\delta\xi(\Delta\mathbf{u}, \Delta\mathbf{f})$ from the load adjustment. When passing a load limit point, i.e. around a load maximum or minimum, the current load increment $\Delta\mathbf{f}$ may vanish, whereby all three vector components become 'horizontal'. This may cause convergence problems. These can be avoided by using the load adjustment term corresponding to the first step of the iteration procedure, indicated as $\delta\xi(\Delta\mathbf{u}_1, \Delta\mathbf{f}_1)$. This corresponds to selecting the displacement sub-increment in the form

$$\delta\mathbf{u} = \delta\mathbf{u}_r + \delta\xi\,\Delta\mathbf{u}_1, \qquad (8.35)$$

where $\Delta\mathbf{u}_1$ follows from the first load increment $\Delta\mathbf{f}_1$ by (8.19). This corresponds to the return direction originally introduced by Riks (1979).

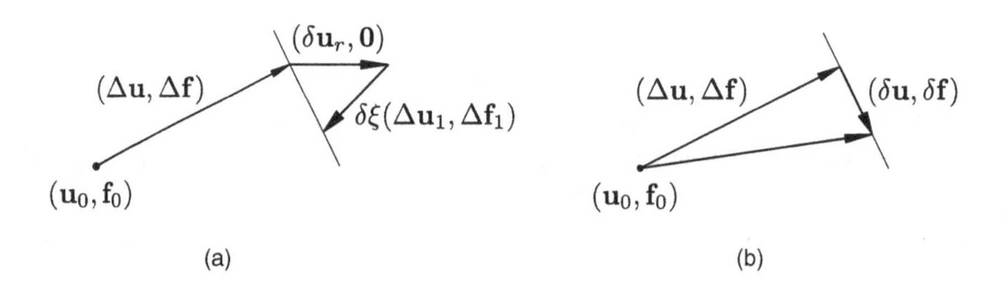

Fig. 8.13. (a) Residual and return components, (b) increment and sub-increment.

The linear constraint is a hyperplane orthogonal to the combined displacement load increment. A fixed hyperplane is obtained by using a hyperplane with normal vector $(\Delta\mathbf{u}_1, \Delta\mathbf{f}_1)$. However, as no additional computational cost is implied by using a hyperplane orthogonal to the current combined displacement load step $(\Delta\mathbf{u}, \Delta\mathbf{f})$, this is illustrated in Fig. 8.13(b) and used in the implementation of Algorithm 8.2. The condition that the sub-increment $(\delta\mathbf{u}, \delta\mathbf{f})$ lies in a hyperplane orthogonal to the current total increment $(\Delta\mathbf{u}, \Delta\mathbf{f})$ is expressed by the condition

$$(\Delta\mathbf{u}, \Delta\mathbf{f}) \cdot (\delta\mathbf{u}, \delta\mathbf{f}) = 0, \tag{8.36}$$

where the dot symbol represents a suitable scalar product in the combined displacement–load space. A proper definition of a scalar product depends on introduction of a suitable metric in the combined displacement–load space. The role of the metric is to ensure that the individual components get appropriate weights in the formation of the scalar product, and thereby that the scalar product is suitable for expressing distance in the combined displacement–load space. The central problem is that the load and displacement increments represent different physical quantities and therefore have different units and scaling. Thus, a scalar product containing load components as well as displacement components should contain an appropriate scaling of load increments relative to displacement increments. In addition to this the displacement vector components may represent different behavior, e.g. translations and rotations, and similarly for the individual load components.

The problem of a suitable metric is often resolved in a pragmatic way by accounting for the different units involved in the displacement vector and in the load vector, but neglecting possible differences in the individual components in each of these vectors. This leads to a scalar product of the form

$$(\Delta\mathbf{u}, \Delta\mathbf{f}) \cdot (\delta\mathbf{u}, \delta\mathbf{f}) = \Delta\mathbf{u}^T \delta\mathbf{u} + \beta^2 \Delta\mathbf{f}^T \delta\mathbf{f}. \tag{8.37}$$

Here β is a flexibility parameter, representing the ratio of a displacement to a corresponding load component. There are several papers discussing various schemes for scaling the terms in (8.37), e.g. Schweizerhof and Wriggers (1986) and Al-Rasby (1991). However, no single method of scaling has yet demonstrated general superiority, and it is interesting to recall the conclusion reached by Crisfield (1981), that *numerical experience has shown that it is preferable to fix the 'incremental length' in n-dimensional space*, i.e. in displacement space. This corresponds to $\beta = 0$, whereby the geometric considerations deal with a projection into the displacement space indicated in Fig. 8.3(b).

The load factor increment $\delta\xi$ is determined from the orthogonality condition (8.36), using the definition (8.37) of the scalar product.

$$\delta\xi = -\frac{\Delta\mathbf{u}^T\delta\mathbf{u}_r}{\Delta\mathbf{u}^T\Delta\mathbf{u}_1 + \beta^2\,\Delta\mathbf{f}^T\Delta\mathbf{f}_1}. \tag{8.38}$$

This formula corresponds to a piecewise linear constraint with return to the constraint along the first increment of the load step $(\Delta\mathbf{u}_1, \Delta\mathbf{f}_1)$. The corresponding formula for the case with return along the current displacement–load vector $(\Delta\mathbf{u}, \Delta\mathbf{f})$ follows by removing the subscript 1 in the denominator of (8.38) and in the sub-increment relation (8.35). Conversely, the relation corresponding to using a fixed hyperplane for the iteration process is obtained by introducing the subscript 1 on the increments shown without subscript.

The linear constraint is a special case of the general constraint condition, and thus the expression (8.38) for the load factor increment $\delta\xi$ is a special case of the general expression (8.34). In the case of a linear constraint the combined displacement–load state is returned to the constraint surface in each iteration step, and therefore $c = 0$. The gradient vector $\mathbf{c}_u = \partial c/\partial\mathbf{u}$ corresponds to the displacement part of the normal to the constraint hyperplane, and thus $\mathbf{c}_u = \Delta\mathbf{u}$ as also seen from the formula (8.38). The last coefficient c_ξ represents the dependence of the constraint on the load level. It is here represented by $\beta^2\Delta\mathbf{f}^T\Delta\mathbf{f}_1$, where the factor β^2 determines the weight of the load contribution. The parameter β may be chosen from two different considerations. One argument is that a certain uniformity in the full displacement–load space is desired, and therefore β should represent a typical flexibility associated with a representative degree of freedom. Alternatively, it has been found that the geometric argument involved in the orthogonality condition can be restricted to apply only to the projection in the displacement part of the space, illustrated by Fig. 8.3(b). Geometrically this corresponds to using 'vertical' hyperplanes in the combined **u**–**f** space. This choice corresponds to selecting $\beta = 0$, whereby the load components are removed from the formula for the load factor increment. In this case the load factor increment is calculated as

$$\delta\xi = -\frac{\Delta\mathbf{u}^T\delta\mathbf{u}_r}{\Delta\mathbf{u}^T\Delta\mathbf{u}_1}. \tag{8.39}$$

This formula bears a considerable similarity to the load factor formula (8.17) of the orthogonal residual method. In fact the main difference appears to be the absence of a stiffness matrix in the scalar products of the present formula for the arc-length method. While this may introduce a certain measure of arbitrariness, when combining displacement components of different di-

ALGORITHM 8.2. Orthogonal arc-length algorithm.

Initial state: $\mathbf{u}_0, \mathbf{f}_0, \Delta\mathbf{u} = \mathbf{0}$

Load increments $n = 1, 2, \ldots, n_{\max}$

$\quad \Delta\mathbf{u}_1 \;=\; \mathbf{K}^{-1}\Delta\mathbf{f}$

$\quad$ *if* $\Delta\mathbf{u}_1^T\Delta\mathbf{u} < 0$ *then* $\Delta\mathbf{u}_1 = -\Delta\mathbf{u}_1, \quad \Delta\mathbf{f} = -\Delta\mathbf{f}$

$\quad \xi \;=\; \ell/\|\Delta\mathbf{u}_1\|$

$\quad \Delta\mathbf{u} \;=\; \xi\,\Delta\mathbf{u}_1$

$\quad$ Iterations $i = 1, 2, \ldots, i_{\max}$

$\qquad \Delta\mathbf{g} \;=\; \mathbf{g}(\mathbf{u}_{n-1} + \Delta\mathbf{u}) \;-\; \mathbf{f}_{n-1}$

$\qquad \mathbf{r} \;=\; \xi\Delta\mathbf{f} \;-\; \Delta\mathbf{g}$

$\qquad \delta\mathbf{u}_r \;=\; \mathbf{K}_n^{-1}\mathbf{r}$

$\qquad \delta\xi \;=\; -\Delta\mathbf{u}^T\delta\mathbf{u}_r/\Delta\mathbf{u}^T\Delta\mathbf{u}_1$

$\qquad \delta\mathbf{u} \;=\; \delta\mathbf{u}_r \;+\; \delta\xi\,\Delta\mathbf{u}_1$

$\qquad \Delta\mathbf{u} \;=\; \Delta\mathbf{u} \;+\; \delta\mathbf{u}$

$\qquad \xi \;=\; \xi \;+\; \delta\xi$

$\quad$ Stop iteration when $\|\mathbf{r}\| \;<\; \varepsilon_f\,\|\Delta\mathbf{f}\|$

$\quad \mathbf{u}_n \;=\; \mathbf{u}_{n-1} \;+\; \Delta\mathbf{u}$

$\quad \mathbf{f}_n \;=\; \mathbf{f}_{n-1} \;+\; \xi\Delta\mathbf{f}$

Stop load incrementation when $\|\mathbf{f}_n\| \;>\; f_{\max}$

---

mension, the arc-length method obtains considerable numerical robustness by avoiding use of the current stiffness matrix that becomes non-positive definite around load limit and bifurcation points.

The arc-length algorithm for the case of piecewise linear constraint is summarized as Algorithm 8.2. The structure is very similar to that of the orthogonal residual algorithm. The algorithm starts from a state of equilibrium $\mathbf{u}_0, \mathbf{f}_0$ and proceeds by application of load increments $\Delta\mathbf{f}_n$, $n = 1, 2, \ldots$ The applied load increments are modified by the algorithm, and thus they do not add up to the total load. Each load step starts with calculation of the first displacement increment $\Delta\mathbf{u}_1$. The initial increments $\Delta\mathbf{f}_n$ and $\Delta\mathbf{u}_1$ are retained as reference, and a scaled increment $\Delta\mathbf{u} = \xi\Delta\mathbf{u}_1$ is used to define the actual first displacement step. The residual force corresponding to this step is calculated, and the load factor increment is determined according to the formula (8.39). In this formula and the following formula for $\delta\mathbf{u}$ the original unscaled form $\Delta\mathbf{u}_1$ is used to simplify the algebra. The algorithm can be changed to use of return to the constraint by the current displacement–load increment simply by omitting the subscript 1 on $\Delta\mathbf{u}_1$ in these two expressions. This corresponds to the method proposed by Ramm (1981). The algorithm is shown with the residual displacement $\delta\mathbf{u}_r$ determined by the first stiffness matrix in the load step $\mathbf{K}_n$ in all iterations. This has been found to increase the robustness of the algorithm and avoids repeated factorization of the stiffness matrix within the iteration loop, similar to the modified Newton–Raphson method.

8.3.3 Hypersphere constraint

In the early formulation of the arc-length method by Crisfield (1981) a quadratic constraint condition was used, and this is still much in use. The idea is that a spherical surface corresponds to a constant distance, and thereby to a constant length of the combined displacement–load increment vector. The corresponding hypersphere is a quadratic function of the vector components, and thus return to the hypersphere will define the load factor ξ or its increment $\delta\xi$ via a quadratic equation. The exact return to the hypersphere constraint in each iteration step makes this algorithm different from the bordering algorithm described in Section 8.3.1, in which a non-linear constraint was only approached during the general iteration process. It is therefore more direct to describe the hypersphere arc-length algorithm as an extension of the linear constraint formulation of the previous section.

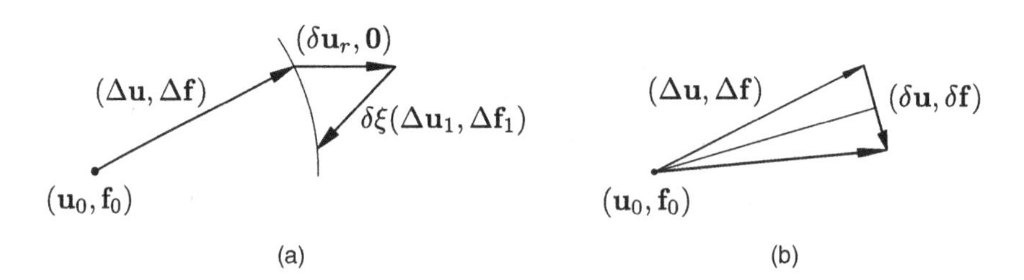

Fig. 8.14. (a) Residual and return components, (b) increment and sub-increment.

The quadratic constraint condition is illustrated in Fig. 8.14(a). The last established state of equilibrium is shown as $(\mathbf{u}_0, \mathbf{f}_0)$, and the current combined displacement and load increment is shown as $(\Delta\mathbf{u}, \Delta\mathbf{f})$. The sub-increment $(\delta\mathbf{u}, \delta\mathbf{f})$ is constructed from the displacement $(\delta\mathbf{u}_r, \mathbf{0})$ from the residual force and the contribution $\delta\xi(\Delta\mathbf{u}, \Delta\mathbf{f})$ from the load adjustment as in the case of linear constraints. Also in this case the return will be shown along the direction of the first increment in the load step as $\delta\xi(\Delta\mathbf{u}_1, \Delta\mathbf{f}_1)$.

This corresponds to selecting the displacement sub-increment in the two-component form (8.35). The condition that the current iteration point lies on a hypersphere corresponds to orthogonality of the sub-increment $(\delta\mathbf{u}, \delta\mathbf{f})$ with the mean value of the current and predicted form of the total increment $(\delta\mathbf{u} + \frac{1}{2}\delta\mathbf{u}, \delta\mathbf{f} + \frac{1}{2}\delta\mathbf{f})$ as illustrated in Fig. 8.14(b). This is expressed by the orthogonality condition

$$(\Delta\mathbf{u} + \tfrac{1}{2}\delta\mathbf{u}, \Delta\mathbf{f} + \tfrac{1}{2}\delta\mathbf{f}) \cdot (\delta\mathbf{u}, \delta\mathbf{f}) = 0. \tag{8.40}$$

When using the definition of the norm in which the load components are scaled by a common flexibility parameter β similar to (8.37), this gives the relation

$$(\Delta\mathbf{u} + \tfrac{1}{2}\delta\mathbf{u})^T\delta\mathbf{u} + \beta^2(\Delta\mathbf{f} + \tfrac{1}{2}\delta\mathbf{f})^T\delta\mathbf{f} = 0. \tag{8.41}$$

The difference from the linear constraint condition is the presence of the terms $\frac{1}{2}\delta\mathbf{u}$ and $\frac{1}{2}\delta\mathbf{f}$ in the quadratic constraint condition. When introducing the displacement sub-increment $\delta\mathbf{u}$ from (8.35) and the load sub-increment $\delta\mathbf{f} = \delta\xi\Delta\mathbf{f}_1$, the constraint equation becomes

$$\begin{aligned}(\Delta\mathbf{u}_1^T\Delta\mathbf{u}_1 + \beta^2\Delta\mathbf{f}_1^T\Delta\mathbf{f}_1)\delta\xi^2 \\ + 2(\Delta\mathbf{u}^T\Delta\mathbf{u}_1 + \Delta\mathbf{u}^T\delta\mathbf{u}_r + \beta^2\Delta\mathbf{f}^T\Delta\mathbf{f}_1)\delta\xi \\ + 2\Delta\mathbf{u}^T\delta\mathbf{u}_r + \delta\mathbf{u}_r^T\delta\mathbf{u}_r = 0.\end{aligned} \tag{8.42}$$

This is a quadratic equation for the load factor increment $\delta\xi$ of the form

$$A\,\delta\xi^2 + 2B\,\delta\xi + C = 0 \tag{8.43}$$

with coefficients

$$\begin{aligned} A &= \Delta\mathbf{u}_1^T\Delta\mathbf{u}_1 + \beta^2\Delta\mathbf{f}_1^T\Delta\mathbf{f}_1, \\ B &= \Delta\mathbf{u}^T\Delta\mathbf{u}_1 + \Delta\mathbf{u}^T\delta\mathbf{u}_r + \beta^2\Delta\mathbf{f}^T\Delta\mathbf{f}_1, \\ C &= 2\Delta\mathbf{u}^T\delta\mathbf{u}_r + \delta\mathbf{u}_r^T\delta\mathbf{u}_r. \end{aligned} \tag{8.44}$$

The solution to this quadratic equation is given by

$$\delta\xi = \Big(-B \pm \sqrt{B^2 - AC}\,\Big)/A. \tag{8.45}$$

Solutions corresponding to points on the hypersphere require that $B \geq \sqrt{AC}$. As seen from Fig. 8.14(a), this condition may be violated if the displacement increment $\delta\mathbf{u}_r$ generated by the residual force is too large. The typical solution to this problem is to decrease the initial load increment $\Delta\mathbf{f}_1$ of the load step. When a tangent stiffness is used the residual is of higher order in the load increment size, thus leading to intersection with the hypersphere for a sufficiently small load increment. For $\beta = 0$ this is the condition introduced by Crisfield (1981, 1991).

When using an appropriate load increment there will be two solutions to the equation (8.43). The root can be chosen in analogy with the argument illustrated in Fig. 8.4. The initial and final increment vectors $(\Delta\mathbf{u}, \Delta\mathbf{f})$ and $(\Delta\mathbf{u} + \delta\mathbf{u}, \Delta\mathbf{f} + \delta\mathbf{f})$ should then have a positive scalar product:

$$(\Delta\mathbf{u}, \Delta\mathbf{f}) \cdot (\Delta\mathbf{u} + \delta\mathbf{u}, \Delta\mathbf{f} + \delta\mathbf{f}) > 0. \tag{8.46}$$

Accordingly the appropriate root $\delta\xi$ is chosen to maximize this scalar product. The method implicitly assumes that the current increment vector $(\Delta\mathbf{u}, \Delta\mathbf{f})$ has a positive scalar product with the increment vector used for return to the constraint condition. Substitution of the appropriate expressions into (8.46) then shows that the appropriate choice is $\max(\delta\xi)$. When $\delta\xi$ is negative this corresponds to selecting the smallest absolute value.

In the arc-length method the iterative update of the current increment vector $(\Delta\mathbf{u}, \Delta\mathbf{f})$ is obtained by using the residual sub-increment $(\delta\mathbf{u}_r, \mathbf{0})$ and the direction of return, here represented by the first increment of the load step $(\Delta\mathbf{u}_1, \Delta\mathbf{f}_1)$. If the current increment is used as return direction, these three vectors become co-planar, and the hypersphere condition can be obtained as a fairly simple modification of the hyperplane constraint result. When using the current increments for the return direction the hyperplane condition (8.38) determines the load factor increment as

$$\delta\xi_p = -\frac{\Delta\mathbf{u}^T\delta\mathbf{u}_r}{\Delta\mathbf{u}^T\Delta\mathbf{u} + \beta^2\,\Delta\mathbf{f}^T\Delta\mathbf{f}}. \tag{8.47}$$

Similarly the subscript 1 is omitted in the quadratic equation (8.42). This equation is divided by $(\Delta\mathbf{u}^T\Delta\mathbf{u} + \beta^2\Delta\mathbf{f}^T\Delta\mathbf{f})$, and the product $\Delta\mathbf{u}^T\delta\mathbf{u}_r$ is expressed in terms of the hyperplane load factor increment $\delta\xi_p$ by use of (8.47). The result is a normalized quadratic equation for the load factor increment $\delta\xi_s$ corresponding to the hypersphere in the form

$$\delta\xi_s^2 + 2(1 - \delta\xi_p)\delta\xi_s - 2\delta\xi_p + \frac{\delta\mathbf{u}_r^T\delta\mathbf{u}_r}{\Delta\mathbf{u}^T\Delta\mathbf{u} + \beta^2\Delta\mathbf{f}^T\Delta\mathbf{f}} = 0. \tag{8.48}$$

The load factor increment $\delta\xi_s$ for the hypersphere is now expressed in terms of the similar load factor increment $\delta\xi_p$ for the hyperplane as

$$\delta\xi_s = \delta\xi_p + \delta\xi_*. \tag{8.49}$$

Substitution of this expression for $\delta\xi_s$ into (8.48) gives the following equation for $\delta\xi_*$:

$$(1 + \delta\xi_*)^2 - (1 + \delta\xi_p^2) + \frac{\delta\mathbf{u}_r^T\delta\mathbf{u}_r}{\Delta\mathbf{u}^T\Delta\mathbf{u} + \beta^2\Delta\mathbf{f}^T\Delta\mathbf{f}} = 0. \tag{8.50}$$

Solution of this equation leads to

$$\delta\xi_* = -1 + \sqrt{1 + \delta\xi_p^2 - \frac{\delta\mathbf{u}_r^T\delta\mathbf{u}_r}{\Delta\mathbf{u}^T\Delta\mathbf{u} + \beta^2\Delta\mathbf{f}^T\Delta\mathbf{f}}}. \tag{8.51}$$

An iterative step with this method starts with calculating the hyperplane load factor increment $\delta\xi_p$ from (8.47), and then proceeds to calculation of the additional load factor increment $\delta\xi_*$ from (8.51). Return to the hypersphere is obtained by using the sum according to (8.49). This procedure is particularly convenient for obtaining a common implementation of the two arc-length conditions in a single routine. As in the previous formulae the special case $\beta = 0$ reduces the geometric considerations to the projection on the displacement space. A different correction to the hyperplane condition has been proposed by Forde and Stiemer (1987), who calculated the additional length $\delta\xi_*$ via the radial distance from the center of the hypersphere to the point on the hyperplane.

EXAMPLE 8.3. 12-BAR TRUSS. In this example the full equilibrium path of the 12-bar space truss structure shown in Fig. 8.15 is calculated by the hyperplane arc-length method as shown in Algorithm 8.2. This structure has been used to illustrate large strain implementation by Yang and Leu (1991) and numerical solution algorithms by Krenk and Hededal (1995).

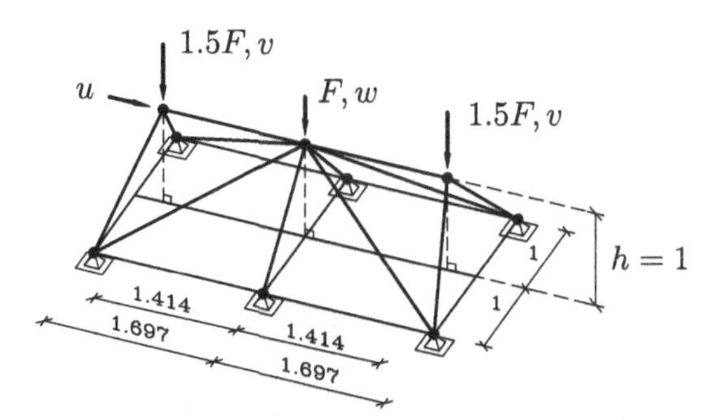

Fig. 8.15. Space truss with 12 bars.

The truss structure consists of 12 bars with identical stiffness EA. The deformation of the structure is characterized by the three displacement components u, v, w shown in the figure. The center top node is loaded by a vertical force of magnitude F, while the two neighboring top nodes carry a vertical force of magnitude $1.5F$. The initial load increment is $\Delta F/EA = 0.03$, and the following load increments are adjusted to keep constant length of the displacement increment.

The results are illustrated in Fig. 8.16. The general behavior is probably best understood with reference to the F–v plot in Fig. 8.16(b). In the first stage of the loading process the side nodes move downwards, finally reaching nearly the level $-h$. At that point the center node has reached the level 0 as seen from Fig. 8.16(c). The side nodes now start moving up, and the structure reaches a prestressed stage in the middle of the load history, where $v = w = h$ and $F/EA = 0$. From then on a mirror image of the first part of the equilibrium path is followed.

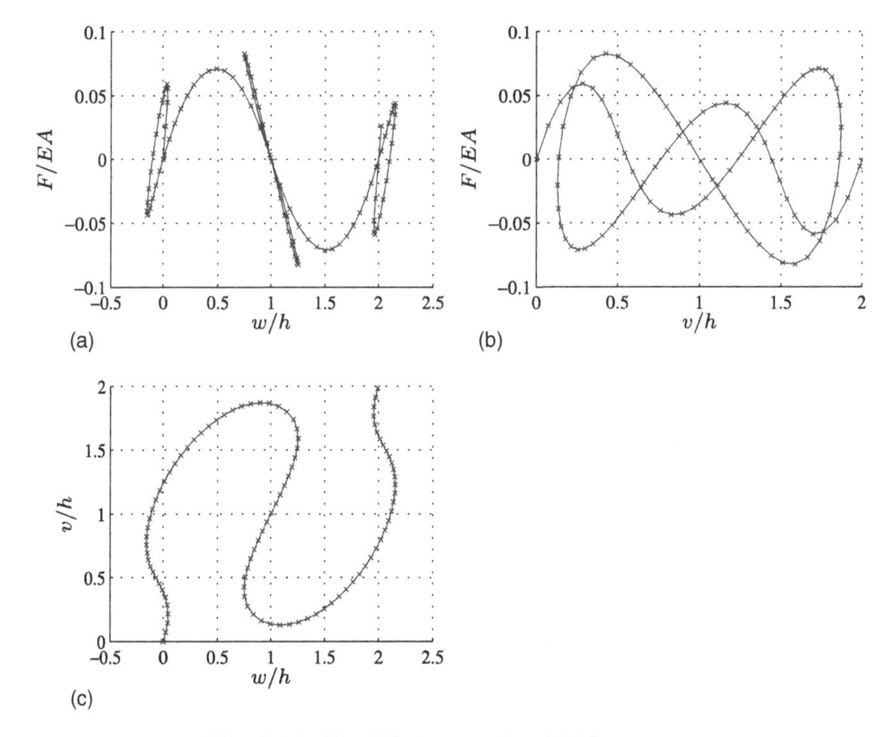

Fig. 8.16. Equilibrium path of 12-bar truss.

The equilibrium curve shown in the figure is traced in $n_{\text{total}} = 95$ steps. The use of a tolerance of $\epsilon_f = 10^{-4}$ leads to a total of $i_{\text{total}} = 389$ iterations, including the initial calculation of the displacement increment $\Delta\mathbf{u}_1$. The maximum number of iterations in any step was $i_{\max} = 6$. The present form of the arc-length algorithm, in which the stiffness matrix is retained during the iterations within a load step, does not experience any convergence problems. However, a full Newton–Raphson implementation is less robust and experiences problems at the two points, where the displacement component v changes direction.

8.4 Quasi-Newton methods

Iterative methods for solution of systems of non-linear equations typically make use of a sequence of solutions of linear systems that represent a local approximation to the non-linear system. In the methods described so far the linear equations have used a tangent stiffness matrix – either the current tangent stiffness or a tangent stiffness from the initial iteration within the load step. However, as soon as a displacement increment has been determined, the corresponding internal forces can be evaluated and this provides additional information on the stiffness of the system. This information can be used to construct systematic modifications to the stiffness matrix, whereby improved convergence properties can often be obtained with a modest computational effort. This class of methods are often called quasi-Newton methods.

Let the non-linear equation to be solved be given in the form

$$\mathbf{g}(\mathbf{u}) = \mathbf{f}, \tag{8.52}$$

where $\mathbf{f}$ is the external load vector and $\mathbf{g}$ is the internal force vector, given in terms of the current displacement vector $\mathbf{u}$. The solution starts from a state of equilibrium, given by the load $\mathbf{f}_0$ and the internal force $\mathbf{g} = \mathbf{g}(\mathbf{u}_0)$. The first step is to impose a load increment $\Delta\mathbf{f}$ and to calculate the corresponding displacement increment $\Delta\mathbf{u}$ from the equation

$$\mathbf{K}_0 \, \Delta\mathbf{u} = \Delta\mathbf{f}. \tag{8.53}$$

In this equation $\mathbf{K}_0$ is a stiffness matrix obtained at the state $\mathbf{u}_0, \mathbf{f}_0$. The internal force is now determined corresponding to the new state of displacement, $\mathbf{g} = \mathbf{g}(\mathbf{u}_0 + \Delta\mathbf{u})$. Hereby both the displacement increment $\Delta\mathbf{u}$ and the corresponding internal force increment $\Delta\mathbf{g} = \mathbf{g} - \mathbf{g}_0$ are known. These vectors are related by a secant stiffness relation of the form

$$\mathbf{K} \, \Delta\mathbf{u} = \Delta\mathbf{g}. \tag{8.54}$$

This so-called quasi-Newton condition is illustrated in Fig. 8.17. Clearly, the two vectors $\Delta\mathbf{u}$ and $\Delta\mathbf{g}$ do not define the matrix $\mathbf{K}$, but rather impose a set of constraints. Thus, the idea of quasi-Newton methods is to use the secant condition (8.54) to modify the already known stiffness matrix $\mathbf{K}_0$.

The basic idea is to introduce corrections to the stiffness matrix $\mathbf{K}_0$ in the form of exterior products of the vectors $\Delta\mathbf{u}$ and $\Delta\mathbf{g}$. This can be done in several different ways, see e.g. Luenberger (1984). The most popular is a symmetric rank two correction, the so-called Broyden–Fletcher–Goldfarb–Shanno (BFGS) update. In the BFGS update the new stiffness matrix is

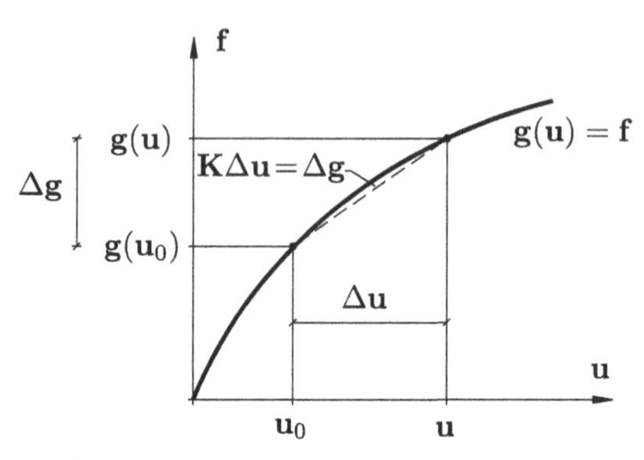

Fig. 8.17. The quasi-Newton condition.

obtained via two additive contributions as

$$\mathbf{K} = \mathbf{K}_0 - \frac{(\mathbf{K}_0\Delta\mathbf{u})(\mathbf{K}_0\Delta\mathbf{u})^T}{\Delta\mathbf{u}^T\mathbf{K}_0\Delta\mathbf{u}} + \frac{\Delta\mathbf{g}\,\Delta\mathbf{g}^T}{\Delta\mathbf{g}^T\Delta\mathbf{u}}. \tag{8.55}$$

It is easily verified that multiplication of $\mathbf{K}$ with $\Delta\mathbf{u}$ leads to the desired force increment $\Delta\mathbf{g}$. The structure of the formula is special, removing the stiffness of $\mathbf{K}_0$ in the direction of $\Delta\mathbf{u}$ and replacing it by an exterior product of $\Delta\mathbf{g}$. In directions orthogonal to $\Delta\mathbf{u}$ no change of stiffness is introduced. The special form of the BFGS update implies that the quasi-Newton condition (8.54) will also be satisfied if the original stiffness matrix $\mathbf{K}_0$ were scaled by a factor β. This possibility will be discussed later.

The inverse of $\mathbf{K}$ is obtained by use of the structure as a modification of $\mathbf{K}_0$ via addition of exterior products. The basic result is the Sherman–Morrison formula for the inverse of a matrix, where an exterior vector product has been added. The Sherman–Morrison formula

$$\left(\mathbf{A} + \frac{\mathbf{a}\mathbf{b}^T}{H}\right)^{-1} = \mathbf{A}^{-1} - \frac{(\mathbf{A}^{-1}\mathbf{a})(\mathbf{A}^{-1}\mathbf{b})^T}{H + \mathbf{b}^T\mathbf{A}^{-1}\mathbf{a}} \tag{8.56}$$

is most easily proved directly by multiplication with the matrix to be inverted. A similar format occurs in the theory of plasticity, discussed in Chapter 7. In the present context this formula is used twice – first to evaluate the inverse of $\mathbf{K}_0 + \Delta\mathbf{g}\Delta\mathbf{g}^T/\Delta\mathbf{g}^T\Delta\mathbf{u}$, and then combining this with the exterior product involving the vector $\mathbf{K}_0\Delta\mathbf{u}$. The result of these manipulations is

$$\mathbf{K}^{-1} = \left(\mathbf{I} - \frac{\Delta\mathbf{u}\Delta\mathbf{g}^T}{\Delta\mathbf{g}^T\Delta\mathbf{u}}\right)\mathbf{K}_0^{-1}\left(\mathbf{I} - \frac{\Delta\mathbf{g}\Delta\mathbf{u}^T}{\Delta\mathbf{g}^T\Delta\mathbf{u}}\right) + \frac{\Delta\mathbf{u}\Delta\mathbf{u}^T}{\Delta\mathbf{g}^T\Delta\mathbf{u}}. \tag{8.57}$$

Note that here multiplication with $\Delta\mathbf{g}$ produces $\Delta\mathbf{u}$ in accordance with inversion of the tangent relation (8.54).

The following sub-increment $\delta\mathbf{u}$ can be determined from the inverse (8.56) of the modified stiffness matrix as $\mathbf{K}^{-1}\mathbf{r}$. However, if consecutive steps are needed, it is convenient to have the inverse in product format. As demonstrated by Matthies and Strang (1979), the product format

$$\mathbf{K}^{-1} = \left(\mathbf{I} - \frac{\Delta\mathbf{u}\,\mathbf{v}^T}{\Delta\mathbf{g}^T\Delta\mathbf{u}}\right)\mathbf{K}_0^{-1}\left(\mathbf{I} - \frac{\mathbf{v}\,\Delta\mathbf{u}^T}{\Delta\mathbf{g}^T\Delta\mathbf{u}}\right) \tag{8.58}$$

is obtained, when the vector $\mathbf{v}$ is defined as

$$\mathbf{v} = \Delta\mathbf{g} - \sqrt{\frac{\Delta\mathbf{g}^T\Delta\mathbf{u}}{\Delta\mathbf{u}^T\mathbf{K}_0\Delta\mathbf{u}}}\,\mathbf{K}_0\Delta\mathbf{u}. \tag{8.59}$$

The result is proved by substitution into (8.57). This factored format is suitable for building a sequence of stiffness matrices $\mathbf{K}_0, \mathbf{K}_1, \mathbf{K}_2, \ldots$, each incorporating the latest displacement and internal force sub-increment $\delta\mathbf{u}_i, \delta\mathbf{g}_i$. The updated inverse matrices $\mathbf{K}_1^{-1}, \mathbf{K}_2^{-1}, \ldots$ need never be formed explicitly. The displacement increments $\mathbf{K}_i^{-1}\mathbf{r}_i$ can be found directly from the original stiffness matrix and the stored vectors $(\mathbf{v}_1, \delta\mathbf{u}_1^T)$, $(\mathbf{v}_2, \delta\mathbf{u}_2^T)$, etc. using scalar vector products. This is a computationally important point as the exterior vector products would otherwise destroy the profile structure of the stiffness matrix.

In its original form the BFGS method removes the stiffness associated with the specific displacement increment $\Delta\mathbf{u}$ and adds a rank one stiffness to obtain the force increment $\Delta\mathbf{g}$. Thus, the stiffness change is related specifically to the directions $\Delta\mathbf{u}_i$ and $\Delta\mathbf{g}$ in displacement and load space. Alternatively a general stiffness reduction β for all finite displacement increments may be introduced by replacing $\mathbf{K}_0$ with $\beta\mathbf{K}_0$. A logical choice for the stiffness reduction factor β_i would be

$$\beta = \frac{\Delta\mathbf{g}^T\Delta\mathbf{u}}{\Delta\mathbf{u}^T\mathbf{K}_0\Delta\mathbf{u}}, \tag{8.60}$$

i.e. the ratio between the actual work $\Delta\mathbf{g}^T\Delta\mathbf{u}$ and the corresponding elastic work $\Delta\mathbf{u}^T\mathbf{K}_0\Delta\mathbf{u}$ using the original stiffness matrix. In most computation strategies the vector $\mathbf{K}_0\Delta\mathbf{u}$ is already available, and thus the denominator of (8.60) only requires a product of two vectors. With this choice for β the inverse of the updated stiffness matrix is

$$\mathbf{K}^{-1} = \beta^{-1}(\,\mathbf{I} - \Delta\mathbf{u}\,\mathbf{w}^T)\mathbf{K}_0^{-1}(\,\mathbf{I} - \mathbf{w}\,\Delta\mathbf{u}^T), \tag{8.61}$$

where the vector $\mathbf{w}$ is defined as

$$\mathbf{w} = \frac{\Delta\mathbf{g}}{\Delta\mathbf{g}^T\Delta\mathbf{u}} - \frac{\mathbf{K}_0\Delta\mathbf{u}}{\Delta\mathbf{u}^T\mathbf{K}_0\Delta\mathbf{u}}. \tag{8.62}$$

This vector is orthogonal to $\Delta\mathbf{u}$, i.e. $\mathbf{w}^T\Delta\mathbf{u} = 0$. Inversion by the Sherman–Morrison formula then gives the updated stiffness matrix in the factored format

$$\mathbf{K} = \beta\big(\,\mathbf{I} + \mathbf{w}\,\Delta\mathbf{u}^T\big)\mathbf{K}_0\big(\,\mathbf{I} + \Delta\mathbf{u}\,\mathbf{w}^T\big). \tag{8.63}$$

The scaled formulation is computationally convenient and has a certain physical logic. The stiffness is reduced by the factor β for all displacement combinations, and the matrix $(\mathbf{I} + \Delta\mathbf{u}\,\mathbf{w}^T)$ only serves to adjust the mutual directions of the force and displacement vectors.

The scaled update has been discussed by Luenberger (1984) in the context of optimization. Luenberger introduced scaling in order to counteract possible ill-conditioning caused by use of a very crude original estimate of the matrix $\mathbf{K}_0$, e.g. the unit matrix. Considerations of optimal condition number for the updated matrices led to the scaling factor β given by (8.60). It was demonstrated by Luenberger that the scaled form was robust and did not accumulate errors when used in connection with inaccurate line searches.

The use of the sequential form of the BFGS method in non-linear finite element analysis has been discussed by Bathe and Cimento (1980) and Bathe (1996). An alternative formulation, a so-called secant-Newton method, has been proposed by Crisfield (1979, 1980, 1991). In the secant-Newton method the current displacement and internal force increments $\delta\mathbf{u}_i$ and $\delta\mathbf{g}_i$ with respect to the last equilibrium state are used to update a fixed stiffness matrix $\mathbf{K}_0$, e.g. the tangent stiffness at the beginning of the load increment. Crisfield gives a format in which the displacement update is expressed in terms of the previous displacement updates by a recurrence relation with two or three terms. This formulation is somewhat simpler to implement than the full BFGS method.

In the orthogonal residual method presented in Section 8.2 the residual force satisfies the orthogonality relation $\mathbf{r}^T\Delta\mathbf{u} = 0$. In that case a BFGS update using the current total increments $\Delta\mathbf{u}, \Delta\mathbf{g}$ leads to a particularly simple form of the inversion formula (8.57). This has been used in the dual orthogonality algorithm by Krenk and Hededal (1993), used for shell stability analysis e.g. by Poulsen and Damkilde (1996). Numerical experience indicates a need for step-size control around limit points, leading to a recent modification by Kouhia (2008).

8.5 Summary

Numerical analysis of non-linear problems of solids and structures usually leads to a need for solution of systems of non-linear equations on the global level, i.e. fair-sized or large systems of non-linear equations. In the case of geometric non-linearity the equations can be formulated directly, while the implicit formulation of elasto-plastic problems also has a non-linear analysis step on the local element level. While the standard Newton–Raphson iteration method using a fixed load increment can be applied to moderately non-linear problems, a central feature of general non-linear solvers is an iteration cycle in which the load level is modified together with the displacement increment. Two essentially different approaches to load level adjustment have been discussed: the orthogonal residual method, and arc-length methods. The difference is in the way the combined load and displacement increments are scaled and adjusted in the iteration cycle.

In the orthogonal residual method the load increment is scaled such that the residual force $\mathbf{r}$ is orthogonal to the current displacement increment $\Delta\mathbf{u}$. The idea has a direct analogue to the mechanics of the problem: the optimal load level is the one that does not have a residual component in the direction of the current displacement increment $\Delta\mathbf{u}$. However, the robustness of the method depends on proper scaling of the first displacement increment of each load step.

In the arc-length methods the central point is the scaling of the combined displacement and load increments $\Delta\mathbf{u}, \Delta\mathbf{f}$. The idea of the arc-length methods is to form the combined load and displacement increment from two components: the increments corresponding to the Newton–Raphson method, plus an extra term pulling the load–displacement point back to an imposed constraint. There are numerous variations of this idea originally introduced into computational mechanics by Riks (1979) and Crisfield (1981). There is a large literature on various generalizations of this class of methods, e.g. to bifurcation analysis, see Kouhia and Mikkola (1999a,b). Arc-length methods are typically implemented using a 'length' defined via simple scalar products of e.g. the displacement vector, $l^2 = \Delta\mathbf{u}^T\Delta\mathbf{u}$. While this may appear somewhat removed from the physics of the original problem, and may even involve addition of terms carrying different physical dimension, the simple form contributes to computational efficiency, and a simple scaling will often be sufficient to render the algorithm robust, even around limit or bifurcation points.

The iterations within each load step require the use of an appropriate stiffness matrix. The algorithms in this chapter show the use of the stiffness matrix corresponding to the first increment within the load step, and in the implementation of the arc-length method shown in Algorithm 8.2 the direction for return to the constraint is along the direction of the first increment. While this choice may not provide optimal convergence properties on monotonic parts of the equilibrium path, it does give added robustness around limit points. The use of the initial stiffness can be modified by BFGS methods as briefly indicated. This type of modification improves performance in regular non-linear regimes, but may lead to reduced robustness around critical points where the stiffness matrix is singular.

8.6 Exercises

Exercise 8.1* Implement the orthogonal residual method, described in Algorithm 8.1, and use it to study the two-bar truss illustrated in Fig. 8.8.

(a) The effect of changing the iteration tolerance ε_f.
(b) The effect of the magnitude of the initial load increment.
(c) The effect of the relative inclination on the equilibrium curves and the number of iterations.

The results should be illustrated in graphs like those in Fig. 8.9.

Exercise 8.2* Implement the hyperplane form of the arc-length method, described in Algorithm 8.2, and use it to study the two-bar truss illustrated in Fig. 8.8 with respect to the influence of the parameters as described in Exercise 8.1.

Exercise 8.3* Introduce the flexibility parameter β in the norm (8.37) in the hyperplane arc-length algorithm. Compare the performance of the algorithm for the 12-bar truss of Example 8.3. Use $\beta = 0$ and $\beta = \|\Delta \mathbf{u}_1\| / \|\Delta \mathbf{f}_1\|$ from the first load step, respectively.

Exercise 8.4* Analyze the shallow dome shown in Fig. 8.10 for the anti-symmetric vertical load case described in Exercise 2.6. Use both the orthogonal residual method and the hyperplane arc-length method, and compare performance characteristics.

Exercise 8.5* Modify the hyperplane implementation of the arc-length method to the hypersphere form by including the extra length $\delta\xi_*$ defined by (8.51). Evaluate the effect by reconsidering the 12-bar truss of Example 8.3 already analyzed with the hyperplane arc-length method in Exercise 8.3.

Exercise 8.6 In the case of a hypersphere constraint the two consecutive increments $(\Delta \mathbf{u}, \Delta \mathbf{f})$ and $(\Delta \mathbf{u} + \delta \mathbf{u}, \Delta \mathbf{f} + \delta \mathbf{f})$ have the same length. Thus, the condition

$$\text{maximize } (\Delta\mathbf{u}, \Delta\mathbf{f}) \cdot (\Delta\mathbf{u} + \delta\mathbf{u}, \Delta\mathbf{f} + \delta\mathbf{f})$$

corresponds to minimizing the angle between the two vectors. Introduce the sub-increments $(\delta\mathbf{u}, \delta\mathbf{f})$ in terms of the increment $\delta\xi$ of the load factor, and find the corresponding condition on $\delta\xi$.

Exercise 8.7 Prove the formula (8.58) giving the factored format of the BFGS inverse of the quasi-Newton stiffness matrix $\mathbf{K}$.

Exercise 8.8 Prove the representation (8.63) for the scaled BFGS matrix $\mathbf{K}$ and the representation (8.61) for its inverse.

9

Dynamic effects and time integration

There are many engineering problems in which dynamic effects are important, e.g. in robotics, manufacturing, transportation and civil engineering structures under environmental loads like wind, waves and earthquakes. Dynamic problems with essential non-linearities typically require a combination of one or more of the non-linear features discussed in the previous chapters, with the dynamic effects arising from the motion. These problems are mostly solved by time integration of the non-linear equations of motion. Traditionally these problems have been solved by so-called collocation-type methods, see e.g. Hughes (1987), Argyris and Mlejnek (1991), Géradin and Rixen (1997). In these methods the equation of motion is matched at selected points in combination with suitable assumptions regarding the relation between displacement, velocity and acceleration. Recent years have seen a rapid development of an alternative type of algorithm based on an integrated form of the equation of motion – the so-called momentum methods. This chapter combines a concise summary of the collocation methods, that are still widely used, with an introduction to the momentum-based methods.

In order to bring out the essential features of the two types of time integration methods the present chapter makes use of the following simple format for the equations of motion:

$$\mathbf{M}\ddot{\mathbf{u}} + \mathbf{g}(\mathbf{u}, \dot{\mathbf{u}}) = \mathbf{f}(t). \tag{9.1}$$

In this equation the first term represents the inertial forces generated by the acceleration $\ddot{\mathbf{u}}$, and the second term represents the internal forces expressed via the displacement $\mathbf{u}$ and the velocity $\dot{\mathbf{u}}$. The external load $\mathbf{f}(t)$ is assumed to be a given function of time. The problem also requires a set of initial conditions of the form

$$\mathbf{u}(0) = \mathbf{u}_0, \qquad \dot{\mathbf{u}}(0) = \dot{\mathbf{u}}_0. \tag{9.2}$$

This basic format with constant mass matrix $\mathbf{M}$ enables a presentation of the central features of collocation and momentum methods, but is not sufficiently general to cover a detailed discussion of finite motion described in terms of rotations, see e.g. Géradin and Cardona (2001).

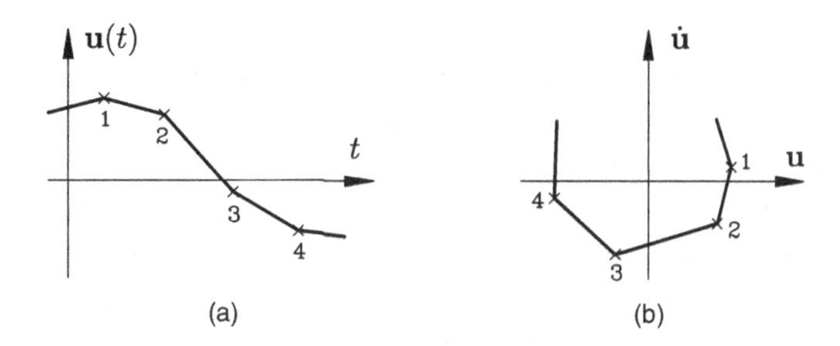

Fig. 9.1. Time history and state-space diagrams.

The basic idea of the two approaches to time integration can be illustrated with reference to Fig. 9.1. The state of the system at any particular time t is described completely by the state vector $[\mathbf{u}(t), \dot{\mathbf{u}}(t)]$ and the load vector $\mathbf{f}(t)$ at time t. Thus, the time development of the system can be represented in state-space as shown in Fig. 9.1(b). The state vector at time t_n can serve as initial conditions for the system response for $t > t_n$. The simplest direct numerical integration procedures for the equation of motion (9.1) are the single-step algorithms, in which the state vector $(\mathbf{u}_{n+1}, \dot{\mathbf{u}}_{n+1})$ at time t_{n+1} is calculated by use of the state vector $(\mathbf{u}_n, \dot{\mathbf{u}}_n)$ at the previous time t_n and the load history between these two states. In the collocation methods the equation of motion (9.1) is satisfied at a selected point in the time interval $[t_n, t_{n+1}]$. This equation must be supplemented by two additional equations, relating the involved displacements, velocities and accelerations. In contrast, the momentum-based methods consider the development of the motion over the time interval $[t_n, t_{n+1}]$. The idea is to integrate the equation of motion over the interval. Hereby the inertial term becomes the finite increment of momentum over the time interval, while the load and internal forces are represented by suitable time averages.

The first part of this chapter deals with single-step collocation methods of the Newmark type, typically used in analysis of the dynamics of structures. First, the algorithm is formulated for a linear system in order to identify the appropriate computational structure, and to enable a detailed analysis of the stability and accuracy, based on consideration of the individual modes of vibration. It is often desirable to introduce artificial damping of high-frequency response components, because they are subject to aliasing by the time discretization and therefore may adversely influence the results. A discretized energy balance equation is set up for the Newmark algorithm, and the influence of the parameters of the algorithm on the representation of damping and stiffness is discussed. The algorithm is then generalized to non-linear systems in Section 9.2.

Some central aspects of momentum-based algorithms are then presented. The basic energy-conserving form is presented in Section 9.3. It is demonstrated that for a system with quadratic non-linearity, like an elastic system with a finite displacement formulation in terms of the quadratic Green strain, energy conservation is obtained by including an additional 'damping term' expressed via the geometric stiffness matrix. It is then shown how damping can be introduced into the algorithm via a balanced combination of a velocity and a displacement term. The chapter concludes with some remarks on ongoing developments in the field of time integration.

9.1 Newmark algorithm for linear systems

The oldest and probably most extensively used algorithm for integration of the equations of structural dynamics is due to Newmark (1959). This algorithm is first discussed in relation to the linear equation of motion

$$\mathbf{M}\,\ddot{\mathbf{u}} + \mathbf{C}\,\dot{\mathbf{u}} + \mathbf{K}\,\mathbf{u} = \mathbf{f}(t), \tag{9.3}$$

where the structure is described by the mass matrix $\mathbf{M}$, the viscous damping matrix $\mathbf{C}$, and the stiffness matrix $\mathbf{K}$.

The objective of the Newmark integration procedure is the computation of the state vector $(\mathbf{u}_{n+1}, \dot{\mathbf{u}}_{n+1})$ at time $t_{n+1} = t_n + h$, given the state vector $(\mathbf{u}_n, \dot{\mathbf{u}}_n)$ at the previous time t_n and the load vector at both these times, $\mathbf{f}_n$ and $\mathbf{f}_{n+1}$. The procedure is established in two steps: first the increments of $\mathbf{u}(t)$ and $\dot{\mathbf{u}}(t)$ are expressed in terms of integrals of the acceleration $\ddot{\mathbf{u}}$ over the time interval $[t_n, t_{n+1}]$, and then the acceleration is evaluated by use of the equations of motion. Thus, the Newmark algorithm rests on two principles: the expression of displacement and velocity increments in terms of acceleration by an approximate representation, asymptotically valid for small time increments, and collocation of the equation of motion at the forward time t_{n+1}. Suitable parameters used in the displacement and velocity representations are determined by considering the linear problem, and the method is then extended to non-linear problems by solving the non-linear equation of motion by iteration.

It follows immediately from integral calculus that the value of the vector $\mathbf{u}(t_{n+1})$ at time t_{n+1} can be expressed as its value $\mathbf{u}(t_n)$ at time t_n plus an increment evaluated as the integral of the time derivative $\dot{\mathbf{u}}(\tau)$ over the time interval:

$$\mathbf{u}(t_n + h) \;=\; \mathbf{u}(t_n) \;+\; \int_{t_n}^{t_{n+1}} \dot{\mathbf{u}}(\tau)\, d\tau. \tag{9.4}$$

Now, a factor 1 can be introduced under the integral sign, and the integral evaluated via integration by parts. When the integral of the factor 1 is selected as $\tau - t_{n+1}$, the result is

$$\mathbf{u}(t_{n+1}) \;=\; \mathbf{u}(t_n) \;-\; [\,(t_{n+1} - \tau)\, \dot{\mathbf{u}}(\tau)\,]_{t_n}^{t_{n+1}} \;+\; \int_{t_n}^{t_{n+1}} (t_{n+1} - \tau)\, \ddot{\mathbf{u}}(\tau)\, d\tau. \tag{9.5}$$

Note that the upper limit in the square brackets does not contribute due to the factor $(t_{n+1} - \tau)$.

In the Newmark procedure a central point is the representation of the velocity and the displacement increments in terms of integrals of the acceleration. The first formula is obtained from (9.4), when replacing $\mathbf{u}$ and $\dot{\mathbf{u}}$ with $\dot{\mathbf{u}}$ and $\ddot{\mathbf{u}}$, respectively. The second formula follows from (9.5) by evaluating the term in the square brackets:

$$\begin{aligned} \dot{\mathbf{u}}_{n+1} \;&=\; \dot{\mathbf{u}}_n \;+\; \int_{t_n}^{t_{n+1}} \ddot{\mathbf{u}}(\tau)\, d\tau, \\ \mathbf{u}_{n+1} \;&=\; \mathbf{u}_n \;+\; h\, \dot{\mathbf{u}}_n \;+\; \int_{t_n}^{t_{n+1}} (t_{n+1} - \tau)\, \ddot{\mathbf{u}}(\tau)\, d\tau, \end{aligned} \tag{9.6}$$

where $h = t_{n+1} - t_n$ is the length of the time interval, and $\mathbf{u}_n = \mathbf{u}(t_n)$.

In the present use of the integration formula (9.6) the acceleration will not be known throughout the interval, and the integrals will have to be evaluated approximately from the value of the acceleration vector at the interval endpoints. The appropriate formulae are given by weighted averages of the acceleration in the form

$$\begin{aligned} \int_{t_n}^{t_{n+1}} \ddot{\mathbf{u}}(\tau)\, d\tau \;&\simeq\; (1 - \gamma)\, h\, \ddot{\mathbf{u}}_n \;+\; \gamma\, h\, \ddot{\mathbf{u}}_{n+1}, \\ \int_{t_n}^{t_{n+1}} (t_{n+1} - \tau)\, \ddot{\mathbf{u}}(\tau)\, d\tau \;&\simeq\; (\tfrac{1}{2} - \beta)\, h^2\, \ddot{\mathbf{u}}_n \;+\; \beta\, h^2\, \ddot{\mathbf{u}}_{n+1}. \end{aligned} \tag{9.7}$$

The parameters $0 < \gamma < 1$ and $0 < \beta < \frac{1}{2}$ determine the degree of forward weighting, with $\gamma = 0, \beta = 0$ corresponding to full backward weighting, and $\gamma = 1, \beta = \frac{1}{2}$ corresponding to full forward weighting. Note that in (9.7a) the sum of the integration weights is h, while in (9.7b) the sum is $\frac{1}{2}h$ due to the linear factor in the integrand. These properties are necessary to ensure convergence for decreasing h, which implies that the formulae must be exact for constant acceleration.

When the approximate integration formulae (9.7) are substituted into (9.6) the following discrete relations are obtained between the displacement, velocity and acceleration vectors at the two interval end-points:

$$\begin{aligned} \dot{\mathbf{u}}_{n+1} &= \dot{\mathbf{u}}_n + (1-\gamma)\,h\,\ddot{\mathbf{u}}_n + \gamma\,h\,\ddot{\mathbf{u}}_{n+1}, \\ \mathbf{u}_{n+1} &= \mathbf{u}_n + h\,\dot{\mathbf{u}}_n + (\tfrac{1}{2}-\beta)\,h^2\,\ddot{\mathbf{u}}_n + \beta\,h^2\,\ddot{\mathbf{u}}_{n+1}. \end{aligned} \tag{9.8}$$

The Newmark integration algorithm is obtained by satisfying the equations of motion at time t_{n+1}, using the representation (9.8a) and (9.8b) for the velocity and displacement vector, respectively. Substitution of these representations into the linear equations of motion (9.3) gives the following equation for the acceleration vector $\ddot{\mathbf{u}}_{n+1}$:

$$\begin{aligned} \Big(\mathbf{M} + \gamma\,h\,\mathbf{C} + \beta\,h^2\,\mathbf{K}\Big)\,\ddot{\mathbf{u}}_{n+1} = \mathbf{f}_{n+1} &- \mathbf{C}\Big(\dot{\mathbf{u}}_n + (1-\gamma)\,h\,\ddot{\mathbf{u}}_n\Big) \\ &- \mathbf{K}\Big(\mathbf{u}_n + h\,\dot{\mathbf{u}}_n + (\tfrac{1}{2}-\beta)\,h^2\,\ddot{\mathbf{u}}_n\Big). \end{aligned} \tag{9.9}$$

This equation permits calculation of the acceleration vector $\ddot{\mathbf{u}}_{n+1}$ at time t_{n+1}, and the velocity and displacement can then be calculated from (9.8).

It is convenient to organize the computation of a time step in the Newmark algorithm in the form of a prediction step followed by a correction step. In the linear problem the definition of the predictor is merely a matter of convenience, while in non-linear problems the predictor serves as the starting point for iterations. In the prediction step preliminary values $\dot{\mathbf{u}}^*_{n+1}$ and $\mathbf{u}^*_{n+1}$ of velocity and displacement are evaluated from (9.8), without the last term containing the as yet unknown acceleration $\ddot{\mathbf{u}}_{n+1}$:

$$\begin{aligned} \dot{\mathbf{u}}^*_{n+1} &= \dot{\mathbf{u}}_n + (1-\gamma)\,h\,\ddot{\mathbf{u}}_n, \\ \mathbf{u}^*_{n+1} &= \mathbf{u}_n + h\,\dot{\mathbf{u}}_n + (\tfrac{1}{2}-\beta)\,h^2\,\ddot{\mathbf{u}}_n. \end{aligned} \tag{9.10}$$

These predicted values appear directly on the right side of the equation (9.9) for the acceleration $\ddot{\mathbf{u}}_{n+1}$. With the notation

$$\mathbf{M}_* = \mathbf{M} + \gamma\,h\,\mathbf{C} + \beta\,h^2\,\mathbf{K} \tag{9.11}$$

for the modified mass matrix, the equation of motion (9.9) takes the simplified form

$$\mathbf{M}_*\,\ddot{\mathbf{u}}_{n+1} = \mathbf{f}_{n+1} - \mathbf{C}\,\dot{\mathbf{u}}^*_{n+1} - \mathbf{K}\,\mathbf{u}^*_{n+1}. \tag{9.12}$$

This equation is solved for the acceleration $\ddot{\mathbf{u}}_{n+1}$. The correction consists

ALGORITHM 9.1. The linear Newmark algorithm.

---

(1) System matrices $\mathbf{K}$, $\mathbf{C}$, $\mathbf{M}$

$$\mathbf{M}_* = \mathbf{M} + \gamma h \mathbf{C} + \beta h^2 \mathbf{K}$$

(2) Initial conditions $\mathbf{u}_0$, $\dot{\mathbf{u}}_0$

$$\ddot{\mathbf{u}}_0 = \mathbf{M}^{-1}(\mathbf{f}_0 - \mathbf{C}\dot{\mathbf{u}}_0 - \mathbf{K}\mathbf{u}_0)$$

(3) Prediction step:

$$\begin{aligned}\dot{\mathbf{u}}^*_{n+1} &= \dot{\mathbf{u}}_n + (1-\gamma) h \ddot{\mathbf{u}}_n \\ \mathbf{u}^*_{n+1} &= \mathbf{u}_n + h\dot{\mathbf{u}}_n + (\tfrac{1}{2}-\beta) h^2 \ddot{\mathbf{u}}_n\end{aligned}$$

(4) Correction step:

$$\begin{aligned}\ddot{\mathbf{u}}_{n+1} &= \mathbf{M}_*^{-1}(\mathbf{f}_{n+1} - \mathbf{C}\dot{\mathbf{u}}^*_{n+1} - \mathbf{K}\mathbf{u}^*_{n+1}) \\ \dot{\mathbf{u}}_{n+1} &= \dot{\mathbf{u}}^*_{n+1} + \gamma h \ddot{\mathbf{u}}_{n+1} \\ \mathbf{u}_{n+1} &= \mathbf{u}^*_{n+1} + \beta h^2 \ddot{\mathbf{u}}_{n+1}\end{aligned}$$

(5) Return to (3) for new time step or stop

---

in adding the last term in (9.8) to the predicted velocity and displacement vector:

$$\begin{aligned}\dot{\mathbf{u}}_{n+1} &= \dot{\mathbf{u}}^*_{n+1} + \gamma h \ddot{\mathbf{u}}_{n+1}, \\ \mathbf{u}_{n+1} &= \mathbf{u}^*_{n+1} + \beta h^2 \ddot{\mathbf{u}}_{n+1},\end{aligned} \tag{9.13}$$

thereby completing the time step.

The implementation of the linear Newmark algorithm is summarized in Algorithm 9.1. For undamped structures the algorithm can be formulated in explicit form if the mass matrix is diagonal and $\beta = 0$. In that case (9.9) can be solved directly for $\ddot{\mathbf{u}}_{n+1}$, without matrix inversion. This particular case is only conditionally stable and requires the time increment h to be within an upper limit, identified in the following subsection.

9.1.1 Energy balance and stability

A time integration method is classified as stable if the free response to a set of finite initial conditions remains bounded. Stability is a necessary condition, and in order to be useful an algorithm must also satisfy some accuracy requirements. Traditionally time integration algorithms have been studied in their linear form by performing a modal decomposition and carrying out a harmonic analysis of the free response. Alternatively some algorithms, including those of Newmark type, lead to an explicit equation for the development of the energy. This equation can be used to identify the stability conditions, and furthermore gives information about the development of the energy of the system as represented by the algorithm.

The energy equation for the linear equation of motion (9.3) is found by pre-multiplication with the velocity $\dot{\mathbf{u}}^T$, followed by a slight rearrangement of terms:

$$\frac{d}{dt}\Big(\tfrac{1}{2}\dot{\mathbf{u}}^T\mathbf{M}\dot{\mathbf{u}} + \tfrac{1}{2}\mathbf{u}^T\mathbf{K}\mathbf{u}\Big) = \dot{\mathbf{u}}^T\mathbf{f} - \dot{\mathbf{u}}^T\mathbf{C}\dot{\mathbf{u}}. \tag{9.14}$$

This equation gives the rate of change of the mechanical energy in terms of the external rate of work minus the rate of energy dissipation in terms of the viscous damping matrix $\mathbf{C}$.

The discrete energy relation similar to the differential energy equation (9.14) is obtained most directly from a phase-space representation of the algorithm. This is a representation in terms of the state-space variables $\mathbf{u}$ and $\dot{\mathbf{u}}$, where the acceleration $\ddot{\mathbf{u}}$ has been eliminated. The state-space representation is obtained by rearranging the Newmark representation formulae (9.8) in terms of increments and mean values, using the notation $\Delta\mathbf{u} = \mathbf{u}_{n+1} - \mathbf{u}_n$ for the displacement increment with similar notation for the velocity and acceleration increments:

$$\begin{aligned}\Delta\dot{\mathbf{u}} &= \tfrac{1}{2}h(\ddot{\mathbf{u}}_{n+1} + \ddot{\mathbf{u}}_n) + (\gamma - \tfrac{1}{2})h\Delta\ddot{\mathbf{u}},\\ \Delta\mathbf{u} &= \tfrac{1}{2}h(\dot{\mathbf{u}}_{n+1} + \dot{\mathbf{u}}_n) + (\beta - \tfrac{1}{2}\gamma)h^2\Delta\ddot{\mathbf{u}}.\end{aligned} \tag{9.15}$$

It is seen that these formulae are fully symmetric for $\gamma = \frac{1}{2}$ and $\beta = \frac{1}{2}\gamma$, while different parameter values introduce a bias.

When these formulae are multiplied by the mass matrix $\mathbf{M}$, the terms containing the acceleration can be eliminated by use of the equation of motion (9.3). The resulting equations can be written in state-space format as

$$\begin{aligned}&\begin{bmatrix}(\gamma-\frac{1}{2})h\mathbf{K} & \mathbf{M}+(\gamma-\frac{1}{2})h\mathbf{C}\\ \mathbf{M}+(\beta-\frac{1}{2}\gamma)h^2\mathbf{K} & (\beta-\frac{1}{2}\gamma)h^2\mathbf{C}\end{bmatrix}\begin{bmatrix}\Delta\mathbf{u}\\ \Delta\dot{\mathbf{u}}\end{bmatrix} +\\ &\qquad\begin{bmatrix}\mathbf{K} & \mathbf{C}\\ \mathbf{0} & -\mathbf{M}\end{bmatrix}\begin{bmatrix}h\bar{\mathbf{u}}\\ h\bar{\dot{\mathbf{u}}}\end{bmatrix} = \begin{bmatrix}h\bar{\mathbf{f}} + (\gamma-\frac{1}{2})h\Delta\mathbf{f}\\ (\beta-\frac{1}{2}\gamma)h^2\Delta\mathbf{f}\end{bmatrix},\end{aligned} \tag{9.16}$$

where the bar symbol denotes the algebraic mean value, e.g. the mean load $\bar{\mathbf{f}} = \frac{1}{2}(\mathbf{f}_{n+1} + \mathbf{f}_n)$. If the parameters are selected as $\gamma = \frac{1}{2}$ and $\beta = \frac{1}{2}\gamma$, these equations simplify considerably and are recognized as the equation of motion and an equation defining the mean velocity in terms of the displacement increment, respectively.

The discrete form of the energy balance equation involves the increment of the mechanical energy over the time interval from t_n to t_{n+1}. This increment can be expressed in terms of mean values and increments of the displacement and velocity by the identity

$$\begin{aligned}\Big[\tfrac{1}{2}\dot{\mathbf{u}}^T\mathbf{M}\dot{\mathbf{u}} + \tfrac{1}{2}\mathbf{u}^T\mathbf{K}\mathbf{u}\Big]_n^{n+1} &= \\ \tfrac{1}{2}(\dot{\mathbf{u}}_{n+1}+\dot{\mathbf{u}}_n)^T\mathbf{M}(\dot{\mathbf{u}}_{n+1}-\dot{\mathbf{u}}_n) &+ \tfrac{1}{2}(\mathbf{u}_{n+1}+\mathbf{u}_n)^T\mathbf{K}(\mathbf{u}_{n+1}-\mathbf{u}_n).\end{aligned} \tag{9.17}$$

It is seen that the energy increment can be obtained from the diagonal of the second matrix when the equations (9.16) are multiplied by the vector $[\Delta\mathbf{u}^T, -\Delta\dot{\mathbf{u}}^T]$. In the present case, where the main purpose is to obtain stability limits, the free response of a system without external damping is sufficient. The complete energy relations for linear systems have been derived by Krenk (2006). For the homogeneous equations without structural damping the energy balance takes the form

$$\Big[\tfrac{1}{2}\dot{\mathbf{u}}^T\mathbf{M}\dot{\mathbf{u}} + \tfrac{1}{2}\mathbf{u}^T\mathbf{K}\mathbf{u}\Big]_n^{n+1} = -(\gamma-\tfrac{1}{2})\Delta\mathbf{u}^T\mathbf{K}\Delta\mathbf{u} + (\beta-\tfrac{1}{2}\gamma)h\,\Delta\dot{\mathbf{u}}^T\mathbf{K}\Delta\mathbf{u}. \tag{9.18}$$

In the last term the velocity increment can be substituted from the first equation of (9.16). This introduces two terms, the first $\Delta\mathbf{u}^T\mathbf{K}\mathbf{M}^{-1}\mathbf{K}\Delta\mathbf{u}$ is a dissipation term, while the second $\frac{1}{2}(\mathbf{u}_{n+1}+\mathbf{u}_n)^T\mathbf{K}\mathbf{M}^{-1}\mathbf{K}\Delta\mathbf{u}$ can be written in increment form and moved to the left side of the equation. The result of these operations is the following energy relation for free response of a linear system without damping as calculated by the Newmark algorithm:

$$\begin{aligned}&\Big[\tfrac{1}{2}\dot{\mathbf{u}}^T\mathbf{M}\dot{\mathbf{u}} + \tfrac{1}{2}\mathbf{u}^T\Big(\mathbf{K} + (\beta-\tfrac{1}{2}\gamma)h^2\mathbf{K}\mathbf{M}^{-1}\mathbf{K}\Big)\mathbf{u}\Big]_n^{n+1} \\ &\qquad = -(\gamma-\tfrac{1}{2})\Delta\mathbf{u}^T\Big(\mathbf{K} + (\beta-\tfrac{1}{2}\gamma)h^2\mathbf{K}\mathbf{M}^{-1}\mathbf{K}\Big)\Delta\mathbf{u}.\end{aligned} \tag{9.19}$$

From this relation it is seen that the algorithm replaces the original stiffness matrix $\mathbf{K}$ by an equivalent stiffness matrix given by

$$\mathbf{K}_{\text{eq}} = \mathbf{K} + (\beta-\tfrac{1}{2}\gamma)h^2\mathbf{K}\mathbf{M}^{-1}\mathbf{K}. \tag{9.20}$$

This equivalent stiffness appears in the definition of the equivalent energy inside the brackets on the left side as well as in the quadratic term on the right side.

For the algorithm to be stable the equivalent energy inside the brackets on the left side must be positive definite and the right side must be zero or provide positive dissipation. It is seen immediately that these conditions are satisfied and full correspondence with the continuous energy equation (9.14) obtained for the parameter values $\gamma = \frac{1}{2}$ and $\beta = \frac{1}{4}$. In general stability requires the right side to be zero or negative, implying that $\gamma \geq \frac{1}{2}$. Taking $\gamma > \frac{1}{2}$ introduces so-called algorithmic damping into the time integration. Stability also requires that the equivalent stiffness matrix $\mathbf{K}_{\rm eq}$ is non-negative definite. The stiffness matrix $\mathbf{K}$ itself is assumed to be positive definite, and for $\beta \geq \frac{1}{2}\gamma$ a positive definite equivalent stiffness is guaranteed, irrespective of the magnitude of the time increment h. This is called unconditional stability, and the corresponding conditions are

$$\gamma \geq \tfrac{1}{2} \qquad \text{and} \qquad \beta \geq \tfrac{1}{2}\gamma. \tag{9.21}$$

These conditions are illustrated in Fig. 9.2.

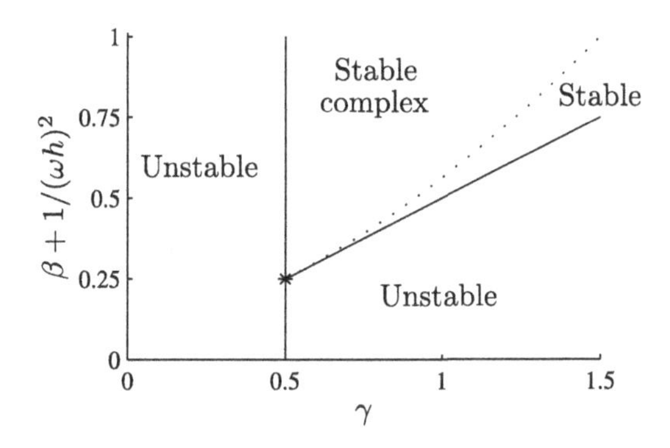

Fig. 9.2. Stability diagram for the Newmark algorithm.

The condition (9.21b) on the parameter β can be relaxed, provided the time increment h is bounded by a limit $h_{\rm max}$ to be calculated from the condition that $\mathbf{K}_{\rm eq}$ is non-negative definite. The time limit $h_{\rm max}$ depends on the behavior of the structure described by the stiffness matrix $\mathbf{K}$ and mass matrix $\mathbf{M}$. For linear structures the response is conveniently represented in terms of the mode shapes $\mathbf{u}_1, \ldots, \mathbf{u}_m$ in the form

$$\mathbf{u}(t) = \sum_{j=1}^{m} u_j(t)\,\mathbf{u}_j. \tag{9.22}$$

The full theory of modal analysis can be found in standard texts on structural dynamics, e.g. Hughes (1987) and Géradin and Rixen (1997), and only a brief summary of the basic facts is included here. In the representation (9.22) the modal coordinates $u_j(t)$ describe the time dependence of components corresponding to the mode shapes $\mathbf{u}_j$. The mode shapes are free vibration solutions of the form

$$\mathbf{u}(t) \;=\; \mathbf{u}_j\,\mathrm{e}^{i\omega_j t}, \tag{9.23}$$

where i is the imaginary unit, ω_j is the angular frequency of the modal vibration, and the result is interpreted as the real part of the equation. Substitution of this representation into the undamped equation of free vibrations gives the generalized eigenvalue problem

$$\big(\mathbf{K} \,-\, \omega_j^2\,\mathbf{M}\big)\,\mathbf{u}_j \;=\; \mathbf{0}. \tag{9.24}$$

The mode shapes satisfy orthogonality conditions with respect to both the mass matrix $\mathbf{M}$ and the stiffness matrix $\mathbf{K}$. These orthogonality relations are conveniently normalized to the form

$$\mathbf{u}_j^T\mathbf{M}\,\mathbf{u}_k \;=\; \delta_{jk}, \qquad \mathbf{u}_j^T\mathbf{K}\,\mathbf{u}_k \;=\; \omega_j^2\,\delta_{jk}. \tag{9.25}$$

These orthogonality relations imply that each mode can be treated separately, and substitution of the modal representation (9.22) into the energy equation (9.19) gives the following energy equation for mode j:

$$\begin{aligned}\Big[\,\tfrac12\dot u_j^2 \,+\, \tfrac12\Big(1 \,+\, (\beta-\tfrac12\gamma)(\omega_j h)^2\Big)\omega_j^2 u_j^2\,\Big]_n^{n+1} \;=&\\ -\,(\gamma-\tfrac12)\Big(1 \,+\, (\beta-\tfrac12\gamma)(\omega_j h)^2\Big)\omega_j^2\,(\Delta u_j)^2.&\end{aligned} \tag{9.26}$$

It is seen from this equation that the modal stiffness is changed from its proper value ω_j^2 to an equivalent modal stiffness $\big[1+(\beta-\frac12\gamma)(\omega_j h)^2\big]\omega_j^2$ on both sides of the equation. This equivalent modal stiffness must be positive for all modes. The condition of positive modal stiffness is conveniently expressed in the form

$$\beta_* \;=\; \beta + (\omega_j h)^{-2} \;\ge\; \tfrac12\gamma \tag{9.27}$$

For a given value of β this condition determines an upper limit on h and the smallest of these is the limit $h_{\max}$ for conditional stability. Conversely, for a given time increment size h the condition determines a lower limit on β. The limits for conditional stability are included in the diagram in Fig. 9.2 simply by replacing β with β_*. The case $\beta = 0$ is of particular interest because the algorithm can then be arranged in explicit form if the mass matrix $\mathbf{M}$ is diagonal, or can be represented as a diagonal matrix. This leads to a conditionally stable algorithm. If no algorithmic damping is imposed $\gamma = \frac12$, and the upper time increment limit is determined from (9.27) as $h \le h_{\max} = 2/\omega_{\max}$. It can be observed from the modal energy equation (9.26) that in this case the equivalent energy is conserved, but for

h close to the stability limit $h_{\max}$ the change in equivalent stiffness leads to fairly large oscillations in the mechanical energy (Krenk, 2006).

9.1.2 Numerical accuracy and damping

The accuracy of an algorithm like the Newmark algorithm is usually determined for linear systems by a spectral analysis. The spectral analysis traces the state-space variables $\mathbf{u}, \dot{\mathbf{u}}$ through a discrete time step and determines the amplification and phase angle increment. Detailed spectral analyses of the linear Newmark algorithm have been given e.g. by Hughes (1987) and Géradin and Rixen (1997). In the present context a concise summary of the spectral properties of the free vibration response of an undamped system is sufficient. The first step is to arrange the state-space equations (9.16) in the form of a recurrence relation. When the first equation is multiplied by $\frac{1}{2}h$ and added to the second, the result for an undamped system without external load is

$$\begin{bmatrix} \gamma h\mathbf{K} & \mathbf{M} \\ \mathbf{M}+\beta h^2\mathbf{K} & \mathbf{0} \end{bmatrix} \begin{bmatrix} \mathbf{u} \\ \dot{\mathbf{u}} \end{bmatrix}_{n+1} = \begin{bmatrix} -(1-\gamma)h\mathbf{K} & \mathbf{M} \\ \mathbf{M}+(\beta-\frac{1}{2})h^2\mathbf{K} & h\mathbf{M} \end{bmatrix} \begin{bmatrix} \mathbf{u} \\ \dot{\mathbf{u}} \end{bmatrix}_{n} . \tag{9.28}$$

These equations are now applied to a single mode with modal coordinates $[\,u(t), \dot{u}(t)\,]$, where the subscript j is left out for convenience. When the corresponding natural frequency is ω, the time increment is characterized by the non-dimensional frequency $\Omega = \omega h$. The equations (9.28) for this particular mode reduce to

$$\begin{bmatrix} \gamma\Omega^2 & 1 \\ 1+\beta\Omega^2 & 0 \end{bmatrix} \begin{bmatrix} u \\ h\dot{u} \end{bmatrix}_{n+1} = \begin{bmatrix} -(1-\gamma)\Omega^2 & 1 \\ 1-(\frac{1}{2}-\beta)\Omega^2 & 1 \end{bmatrix} \begin{bmatrix} u \\ h\dot{u} \end{bmatrix}_{n} . \tag{9.29}$$

The spectral properties of the algorithm follow from analyzing the modal response characterized by an amplification factor λ. This means that the modal response is assumed to be of the form

$$\begin{bmatrix} u \\ h\dot{u} \end{bmatrix}_{n+1} = \lambda \begin{bmatrix} u \\ h\dot{u} \end{bmatrix}_{n} . \tag{9.30}$$

The algorithm is then characterized by the – generally complex – amplification factor λ. The magnitude $|\lambda|$, called the spectral radius, determines the amplification in a single time increment, while the argument of λ determines the development of the phase angle, and thereby the period of the response.

The amplification factor λ is determined by substituting the representation (9.30) into the recurrence equations (9.29). This gives the generalized eigenvalue problem

$$\left(\begin{bmatrix} -(1-\gamma)\Omega^2 & 1 \\ 1-(\frac{1}{2}-\beta)\Omega^2 & 1 \end{bmatrix} - \lambda \begin{bmatrix} \gamma\Omega^2 & 1 \\ 1+\beta\Omega^2 & 0 \end{bmatrix}\right) \begin{bmatrix} u \\ h\dot{u} \end{bmatrix} = \begin{bmatrix} 0 \\ 0 \end{bmatrix}. \tag{9.31}$$

The eigenvalue λ is determined by the characteristic equation. After division by Ω^2 the characteristic equation is

$$\beta_*\lambda^2 - \left[2\beta_* - (\gamma+\tfrac{1}{2})\right]\lambda + \left[\beta_* - (\gamma-\tfrac{1}{2})\right] = 0, \tag{9.32}$$

where β_* was introduced in (9.27). The two roots λ_1 and λ_2 of this equation characterize the algorithm.

In the low frequency limit $\lambda_1 = \lambda_2 = 1$. With increasing frequency the roots form a complex conjugate pair. The condition of a complex conjugate pair follows from the discriminant of (9.32) in the form

$$\beta_* = \beta + \Omega^{-2} > \tfrac{1}{4}(\gamma+\tfrac{1}{2})^2. \tag{9.33}$$

This condition is shown by a dashed curve in Fig. 9.2. It is seen that roots in the form of a complex conjugate pair can be guaranteed for any set of parameters γ and β if the non-dimensional frequency Ω is taken to be sufficiently small. In structural dynamics high frequencies are often present due to the modeling procedure, and there is a particular interest in algorithms that do not impose restrictions on Ω. If complex conjugate roots are required for all frequencies the parameters must satisfy the inequality

$$\beta \geq \tfrac{1}{4}(\gamma+\tfrac{1}{2})^2, \tag{9.34}$$

where the equality will produce a real double root in the limit of infinite frequency.

The complex conjugate roots can be expressed as $\lambda_{1,2} = |\lambda|\exp(\pm i\varphi)$, where $|\lambda|$ is the spectral radius and φ is the phase angle. The spectral radius and phase angle are most easily extracted by expressing the characteristic equation in the normalized form

$$\lambda^2 - 2\cos\varphi|\lambda_{1,2}|\,\lambda + |\lambda_{1,2}|^2 = 0. \tag{9.35}$$

The spectral radius then follows immediately from normalizing the constant term in (9.32):

$$|\lambda| = \left(1 - \frac{\gamma-\frac{1}{2}}{\beta_*}\right)^{1/2} = 1 - \tfrac{1}{2}(\gamma-\tfrac{1}{2})\Omega^2\left(1+\mathrm{O}(\Omega^2)\right). \tag{9.36}$$

Thus, stability requires $\gamma \geq \frac{1}{2}$, and unconditional stability with complex conjugate roots imposes the additional condition (9.34) on β.

In some applications it is advantageous to use a numerical integration algorithm that retains the full response below a certain cut-off frequency, but introduces numerical damping at higher frequencies. There are two common causes for this. Often the discretized model may contain an erroneous representation of the underlying physical system in the high-frequency domain, and it may therefore be desirable to have damping of the high-frequency components to prevent them from retaining or even building up their part of the system energy, thereby adding noise to the solution and possibly creating convergence problems in the iterative solution of non-linear problems. A second class of problems are those with very high-frequency components – e.g. in the modeling of mechanisms, where rigid links may be associated with infinite frequencies. These high-frequency components are misrepresented by discretized collocation-type time integration algorithms, and numerical damping can be used to remove spurious high-frequency oscillations, see e.g. Cardona and Géradin (1989). In contrast, energy-conserving algorithms of the type developed subsequently can typically accommodate constraints without the need for special algorithmic damping measures.

In linear vibrations the attenuation produced by damping is characterized by the damping ratio ζ, and the attenuation over a small time step h is $e^{-\zeta\Omega} \simeq 1 - \zeta\Omega$. It follows from comparison with the expression (9.36) that for low frequencies the damping imposed by the Newmark algorithm can be characterized by the damping ratio $\zeta \simeq \frac{1}{2}(\gamma - \frac{1}{2})\Omega$. This means that if damping is introduced by selecting $\gamma > \frac{1}{2}$, the low-frequency regime will experience an algorithmic damping ratio that is proportional to ωh.

For an ideal algorithm the phase angle φ would be equal to Ω. However, discretization prevents this, and in fact for algorithms of Newmark type the phase angle is mapped on the interval $[0, \pi[$. The phase angle follows by normalizing the characteristic equation (9.32) to the form (9.35),

$$\cos\varphi = \frac{1 - \frac{1}{2}(\gamma + \frac{1}{2})/\beta_*}{\left(1 - (\gamma - \frac{1}{2})/\beta_*\right)^{1/2}}. \tag{9.37}$$

A low-frequency expansion of the phase angle is found by substitution of a series representation of φ into (9.37) and expanding both sides in powers of Ω. The result is

$$\frac{\varphi}{\Omega} \simeq 1 + \left[\tfrac{1}{24} - \tfrac{1}{2}\beta + \tfrac{1}{8}(\gamma - \tfrac{1}{2})\big(2 - (\gamma - \tfrac{1}{2})\big)\right]\Omega^2. \tag{9.38}$$

This result can also be stated as an expression for the relative period error

$$\frac{\Delta T}{T} = \frac{\Omega}{\varphi} - 1 \simeq \tfrac{1}{2}\left[\beta - \tfrac{1}{12} - \tfrac{1}{4}(\gamma - \tfrac{1}{2})(\tfrac{5}{2} - \gamma)\right]\Omega^2. \tag{9.39}$$

For the undamped algorithm, $\gamma = \frac{1}{2}$, the optimal phase behavior is obtained for $\beta = \frac{1}{12}$. However, unconditional stability requires $\beta \geq \frac{1}{4}$. It is concluded that for parameters in this neighborhood the value of β should be chosen as small as possible, while satisfying the conjugate root condition (9.34).

From these considerations it follows that the parameters of the Newmark algorithm with unconditional stability and complex conjugate roots can be expressed in terms of a parameter α as

$$\gamma = \tfrac{1}{2} + \alpha, \qquad \beta = \tfrac{1}{4}(1+\alpha)^2. \tag{9.40}$$

The parameter $\alpha \geq 0$ serves to introduce algorithmic damping, characterized by the spectral radius. The low-frequency spectral radius follows from (9.36)

$$|\lambda| = 1 - \tfrac{1}{2}\alpha\Omega^2\big[\,1 + \mathrm{O}(\Omega^2)\,\big], \tag{9.41}$$

corresponding to the algorithmic damping ratio

$$\zeta = \tfrac{1}{2}\alpha\Omega\,\big[\,1 + \mathrm{O}(\Omega^2)\,\big]. \tag{9.42}$$

For finite frequencies the eigenvalues λ_1 and λ_2 form a complex conjugate pair, meeting in the double root $\lambda_{1,2} = -\lambda_\infty$ for infinite frequency, with

$$\lambda_\infty = \frac{1-\alpha}{1+\alpha}, \qquad \alpha = \frac{1-\lambda_\infty}{1+\lambda_\infty}. \tag{9.43}$$

The low-frequency period error follows from (9.39) as

$$\frac{\Delta T}{T} \simeq \tfrac{1}{12}(1+3\alpha^2)\,\Omega^2. \tag{9.44}$$

It is seen that increasing α leads to increased period error, but the effect is modest for the fairly small values of α needed in practice.

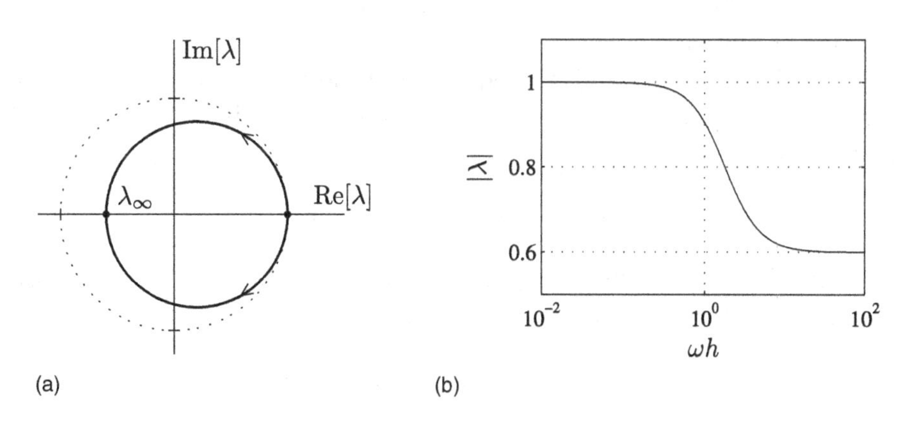

Fig. 9.3. Newmark algorithm with $\lambda_\infty = 0.6$. (a) Locus of the amplification factor λ, (b) spectral radius $|\lambda|$ as a function of ωh.

The complex locus of the amplification factor λ is illustrated in Fig. 9.3(a). In the case of $\alpha = 0$ the two roots trace the upper and lower half of the unit circle, respectively, and the error consists only of period distortion. For $\alpha > 0$ the spectral radius $|\lambda|$ decreases with increasing frequency, with minimum value λ_∞ approached at infinite frequency. The development of the spectral radius with frequency is illustrated in Fig. 9.3(b).

Many problems of dynamics are formulated by models where e.g. the need for geometric detail leads to many high-frequency vibration modes that are not accurately determined by the discrete model. It is therefore desirable to introduce artificial energy dissipation with filter characteristics such that only the high-frequency modes are damped. This kind of damping has been introduced in the Newmark algorithm in the form of various averages of the individual terms in the equation of motion by Hilber *et al.* (1977), Wood *et al.* (1981), Chung and Hulbert (1993), and recently a general procedure containing these as special cases was developed by Krenk and Høgsberg (2005). These procedures are known as α-modifications of the Newmark procedure, because of the α-parameters used to form the averages. A detailed analysis of energy conservation and dissipation in linear Newmark-type algorithms and their α-modifications is given in Krenk (2006).

9.2 Non-linear Newmark algorithm

In non-linear structural mechanics problems the non-linearity is often associated with the displacement, e.g. via a non-linear strain or non-linear material behavior. It is therefore convenient to rearrange the solution method in such a way that the prediction concerns the velocity $\dot{\mathbf{u}}$ and the acceleration $\ddot{\mathbf{u}}$, while the displacement $\mathbf{u}$ is solved for in the iterative solution of the equation of motion.

In the Newmark method, the equation of motion is satisfied at the time increments $\ldots, t_n, t_{n+1}$. Thus, the solution at t_{n+1} is obtained from the equation of motion

$$\mathbf{M}\,\ddot{\mathbf{u}}_{n+1} + \mathbf{g}(\mathbf{u}_{n+1}, \dot{\mathbf{u}}_{n+1}) = \mathbf{f}_{n+1} \tag{9.45}$$

together with the Newmark representation formulae (9.8). The solution is obtained by Newton iterations on the residual

$$\mathbf{r} = \mathbf{f}_{n+1} - \mathbf{M}\,\ddot{\mathbf{u}}_{n+1} - \mathbf{g}(\mathbf{u}_{n+1}, \dot{\mathbf{u}}_{n+1}). \tag{9.46}$$

The idea of Newton iterations is to consider a linearized increment $\delta\mathbf{r}$ to the current value of the residual $\mathbf{r}$. The non-linear equation for the residual

ALGORITHM 9.2. Non-linear Newmark algorithm.

---

(1) Initial conditions $\mathbf{u}_0$, $\dot{\mathbf{u}}_0$

$$\ddot{\mathbf{u}}_0 = \mathbf{M}^{-1}(\mathbf{f}_0 - \mathbf{g}(\mathbf{u}_0, \dot{\mathbf{u}}_0))$$

(2) Prediction step:

$$\begin{aligned} \ddot{\mathbf{u}}_{n+1} &= \ddot{\mathbf{u}}_n \\ \dot{\mathbf{u}}_{n+1} &= \dot{\mathbf{u}}_n + h\,\ddot{\mathbf{u}}_n \\ \mathbf{u}_{n+1} &= \mathbf{u}_n + h\,\dot{\mathbf{u}}_n + \tfrac{1}{2}h^2\,\ddot{\mathbf{u}}_n \end{aligned}$$

(3) Residual calculation:

$$\mathbf{r} = \mathbf{f}_{n+1} - \mathbf{M}\ddot{\mathbf{u}}_{n+1} - \mathbf{g}(\mathbf{u}_{n+1}, \dot{\mathbf{u}}_{n+1})$$

(4) System matrices and increment correction:

$$\begin{aligned} \mathbf{K} &= \partial\mathbf{g}/\partial\mathbf{u}, \quad \mathbf{C} = \partial\mathbf{g}/\partial\dot{\mathbf{u}} \\ \mathbf{K}_* &= \mathbf{K} + \frac{\gamma\,h}{\beta\,h^2}\mathbf{C} + \frac{1}{\beta\,h^2}\mathbf{M} \\ \delta\mathbf{u} &= \mathbf{K}_*^{-1}\,\mathbf{r} \\ \mathbf{u}_{n+1} &= \mathbf{u}_{n+1} + \delta\mathbf{u} \\ \dot{\mathbf{u}}_{n+1} &= \dot{\mathbf{u}}_{n+1} + \frac{\gamma h}{\beta h^2}\,\delta\mathbf{u} \\ \ddot{\mathbf{u}}_{n+1} &= \ddot{\mathbf{u}}_{n+1} + \frac{1}{\beta h^2}\,\delta\mathbf{u} \end{aligned}$$

If $\mathbf{r} > \varepsilon_{\mathbf{r}}$ or $\delta\mathbf{u} > \varepsilon_{\mathbf{u}}$, return to (3) for new iteration

(5) Return to (2) for new time step or stop

---

increment then is

$$\mathbf{r} + \delta\mathbf{r} + \cdots = \mathbf{0}, \tag{9.47}$$

where the dots indicate non-linear contributions to the increment.

In the present case the residual $\mathbf{r}$ depends on $\mathbf{u}$, $\dot{\mathbf{u}}$ and $\ddot{\mathbf{u}}$. Assume that an estimate of these vectors is available. Corrections $\delta\mathbf{u}$, $\delta\dot{\mathbf{u}}$ and $\delta\ddot{\mathbf{u}}$ are then determined by using the linearized system of equations

$$\mathbf{r} + \left(\frac{\partial\mathbf{r}}{\partial\mathbf{u}}\delta\mathbf{u} + \frac{\partial\mathbf{r}}{\partial\dot{\mathbf{u}}}\delta\dot{\mathbf{u}} + \frac{\partial\mathbf{r}}{\partial\ddot{\mathbf{u}}}\delta\ddot{\mathbf{u}}\right) + \cdots = \mathbf{0}. \tag{9.48}$$

These equations are supplemented with the incremental form of the Newmark representation formulae (9.8), taking the form

$$\delta\dot{\mathbf{u}} = \gamma\,h\,\delta\ddot{\mathbf{u}}, \qquad \delta\mathbf{u} = \beta\,h^2\,\delta\ddot{\mathbf{u}}. \tag{9.49}$$

These linear equations enable the elimination of the increments $\delta\dot{\mathbf{u}}$ and $\delta\ddot{\mathbf{u}}$ from the linearized residual equation (9.48). By elimination the linear term becomes

$$\begin{aligned}\delta\mathbf{r} &= \frac{d\mathbf{r}}{d\mathbf{u}}\,\delta\mathbf{u} = \frac{\partial\mathbf{r}}{\partial\mathbf{u}}\delta\mathbf{u} + \frac{\partial\mathbf{r}}{\partial\dot{\mathbf{u}}}\delta\dot{\mathbf{u}} + \frac{\partial\mathbf{r}}{\partial\ddot{\mathbf{u}}}\delta\ddot{\mathbf{u}} \\ &= \Big(\frac{\partial\mathbf{r}}{\partial\mathbf{u}} + \frac{\gamma\,h}{\beta\,h^2}\frac{\partial\mathbf{r}}{\partial\dot{\mathbf{u}}} + \frac{1}{\beta\,h^2}\frac{\partial\mathbf{r}}{\partial\ddot{\mathbf{u}}}\Big)\,\delta\mathbf{u}.\end{aligned} \tag{9.50}$$

The partial derivatives of the residual vector are expressed in terms of the mass matrix and the tangent stiffness and damping matrices, defined as

$$\mathbf{K} = \partial\mathbf{g}/\partial\mathbf{u}, \qquad \mathbf{C} = \partial\mathbf{g}/\partial\dot{\mathbf{u}}. \tag{9.51}$$

The total derivative of the residual vector with respect to the displacement vector $\mathbf{u}$ defines the modified tangent stiffness matrix

$$\mathbf{K}_* = -\frac{d\mathbf{r}}{d\mathbf{u}} = \mathbf{K} + \frac{\gamma}{\beta\,h}\mathbf{C} + \frac{1}{\beta\,h^2}\mathbf{M}. \tag{9.52}$$

When formulated in terms of the displacement increments $\delta\mathbf{u}$, the residual equation (9.48) takes the more familiar form

$$\mathbf{K}_*\,\delta\mathbf{u} = \mathbf{r}, \tag{9.53}$$

where $\mathbf{K}_*$ is the modified tangent stiffness matrix (9.52). The non-linear Newmark algorithm is shown as pseudo-code in Algorithm 9.2. The system matrices $\mathbf{K}$ and $\mathbf{C}$ representing the non-linear properties are now moved inside the iteration loop.

EXAMPLE 9.1. THE ELASTIC PENDULUM. The pendulum shown in Fig. 9.4 consists of a concentrated mass m suspended in a hinged elastic bar with negligible mass and stiffness EA and length l_0 in the unloaded state.

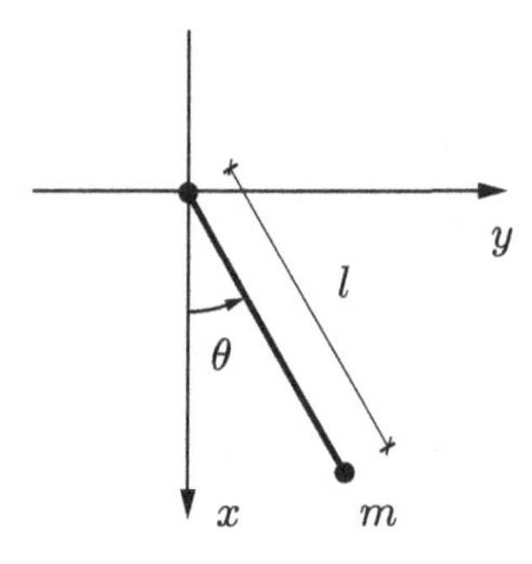

Fig. 9.4. Elastic pendulum with concentrated mass.

Gravitation gives the force mg in the downward x-direction. The position of the mass is described by the coordinates (x, y) and the velocity is $(\dot{x}, \dot{y})$. The kinetic and potential energy are

$$T = \tfrac{1}{2}m(\dot{x}^2 + \dot{y}^2), \qquad U = \tfrac{1}{2}l_0\, EA\, \varepsilon^2 - mg\, x,$$

where ε is the strain in the bar, corresponding to the force $N = EA\varepsilon$. In this example the Green strain $\varepsilon_G = (l^2 - l_0^2)/2l_0^2$ is used, while engineering strain is the subject of Exercises 9.3 and 9.4.

The equations of motion follow from the time derivatives of the total energy,

$$\frac{d}{dt}\big(T + U\big) = \Big(m\ddot{x} + \frac{\partial U}{\partial x}\Big)\,\dot{x} + \Big(m\ddot{y} + \frac{\partial U}{\partial y}\Big)\,\dot{y} = 0.$$

The terms in parentheses define the components of the equations of motion. Thus, the pendulum is described by the mass matrix, the internal and the external forces

$$\mathbf{M} = \begin{bmatrix} m & 0 \\ 0 & m \end{bmatrix}, \quad \mathbf{g} = \frac{N}{l_0}\begin{bmatrix} x \\ y \end{bmatrix}, \quad \mathbf{f} = \begin{bmatrix} mg \\ 0 \end{bmatrix}.$$

The tangent stiffness matrix $\mathbf{K} = \partial\mathbf{g}/\partial\mathbf{x}$ consists of a geometric and a constitutive component

$$\mathbf{K} = \frac{N}{l_0}\begin{bmatrix} 1 & 0 \\ 0 & 1 \end{bmatrix} + \frac{EA}{l_0^3}\begin{bmatrix} x^2 & xy \\ yx & y^2 \end{bmatrix}.$$

These system properties are now used in the Newmark algorithm.

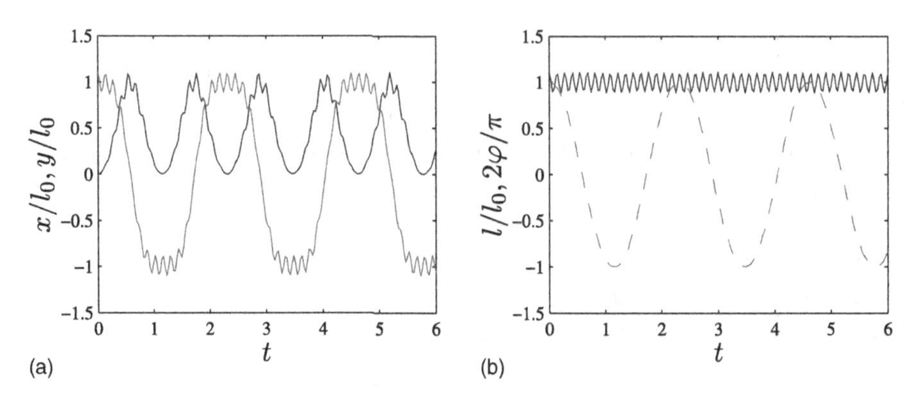

Fig. 9.5. Newmark integration of pendulum vibrations with $h = 0.03$. (a) Coordinates $x(t)$ and $y(t)$, (b) length l/l_0 and angle $2\varphi/\pi$.

In the present example $m = 1$, $l_0 = 1$, $g = 10$ and $EA = 3000$. The time scales of the problem are the period pendulum vibrations $T_p = 2\pi/\omega_p =$

$2\pi/\sqrt{g/l_0} = 1.987$ and the vibration period of the bar $T_b = 2\pi/\omega_b = 2\pi/\sqrt{EA/ml_0} = 0.1147$. The ratio of these time scales is $T_p/T_b = 17.3$, and thus the oscillations of the bar are rapid relative to the swinging pendulum motion. The integration time increment $h = 0.03$ is chosen to enable visualization of the rapid oscillations without attempting an accurate representation of these. In the initial state the pendulum is horizontal with 10 pct. elongation, $(x_0, y_0) = (0, 1.1\,l_0)$. The coordinates $x(t)$ and $y(t)$ are integrated with the non-linear Newmark Algorithm 9.2 with $\gamma = \frac{1}{2}$ and $\beta = \frac{1}{4}$ and shown in Fig. 9.5(a), while the alternative representation of the motion in terms of length $l(t)$ and angle $\varphi(t)$ is shown in Fig. 9.5(b). Both figures clearly illustrate the slow pendulum motion superposed by the rapid oscillations of the elastic bar.

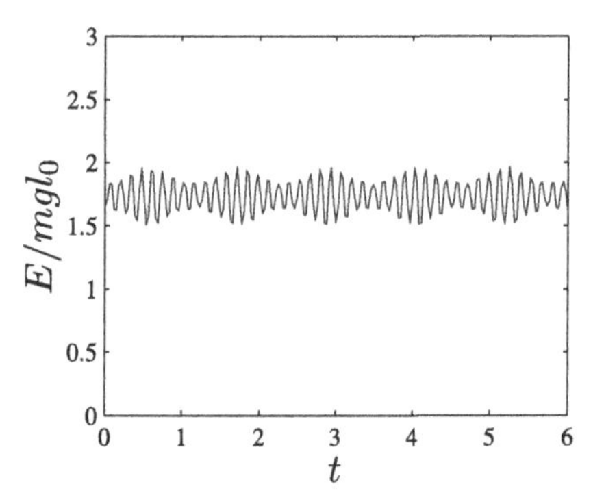

Fig. 9.6. Energy development in Newmark integration of pendulum vibrations for $h = 0.03$ with $\gamma = \frac{1}{2}$, $\beta = \frac{1}{4}$.

The development of the total energy $E = U + T$ with time is shown in Fig 9.6. It is seen that in contrast to the theory for the continuous problem and the linear algorithm, that both predict energy conservation, the total energy of the numerically integrated solution fluctuates by about ± 10 pct. of the energy level. The figures reveal that the fluctuations of the energy follow the period of the oscillations in the bar and are in opposite phase to the elongation. This behavior will be explained in the following section, in which the energy-conserving algorithm is described.

A high-frequency damping scheme for the non-linear Newmark method has been developed by Kuhl and Crisfield (1999), inspired by the generalized alpha methods developed for linear systems. However, numerical results and an analysis of the algorithm by Erlicher *et al.* (2002) indicate that algorithmic damping may lead to undesirable energy fluctuations similar to those illustrated in the example.

9.3 Energy-conserving integration

The Newmark method is based on collocation of the equation of motion at selected points in time. It was demonstrated in Section 9.1.1 that in the linear case energy is only conserved for the particular choice of parameters $\beta = \frac{1}{2}\gamma = \frac{1}{4}$ and Example 9.1 demonstrated that even for this set of parameters energy is not conserved for non-linear systems. It is desirable that time integration algorithms conserve basic quantities like momentum and energy, or dissipate energy in a controlled manner. In order to derive algorithms with conservation properties the idea of collocation at selected times must be replaced with algorithms that are based on the development of the system properties over a time step, e.g. from t_n to t_{n+1}. Time integration methods with conservation properties have been the subject of intensive research since the papers of Simo and Wong (1991) on finite rotations and Simo and Tarnow (1992) on finite elastic deformation.

The basic principles of conservation methods will here be illustrated by considering systems with non-linear kinematics described in terms of non-linear strain. The mass and damping matrices are assumed to be constant, giving the equation of motion in the form

$$\mathbf{M}\ddot{\mathbf{u}}(t) + \mathbf{C}\dot{\mathbf{u}}(t) + \boldsymbol{\nabla}_{\mathbf{u}} G(\mathbf{u}) \;=\; \mathbf{f}(t). \tag{9.54}$$

$\mathbf{M}$ and $\mathbf{C}$ are the mass and damping matrices as before, while the internal forces $\mathbf{g}(\mathbf{u}) = \boldsymbol{\nabla}_{\mathbf{u}} G(\mathbf{u})$ are the displacement derivatives of a potential function $G(\mathbf{u})$. With this definition of the internal forces, multiplication by the velocity $\dot{\mathbf{u}}^T$ leads to the non-linear energy balance equation

$$\frac{d}{dt}\Big(\tfrac{1}{2}\dot{\mathbf{u}}^T\mathbf{M}\dot{\mathbf{u}} + G(\mathbf{u}) \Big) \;=\; \dot{\mathbf{u}}^T\mathbf{f} - \dot{\mathbf{u}}^T\mathbf{C}\dot{\mathbf{u}}. \tag{9.55}$$

This result follows from the observation that the time derivative is obtained from the 'chain rule' as

$$\frac{d}{dt} G(\mathbf{u}) \;=\; \dot{\mathbf{u}}^T\, \boldsymbol{\nabla}_{\mathbf{u}} G(\mathbf{u}). \tag{9.56}$$

As demonstrated in the following, the key to energy conservation is to develop an algorithm that contains the equivalent relation for a finite time interval and thereby a finite displacement increment $\Delta\mathbf{u}$.

9.3.1 State-space formulation

It is often advantageous to recast the second-order differential equations of dynamics into an augmented set of first-order differential equations by introduction of a new independent variable representing the velocity. The simplest form is the introduction of a new vector $\mathbf{v} = \dot{\mathbf{u}}$. This is changed to a momentum relation by multiplication with the mass matrix $\mathbf{M}$. The augmented system then takes the symmetric form

$$\begin{bmatrix} \mathbf{C} & \mathbf{M} \\ \mathbf{M} & \mathbf{0} \end{bmatrix} \begin{bmatrix} \dot{\mathbf{u}} \\ \dot{\mathbf{v}} \end{bmatrix} + \begin{bmatrix} \nabla_{\mathbf{u}} G(\mathbf{u}) \\ -\mathbf{M}\,\mathbf{v} \end{bmatrix} = \begin{bmatrix} \mathbf{f}(t) \\ \mathbf{0} \end{bmatrix}. \tag{9.57}$$

The purpose of a single-step time integration is to advance the state-space variables $\mathbf{u}$ and $\mathbf{v}$ from time t_n to t_{n+1}. Therefore, the equations are integrated over the time interval $[t_n, t_{n+1}]$ as

$$\begin{bmatrix} \mathbf{C} & \mathbf{M} \\ \mathbf{M} & \mathbf{0} \end{bmatrix} \begin{bmatrix} \Delta\mathbf{u} \\ \Delta\mathbf{v} \end{bmatrix} + \begin{bmatrix} \int \nabla_{\mathbf{u}} G(\mathbf{u})\, dt \\ -\mathbf{M} \int \mathbf{v}\, dt \end{bmatrix} = \begin{bmatrix} \int \mathbf{f}(t)\, dt \\ \mathbf{0} \end{bmatrix}. \tag{9.58}$$

The time integral of the external force is represented as the interval length h times the algebraic mean of the end-point values $\bar{\mathbf{f}}$, and the velocity integral is similarly represented as $h\bar{\mathbf{v}}$. The potential term $\nabla_{\mathbf{u}}^T G(\mathbf{u})$ is generally non-linear in $\mathbf{u}$, and the integral is first expressed in terms of a representative value in the form

$$\int_n^{n+1} \nabla_{\mathbf{u}} G(\mathbf{u})\; dt \;\simeq\; h\, \nabla_{\mathbf{u}} G_*. \tag{9.59}$$

The representative 'mean' value $\nabla_{\mathbf{u}} G_*$ is determined later from an energy conservation condition. In terms of this notation the discretized equations of motion take the form

$$\begin{bmatrix} \mathbf{C} & \mathbf{M} \\ \mathbf{M} & \mathbf{0} \end{bmatrix} \begin{bmatrix} \Delta\mathbf{u} \\ \Delta\mathbf{v} \end{bmatrix} + \begin{bmatrix} h\, \nabla_{\mathbf{u}} G_* \\ -h\,\mathbf{M}\,\bar{\mathbf{v}} \end{bmatrix} = \begin{bmatrix} h\,\bar{\mathbf{f}} \\ \mathbf{0} \end{bmatrix}. \tag{9.60}$$

It is observed that the damping term is here represented as $\mathbf{C}\Delta\mathbf{u}$ instead of $\mathbf{C}h\bar{\mathbf{v}}$ as in collocation methods.

The energy relation for the discrete algorithm corresponding to (9.55) follows from the difference form (9.60) by pre-multiplication with $[\Delta\mathbf{u}^T, -\Delta\mathbf{v}^T]$:

$$[\Delta\mathbf{u}^T, -\Delta\mathbf{v}^T] \left(\begin{bmatrix} \mathbf{C} & \mathbf{M} \\ \mathbf{M} & \mathbf{0} \end{bmatrix} \begin{bmatrix} \Delta\mathbf{u} \\ \Delta\mathbf{v} \end{bmatrix} + \begin{bmatrix} h\, \nabla_{\mathbf{u}} G_* \\ -h\,\mathbf{M}\,\bar{\mathbf{v}} \end{bmatrix} \right) = h\,\Delta\mathbf{u}^T\,\bar{\mathbf{f}}. \tag{9.61}$$

The contributions from the off-diagonal sub-matrices of the first matrix cancel, while the diagonal term represents the dissipation by damping. The

$\mathbf{v}$-contributions from the second matrix are rewritten by use of the relation

$$(\mathbf{v}_{n+1}^T - \mathbf{v}_n^T)\,\mathbf{M}\,(\mathbf{v}_{n+1} + \mathbf{v}_n) \;=\; \big[\,\mathbf{v}^T\mathbf{M}\,\mathbf{v}\,\big]_n^{n+1}. \qquad (9.62)$$

After division by h the energy equation then takes the form

$$\big[\,\tfrac{1}{2}\mathbf{v}^T\mathbf{M}\,\mathbf{v}\,\big]_n^{n+1} + \Delta\mathbf{u}^T\boldsymbol{\nabla}_{\mathbf{u}}G_* \;=\; \Delta\mathbf{u}^T\bar{\mathbf{f}} \;-\; \frac{1}{h}\Delta\mathbf{u}^T\mathbf{C}\,\Delta\mathbf{u}. \qquad (9.63)$$

This equation will give the correct energy balance equation if the representative value of the internal force $\boldsymbol{\nabla}_{\mathbf{u}}^T G_*$ is chosen such that the corresponding term in (9.63) represents the increment of the energy potential $G(\mathbf{u})$,

$$\big[\,G(\mathbf{u})\,\big]_n^{n+1} \;=\; \Delta\mathbf{u}^T\boldsymbol{\nabla}_{\mathbf{u}}G_*. \qquad (9.64)$$

This is recognized as a finite increment form of the differentiation formula (9.56). The general question of finite increment formulation of non-linear internal forces has been discussed by e.g. Simo and Tarnow (1992) and Gonzalez (2000). The main points will be illustrated in the following with reference to a linear elastic material with non-linear kinematics described by the Green strain following Krenk (2007a). More general material models are discussed in Section 9.5.

9.3.2 Non-linear kinematics for Green strain

The material is now assumed to be linear elastic in terms of the Green strain tensor $\mathbf{E}$ and the second Piola–Kirchhoff stress tensor $\mathbf{S}$. This implies the linear relation

$$\mathbf{S} \;=\; \mathbb{C}\,\mathbf{E}, \qquad (9.65)$$

where $\mathbb{C}$ is the constant stiffness tensor. In the following, no specific distinction will be made between the formulation in terms of tensor components and the equivalent reduced formulation in terms of vector–matrix components, where stresses and strains are one-dimensional arrays. The transpose symbol is therefore introduced, when needed in the vector–matrix formalism. The Green strain formulation makes use of an initial configuration $\mathbf{x}_0$ with the initial volume element dV_0. The elastic potential of a model consisting of linear elastic elements with Green strain $\mathbf{E}$ is given by

$$G(\mathbf{u}) \;=\; \int_{V_0} \tfrac{1}{2}\mathbf{E}^T\mathbb{C}\,\mathbf{E}\; dV_0, \qquad (9.66)$$

where dV_0 denotes integration with respect to initial volume. The transpose is introduced to make the formulation valid also in matrix notation. The internal force in a displacement state $\mathbf{u}$ follows from the elastic potential (9.66) by differentiation:

$$\mathbf{g}(\mathbf{u}) = \nabla_{\mathbf{u}} G(\mathbf{u}) = \int_{V_0} \nabla_{\mathbf{u}} \mathbf{E}^T \, \mathbf{S} \, dV_0, \tag{9.67}$$

where strain derivatives and stress are evaluated corresponding to the displacement state $\mathbf{u}$.

A key role in conservative integration is played by the increment of the elastic potential. It follows from an argument similar to (9.62) that the increment of the elastic potential (9.66) can be expressed in the factored form

$$\big[G(\mathbf{u}) \big]_n^{n+1} = \int_{V_0} \Delta\mathbf{E}^T \mathbb{C} \, \bar{\mathbf{E}} \, dV_0 = \int_{V_0} \Delta\mathbf{E}^T \bar{\mathbf{S}} \, dV_0. \tag{9.68}$$

This is the form used by Simo and Tarnow (1992). In the present formulation the integral is factored in the form (9.64) to identify the representative 'mean' value $\mathbf{g}_* = \nabla_{\mathbf{u}} G_*$ to be used in the algorithm. While this in principle can be done for any definition of the strain measure, the result takes a particularly simple and convenient form for the Green strain.

Special properties of Green strain

The Green strain tensor is obtained from a product of the deformation gradient tensor as discussed in Chapters 6 and 7,

$$\mathbf{E} = \tfrac{1}{2}\big(\mathbf{F}^T \mathbf{F} - \mathbf{I} \big). \tag{9.69}$$

It follows from this definition that the Green strain tensor $\mathbf{E}$ is a quadratic function of the displacement vector $\mathbf{u}$. Quadratic functions have the property that their increment can be factored as the product of a difference and a mean value. Furthermore, the term in the mean value is the partial derivative vector of the quadratic form. In the case of the Green strain this implies that

$$\Delta\mathbf{E}^T = \Delta\mathbf{u}^T \big(\overline{\nabla_{\mathbf{u}} \mathbf{E}^T} \big). \tag{9.70}$$

It is important to notice that the mean value applies to the derivatives of the Green strain $\mathbf{E}$, i.e. to a linear function of $\mathbf{u}$. While a factored form with the factor $\delta\mathbf{u}^T$ can in principle be obtained for any strain, the property of the second factor as a mean value of the gradient at the end-points is a consequence of the quadratic form of the Green strain, and therefore particular to this strain measure.

When the factored form (9.70) is substituted, the elastic potential increment (9.68) takes the form

$$\big[\,G(\mathbf{u})\,\big]_n^{n+1} \;=\; \Delta\mathbf{u}^T \int_{V_0} (\overline{\boldsymbol{\nabla}_{\mathbf{u}}\mathbf{E}^T})\;\bar{\mathbf{S}}\;dV_0. \tag{9.71}$$

Comparison with (9.64) identifies the representative internal force as

$$\mathbf{g}_* \;=\; \boldsymbol{\nabla}_{\mathbf{u}}G_* \;=\; \int_{V_0} (\overline{\boldsymbol{\nabla}_{\mathbf{u}}\mathbf{E}^T})\;\bar{\mathbf{S}}\;dV_0. \tag{9.72}$$

Thus, the representative mean of the internal force is the product of the arithmetic mean of the strain derivatives, and the arithmetic mean of the stress tensor at the ends of the current time step.

The integrand can also be interpreted as the product of the derivatives of the strain $\boldsymbol{\nabla}_{\mathbf{u}}\mathbf{E}^T(\bar{\mathbf{u}})$ evaluated at the mean configuration described by the mean displacement $\bar{\mathbf{u}}$ with the mean stress $\bar{\mathbf{S}}$. This interpretation follows from the fact that as the strain is quadratic in $\mathbf{u}$, the strain derivatives are linear, whereby

$$\overline{\boldsymbol{\nabla}_{\mathbf{u}}\mathbf{E}^T} \;=\; \tfrac{1}{2}\big[\,\boldsymbol{\nabla}_{\mathbf{u}}\mathbf{E}^T(\mathbf{u}_{n+1}) + \boldsymbol{\nabla}_{\mathbf{u}}\mathbf{E}^T(\mathbf{u}_n)\,\big] \;=\; \boldsymbol{\nabla}_{\mathbf{u}}\mathbf{E}^T(\bar{\mathbf{u}}). \tag{9.73}$$

Substitution of this expression into (9.72) gives the following integral:

$$\mathbf{g}_* \;=\; \boldsymbol{\nabla}_{\mathbf{u}}G(\bar{\mathbf{u}}) \;=\; \int_{V_0} \boldsymbol{\nabla}_{\mathbf{u}}\mathbf{E}^T(\bar{\mathbf{u}})\;\bar{\mathbf{S}}\;dV_0, \tag{9.74}$$

where the strain derivatives refer to the mean state.

Geometric and constitutive stiffness matrices

It is advantageous to reformulate the integral in (9.72) in such a way that the representative internal force $\mathbf{g}_*$ is expressed by the internal forces at the end-points of the time step plus any additional terms. The integrand consists of the product

$$\overline{\boldsymbol{\nabla}_{\mathbf{u}}\mathbf{E}^T}\;\bar{\mathbf{S}} \;=\; \tfrac{1}{4}\Big[\,\boldsymbol{\nabla}_{\mathbf{u}}\mathbf{E}^T_{n+1} + \boldsymbol{\nabla}_{\mathbf{u}}\mathbf{E}^T_n\,\Big]\Big[\,\mathbf{S}_{n+1} + \mathbf{S}_n\,\Big]. \tag{9.75}$$

The products of the individual terms can be grouped to form products of corresponding end-point values and a remainder term:

$$\begin{aligned}\overline{\boldsymbol{\nabla}_{\mathbf{u}}\mathbf{E}^T}\;\bar{\mathbf{S}} \;=\; &\tfrac{1}{2}\Big[\,(\boldsymbol{\nabla}_{\mathbf{u}}\mathbf{E}^T_{n+1})\mathbf{S}_{n+1} + (\boldsymbol{\nabla}_{\mathbf{u}}\mathbf{E}^T_n)\mathbf{S}_n\,\Big]\\ &-\tfrac{1}{4}\Big[\,\boldsymbol{\nabla}_{\mathbf{u}}\mathbf{E}^T_{n+1} - \boldsymbol{\nabla}_{\mathbf{u}}\mathbf{E}^T_n\,\Big]\Big[\,\mathbf{S}_{n+1} - \mathbf{S}_n\,\Big].\end{aligned} \tag{9.76}$$

The first term is the arithmetic mean of the end-point values and the second term is a product of increments,

$$\overline{\nabla_{\mathbf{u}}\mathbf{E}^T}\ \bar{\mathbf{S}} = \overline{(\nabla_{\mathbf{u}}\mathbf{E}^T)\mathbf{S}} - \tfrac{1}{4}\Delta\big(\nabla_{\mathbf{u}}\mathbf{E}^T\big)\,\Delta\mathbf{S}. \tag{9.77}$$

The last term can be reformulated by use of the fact that the strain derivatives $\nabla_{\mathbf{u}}\mathbf{E}^T$ are linear functions of the displacement vector $\mathbf{u}$. This implies that the second derivatives of the strain $\nabla_{\mathbf{u}}\nabla_{\mathbf{u}}^T\mathbf{E}^T$ are independent of $\mathbf{u}$. The strain derivative increment in the last term can therefore be expressed in factored form in terms of the displacement increment $\Delta\mathbf{u}$ as

$$\Delta\big(\nabla_{\mathbf{u}}\mathbf{E}^T\big) = \big(\nabla_{\mathbf{u}}\nabla_{\mathbf{u}}^T\mathbf{E}^T\big)\,\Delta\mathbf{u}. \tag{9.78}$$

This expression is to be interpreted on a strain component basis, i.e. as an equation of each component in the strain tensor $\mathbf{E}$.

Substitution of (9.75) and (9.78) into (9.72) gives the following expression for the representative internal force:

$$\mathbf{g}_* = \tfrac{1}{2}\big[\,\mathbf{g}_{n+1} + \mathbf{g}_n\,\big] - \tfrac{1}{4}\Big[\int_{V_0}\big(\nabla_{\mathbf{u}}\nabla_{\mathbf{u}}^T\mathbf{E}^T\big)\Delta\mathbf{S}\;dV_0\Big]\,\Delta\mathbf{u}. \tag{9.79}$$

This establishes the representative internal force $\mathbf{g}_*$ as the mean value of the internal forces at the end-points of the time step, minus an integral accounting for the effect of non-linearity.

The integral term is closely related to the geometric stiffness matrix of the structure. The internal force $\mathbf{g}(\mathbf{u})$ corresponding to the state of displacement is given by (9.67). The infinitesimal increment of the internal force corresponding to an infinitesimal change of the displacement $d\mathbf{u}$ follows by differentiation. The integral is over the initial volume, which remains constant under differentiation, and the increment $d\mathbf{g}$ is therefore obtained as the sum of the contributions from the increments of the strain and the stress, respectively:

$$d\mathbf{g}(\mathbf{u}) = \Big[\int_{V_0}\big(\nabla_{\mathbf{u}}\nabla_{\mathbf{u}}^T\mathbf{E}^T\big)\,\mathbf{S}\;dV_0 + \int_{V_0}(\nabla_{\mathbf{u}}\mathbf{E}^T)\mathbb{C}(\nabla_{\mathbf{u}}\mathbf{E}^T)^T\,dV_0\Big]\,d\mathbf{u}. \tag{9.80}$$

This is a matrix relation for the infinitesimal internal force increment in the form

$$d\mathbf{g}(\mathbf{u}) = \big[\,\mathbf{K}^g + \mathbf{K}^c\,\big]\,d\mathbf{u}, \tag{9.81}$$

where

$$\mathbf{K}^g = \int_{V_0}\big(\nabla_{\mathbf{u}}\nabla_{\mathbf{u}}^T\mathbf{E}^T\big)\,\mathbf{S}\;dV_0 \tag{9.82}$$

is the geometric tangent stiffness matrix and

$$\mathbf{K}^c = \int_{V_0} (\boldsymbol{\nabla}_{\mathbf{u}}\mathbf{E}^T)\mathbb{C}(\boldsymbol{\nabla}_{\mathbf{u}}\mathbf{E}^T)^T \, dV_0 \tag{9.83}$$

is the constitutive tangent stiffness matrix. These matrices are typically available in a standard finite element implementation.

It is seen that the integral term in the representative internal force $\mathbf{g}_*$ can be expressed directly in terms of the finite increment of the geometric stiffness matrix over the interval of integration,

$$\Delta\mathbf{K}^g = \mathbf{K}^g_{n+1} - \mathbf{K}^g_n = \int_{V_0} (\boldsymbol{\nabla}_{\mathbf{u}}\boldsymbol{\nabla}^T_{\mathbf{u}}\mathbf{E}^T)\,\Delta\mathbf{S}\, dV_0. \tag{9.84}$$

When this expression is introduced into (9.72), the following matrix expression is obtained for the representative internal force:

$$\mathbf{g}_* = \tfrac{1}{2}\big[\,\mathbf{g}_{n+1} + \mathbf{g}_n\,\big] - \tfrac{1}{4}\Delta\mathbf{K}^g\,\Delta\mathbf{u}. \tag{9.85}$$

This form of the representative internal force demonstrates that it consists of a direct arithmetic average plus a term containing the increment of the geometric stiffness matrix and the displacement increment. The additive format presents two advantages: the increment of the geometric stiffness is usually directly available, without implementation of special procedures on the element level, and the presence of the factor $\Delta\mathbf{u}$ allocates a special position of the term in the integration algorithm together with the damping term, thereby providing a physical interpretation of this algorithmic term.

9.3.3 Energy-conserving algorithm

The representative internal force can now be substituted into the state-space form (9.60) of the algorithm. In this substitution the incremental term is moved to the first matrix,

$$\begin{bmatrix} \mathbf{C} - \frac{1}{4}h\Delta\mathbf{K}^g & \mathbf{M} \\ \mathbf{M} & \mathbf{0} \end{bmatrix}\begin{bmatrix} \Delta\mathbf{u} \\ \Delta\mathbf{v} \end{bmatrix} + \frac{h}{2}\begin{bmatrix} \mathbf{g}(\mathbf{u}_{n+1}) + \mathbf{g}(\mathbf{u}_n) \\ -\mathbf{M}\,\mathbf{v}_{n+1} - \mathbf{M}\,\mathbf{v}_n \end{bmatrix} = \begin{bmatrix} h\,\bar{\mathbf{f}} \\ \mathbf{0} \end{bmatrix}. \tag{9.86}$$

This formula gives an energy-conserving discretized form of the equations of motion. It has the form of a simple mean value based algorithm with the exception of a single extra term containing the incremental geometric stiffness matrix $\Delta\mathbf{K}^g$. This term appears in the same position as the viscous damping matrix $\mathbf{C}$. Thus, it is seen that the use of a simple mean value based integration algorithm, in which the incremental geometric stiffness term is omitted, corresponds to introduction of an algorithmic viscous damping of magnitude $\frac{1}{4}h\Delta\mathbf{K}^g$. Typically the sign of the stress increments will change during a response analysis, and thus the geometric non-linearity will result in periods with negative damping if the incremental geometric stiffness term is omitted.

Reduction to displacement format

The full state-space format was introduced to identify the structure of the algorithm and to provide a clear interpretation of the additional term required for energy conservation. In the actual computation it is advantageous to eliminate the explicit dependence on the velocity components $\mathbf{v}_{n+1}$ in the matrix equations. The second equation in (9.86) is just the central difference form of the displacement velocity relation,

$$\Delta\mathbf{u} = \tfrac{1}{2}h(\mathbf{v}_{n+1} + \mathbf{v}_n). \tag{9.87}$$

This equation is used to express the velocity increment,

$$\Delta\mathbf{v} = \frac{2}{h}\Delta\mathbf{u} - 2\mathbf{v}_n. \tag{9.88}$$

When this is introduced into the first of the equations (9.86), the following equation is obtained for the displacement increment:

$$\Big[\frac{4}{h^2}\mathbf{M} + \frac{2}{h}\mathbf{C} - \frac{1}{2}\Delta\mathbf{K}^g\Big]\Delta\mathbf{u} + \mathbf{g}(\mathbf{u}_{n+1}) = \mathbf{f}_{n+1} + \Big[\mathbf{f}_n - \mathbf{g}(\mathbf{u}_n) + \frac{4}{h}\mathbf{M}\mathbf{v}_n\Big]. \tag{9.89}$$

Thus, advancing the solution one time step involves iterative solution of the discretized equation of motion (9.89) for $\Delta\mathbf{u}$, followed by evaluation of the velocity increment $\Delta\mathbf{v}$ by use of (9.88).

The iteration process

The residual is chosen as the difference between the right- and the left-hand sides of (9.89),

$$\mathbf{r} = \mathbf{f}_{n+1}+\mathbf{f}_n - \Big[\mathbf{g}(\mathbf{u}_{n+1})+\mathbf{g}(\mathbf{u}_n)\Big] - \Big[\frac{4}{h^2}\mathbf{M}+\frac{2}{h}\mathbf{C}-\frac{1}{2}\Delta\mathbf{K}^g\Big]\Delta\mathbf{u} + \frac{4}{h}\mathbf{M}\mathbf{v}_n. \tag{9.90}$$

Iteration by the Newton–Raphson procedure essentially amounts to calculating the value of the residual $\mathbf{r}$, and if this is not sufficiently close to zero, a linearized increment $\delta\mathbf{r}$ is calculated in order to make the residual vanish, i.e.

$$\mathbf{r} + \delta\mathbf{r} = \mathbf{0}. \tag{9.91}$$

When obtaining the linearized increment $\delta\mathbf{r}$ it is advantageous to recombine the internal forces and the incremental geometric stiffness matrix by use of (9.85),

$$\delta\mathbf{r} = -\,\delta\big(2\,\mathbf{g}_*\big) - \Big[\frac{4}{h^2}\mathbf{M} + \frac{2}{h}\mathbf{C}\Big]\delta\mathbf{u}. \tag{9.92}$$

The first term can be evaluated from the definition (9.72) of the representative internal force $\mathbf{g}_*$, whereby derivatives of the stiffness matrices are avoided,

$$\delta\big(2\,\mathbf{g}_*\big) = \int_{V_0} \delta(\boldsymbol{\nabla}_{\mathbf{u}}\mathbf{E}^T)\;\bar{\mathbf{S}}\;dV_0 + \int_{V_0} (\overline{\boldsymbol{\nabla}_{\mathbf{u}}\mathbf{E}^T})\;\delta\mathbf{S}\;dV_0. \tag{9.93}$$

When the variations in the integrals are expressed in terms of $\delta\mathbf{u}$, the integrals define a relation of the form

$$\delta\big(2\,\mathbf{g}_*\big) = \big(\mathbf{K}_*^g + \mathbf{K}_*^c\big)\,\delta\mathbf{u}. \tag{9.94}$$

with a geometric iteration matrix defined by

$$\mathbf{K}_*^g = \int_{V_0} (\boldsymbol{\nabla}_{\mathbf{u}}\boldsymbol{\nabla}_{\mathbf{u}}^T\mathbf{E}^T)\;\bar{\mathbf{S}}\;dV_0 \tag{9.95}$$

and a constitutive iteration matrix defined by

$$\mathbf{K}_*^c = \int_{V_0} (\overline{\boldsymbol{\nabla}_{\mathbf{u}}\mathbf{E}^T})\,\mathbb{C}\,(\boldsymbol{\nabla}_{\mathbf{u}}\mathbf{E}^T)^T\,dV_0. \tag{9.96}$$

With these definitions equation (9.91) for the sub-increment $\delta\mathbf{u}$ takes the form

$$\mathbf{K}_*\,\delta\mathbf{u} = \mathbf{r} \tag{9.97}$$

with the effective stiffness matrix

$$\mathbf{K}_* = \mathbf{K}_*^c + \mathbf{K}_*^g + \Big[\frac{4}{h^2}\mathbf{M} + \frac{2}{h}\mathbf{C}\Big]. \tag{9.98}$$

The residual $\mathbf{r}$ was defined in (9.90). In essence the residual is the sum of the unbalance of the equation of motion at times t_n and t_{n+1}. The sum is here chosen instead of the mean value to make the iteration matrices similar to those of the corresponding quasi-static formulation. Use of the mean value would have introduced the factor $\frac{1}{2}$.

It follows from the definition (9.95) that the geometric iteration stiffness matrix is symmetric and given by the mean value of the tangent geometric stiffness matrix at the end-points of the current interval,

$$\mathbf{K}_*^g = \tfrac{1}{2}\big(\mathbf{K}_{n+1}^g + \mathbf{K}_n^g\big). \tag{9.99}$$

The iteration constitutive stiffness matrix is in general non-symmetric, and

ALGORITHM 9.3. Non-linear energy-conserving algorithm.

---

(1) Initial conditions $\mathbf{u}_0$, $\mathbf{v}_0$

(2) Prediction step:

$$\Delta\mathbf{u} = h\,\mathbf{v}_n$$

(3) Residual calculation:

$$\begin{aligned}\mathbf{u}_{n+1} &= \mathbf{u}_n + \Delta\mathbf{u}\\ \mathbf{r} &= \mathbf{f}_{n+1} + \mathbf{f}_n - (\mathbf{g}_{n+1} + \mathbf{g}_n) + (4/h)\mathbf{M}\,\mathbf{v}_n\\ &\quad -\big[(2/h)^2\mathbf{M} + (2/h)\mathbf{C} - \tfrac{1}{2}\Delta\mathbf{K}^g\big]\Delta\mathbf{u}\end{aligned}$$

(4) Displacement sub-increment:

$$\begin{aligned}\mathbf{K}_* &= \big[\mathbf{K}_*^c + \mathbf{K}_*^g + (2/h)^2\mathbf{M} + (2/h)\mathbf{C}\big]\\ \delta\mathbf{u} &= \mathbf{K}_*^{-1}\,\mathbf{r}\\ \Delta\mathbf{u} &= \Delta\mathbf{u} + \delta\mathbf{u}\end{aligned}$$

If $\mathbf{r} > \varepsilon_{\mathbf{r}}$ or $\delta\mathbf{u} > \varepsilon_{\mathbf{u}}$ repeat from (3)

(5) State vector update:

$$\begin{aligned}\mathbf{u}_{n+1} &= \mathbf{u}_n + \Delta\mathbf{u}\\ \mathbf{v}_{n+1} &= (2/h)\Delta\mathbf{u} - \mathbf{v}_n\end{aligned}$$

(6) Return to (2) for new time step, or stop

---

there is no exact expression in terms of the tangent constitutive stiffness matrix. However, the non-symmetry is typically modest, and the iteration matrix can be replaced by a linear combination of the tangent constitutive stiffness matrix at the interval end-points. The asymptotic optimal combination has been given in Krenk (2007a), but numerical experience indicates that a direct mean value like for the geometric stiffness works equally well,

$$\mathbf{K}_*^c \simeq \tfrac{1}{2}\big(\mathbf{K}_{n+1}^c + \mathbf{K}_n^c\big). \tag{9.100}$$

If these mean value expressions are used for both iteration stiffness matrices no additional programming is needed at the element level to implement the energy-conserving time integration algorithm.

The implementation of the algorithm is illustrated as pseudo-code in Algorithm 9.3. The energy conservation algorithm deals specifically with the interval $[t_n, t_{n+1}]$, and the equation of motion is therefore not matched at specific points. This leads to a residual including the extra term $\Delta\mathbf{K}^g\Delta\mathbf{u}$, and combines the equations of motion at t_n and t_{n+1}. The method is formulated entirely in the state-space variables $\mathbf{u}$ and $\mathbf{v}$, and thus the acceleration is not evaluated explicitly in the algorithm. The absence of an explicit value of the acceleration leads to the use of a simple predictor. If needed, the acceleration can be evaluated from the equation of motion at the corresponding time or by a suitable interpolation formula.

EXAMPLE 9.2. THE ELASTIC PENDULUM – ENERGY CONSERVATION. In this example the motion of the elastic pendulum introduced in Example 9.1 is integrated by the energy-conserving algorithm. The mass matrix $\mathbf{M}$, the internal force vector $\mathbf{g}$ and the external load vector $\mathbf{f}$ are as defined before. However, the stiffness matrix now appears in two roles: as the incremental geometric stiffness

$$\Delta\mathbf{K}^g = \frac{\Delta N}{l_0}\begin{bmatrix} 1 & 0 \\ 0 & 1 \end{bmatrix}$$

and as the stiffness matrix needed in the Newton iterations

$$\mathbf{K}_*^g + \mathbf{K}_*^c = \frac{\bar{N}}{l_0}\begin{bmatrix} 1 & 0 \\ 0 & 1 \end{bmatrix} + \frac{EA}{l_0^3}\begin{bmatrix} \bar{x}\,x & \bar{x}\,y \\ \bar{y}\,x & \bar{y}\,y \end{bmatrix}.$$

These matrices are now used in the energy-conserving Algorithm 9.3.

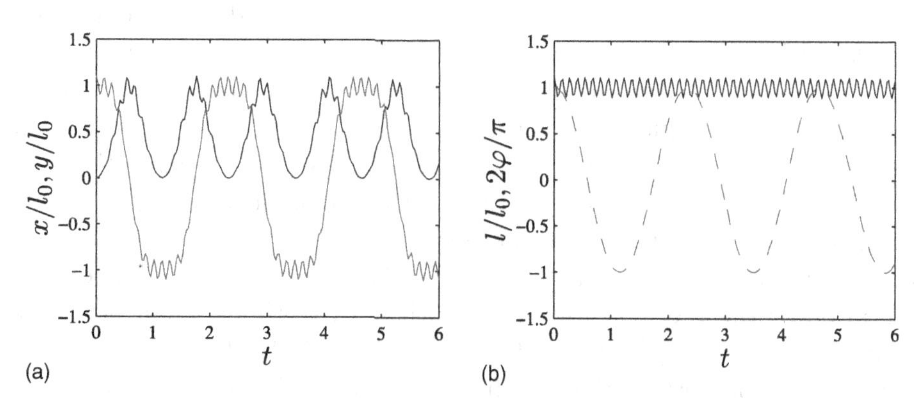

Fig. 9.7. Energy-conserving integration of pendulum vibrations with $h = 0.03$. (a) Coordinates $x(t)$ and $y(t)$, (b) length l/l_0 and angle $2\varphi/\pi$.

The parameters of the elastic pendulum are chosen as in Example 9.1: $m = 1$, $l_0 = 1$, $g = 10$ and $EA = 3000$ with the period of pendulum vibrations $T_p = 1.987$ and the vibration period of the bar $T_b = 0.1147$. The coordinates $x(t)$ and $y(t)$ obtained by the energy-conserving algorithm with $h = 0.03$ are shown in Fig. 9.7(a), and the alternative representation of the motion in terms of $l(t)$ and $\varphi(t)$ is shown in Fig. 9.7(b). The results are very similar to those obtained by the non-linear Newmark method in Example 9.1 and shown in Fig. 9.5. However, small differences lead to differences in the energy, and in the present algorithm the total energy is constant, $E/mgl_0 = 1.6357$, corresponding to the elastic energy of the bar due to the initial elongation.

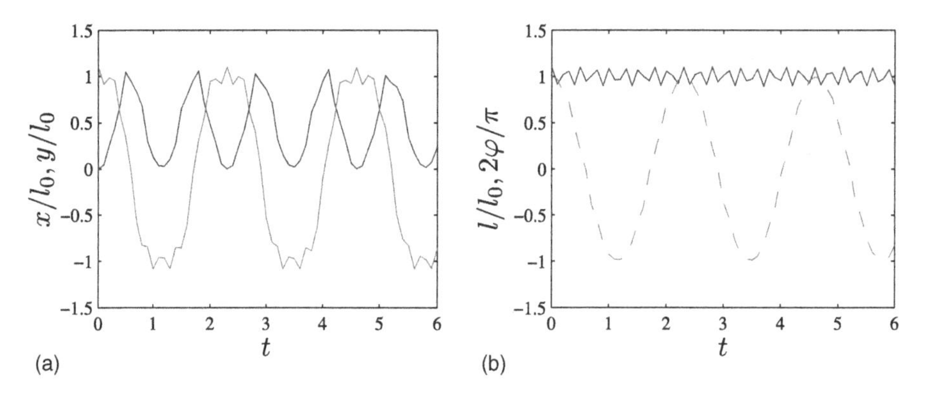

Fig. 9.8. Energy-conserving integration of pendulum vibrations with $h = 0.1$. (a) Coordinates $x(t)$ and $y(t)$, (b) length l/l_0 and angle $2\varphi/\pi$.

The constant energy of Algorithm 9.3 is important because it leads to more accurate results when integrated with an intermediate time increment like $h = 0.03$, but also because it permits the use of a considerably larger time increment. If the high-frequency oscillation is considered as an example of a high-frequency disturbance, which is not a part of the response to be identified, it is desirable to be able to use a fairly large time step, that permits accurate representation of the low-frequency response, while accepting that the high-frequency response will be misrepresented due to under-sampling. This is illustrated in Fig. 9.8, showing the results of an integration by the energy-conserving algorithm with time increment $h = 0.1$. This time increment is close to the period $T_b = 0.1147$ of the bar vibrations, and the time history of the length of the bar $l(t)$ looks erratic due to the crude sampling. However, it is important that the low-frequency pendulum vibration is still accurately represented with approximately 20 points per period, and the energy of the system is constant with its value defined by the initial conditions as before.

In the non-linear Newmark algorithm the energy fluctuations increase with increasing length of the time increment h, and numerical experiments indicate that in the present example the non-linear Newmark algorithm becomes unstable for $h \gtrsim 0.5\,T_b$. This demonstrates the importance of using the energy-conserving form (9.79) for correct evaluation of the internal forces, conveniently implemented as an additional term containing the incremental geometric stiffness matrix $\Delta\mathbf{K}^g$, as shown in Algorithm 9.3.

EXAMPLE 9.3. FREE MOTION OF A SOLID FRAME. This example is concerned with the free motion of a solid frame, illustrated in Fig. 9.9. The frame is quadratic, consisting of four identical sides with interior length $L = 1.0\,\mathrm{m}$ and square cross-section with dimension $b = 0.1\,\mathrm{m}$. The frame is linear elastic in the Green strains with modulus of elasticity $E = 10^6\,\mathrm{Pa}$ and Poisson ratio $\nu = 0.3$. The mass density is $\rho = 10^3\,\mathrm{kg/m^3}$, giving the frame a total mass of $m = 44\,\mathrm{kg}$.

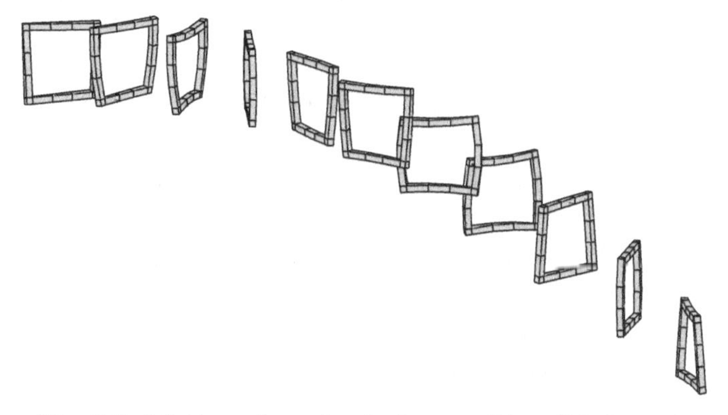

Fig. 9.9. Motion of quadratic frame with solid elements.

The frame is modeled by 16 8-node elements, giving a total of 64 nodes and 192 DOF. In small amplitude vibrations this corresponds to 6 modes of free motion plus 186 vibration modes. The angular frequency range is $0\,\mathrm{rad/s} \le \omega_i \le 1530\,\mathrm{rad/s}$. The vibration modes of most structural models consist of low-frequency modes describing overall deformation of the structure, plus high-frequency modes representing local vibrations on the element level. Typically, the high-frequency modes do not represent the dynamic vibration of the continuous body well, and it is of interest to perform the time integration with a time step that reflects interest in the lower modes.

The modes 7, 9 and 30, representative of the frame-like behavior, are shown in Fig. 9.10. These three modes correspond to warping, in-plane bending and out-of-plane bending, respectively. The modes 70, 120 and 170, representative of element-level vibrations, are shown in Fig. 9.11. Broadly speaking, the frame behavior is described by the modes with natural angular frequency $\omega_i \le 200\,\mathrm{rad/s}$.

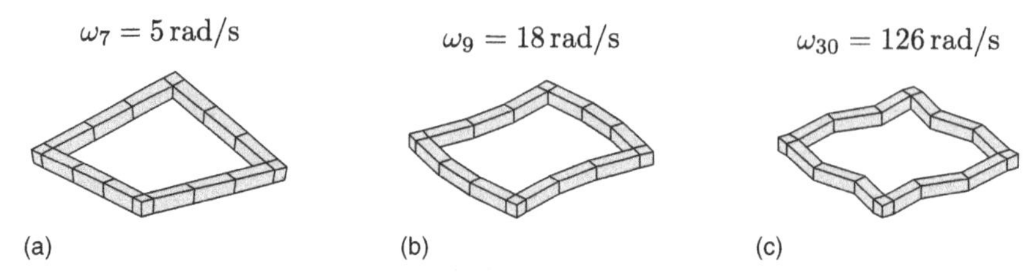

Fig. 9.10. Low-frequency 'frame-like' modes.

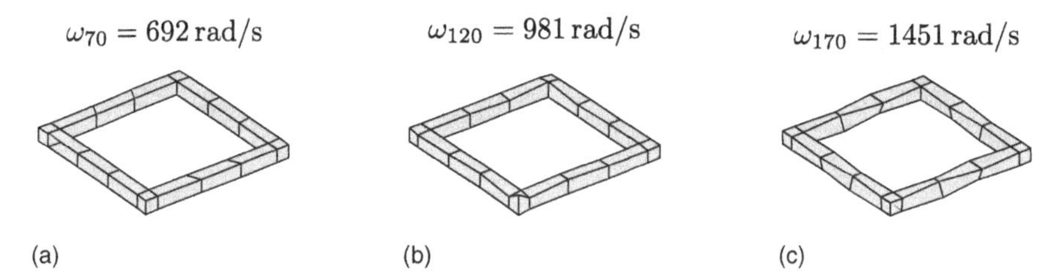

Fig. 9.11. High-frequency 'local vibration' modes.

The frame is moving in a homogeneous gravitational field with acceleration $g = 10\,\mathrm{m/s^2}$. The initial conditions specify the velocity of the center of mass $\mathbf{v}_0 = [10, 0, 0]\,\mathrm{m/s}$ and a rigid body rotation around the center of mass with angular velocity $\boldsymbol{\omega}_0 = [0, 0, 5]\,\mathrm{rad/s}$. In addition, the frame has a deformation in the lowest vibration mode shown in Fig. 9.10(a) corresponding to warping out of the plane of the frame with angular frequency $\omega_7 = 4.52\,\mathrm{rad/s}$.

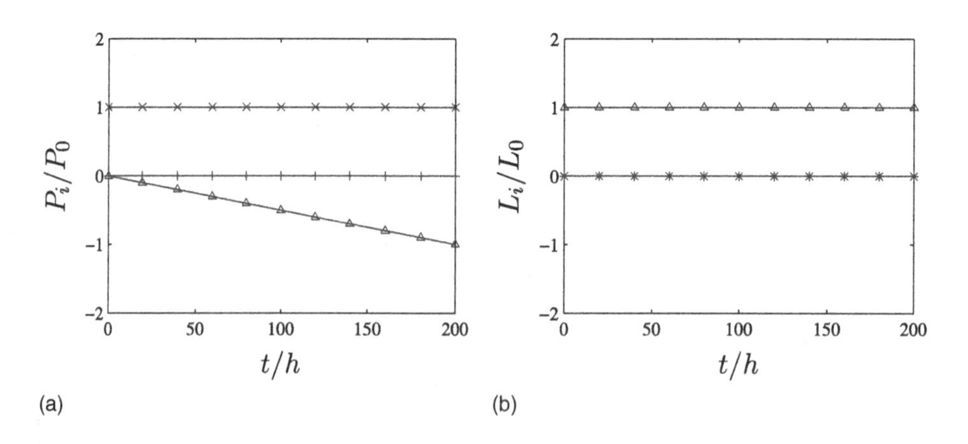

Fig. 9.12. Solid frame. (a) Linear momentum components P_i, (b) angular momentum components L_i.

A time integration was performed by Algorithm 9.3 with time step $h = 0.005\,\mathrm{s}$. This time step corresponds to about 3 points per period at the angular frequency 200 rad/s representing the upper frequency limit of the frame-like modes. The tolerances were set to $\varepsilon_{\mathbf{r}} = 10^{-3}\,$N and $\varepsilon_{\mathbf{u}} = 10^{-12}\,$m. The motion is shown at 20 step intervals in Fig. 9.9. The components of the linear momentum P_i and the angular momentum L_i are shown in Fig. 9.12. The components of the angular momentum are constant to within 12 digits, and the linear momentum is described exactly within the 14 digits of the computed format. The total energy is conserved to a relative accuracy of 14 digits.

9.4 Algorithmic energy dissipation

It is desirable that time integration algorithms used for structural dynamics can introduce energy dissipation in high-frequency components of the response that may occur in the system as a consequence of the discretization of the problem. The Newmark algorithm permits introduction of this so-called algorithmic damping by selecting $\alpha = \gamma - \frac{1}{2} > 0$ and $\beta = \frac{1}{4}(1+\alpha)^2$ as discussed in Section 9.1. However, it is seen from the energy balance equation (9.19) that this way of introducing algorithmic damping also leads to an undesirable reinterpretation of the energy. In view of the importance of precise energy relations in kinematically non-linear problems it is of interest to develop a method that introduces algorithmic damping into the energy balance in the form of energy dissipation terms without influencing the definition of the mechanical energy. The simplest way to do this is to introduce damping terms of order $O(h)$ directly into the state-space form (9.60) of the energy-conserving algorithm. In this process it is important to ensure that the damping terms are balanced to produce a complex amplification factor for all frequencies (Armero and Romero, 2001b; Krenk, 2007a). Therefore the procedure is first developed via a linear spectral analysis, and then extended to non-linear systems in the following two subsections. A high-frequency modification of the algorithmic damping has been developed in Krenk (2008), but is currently not extended to non-linear problems.

9.4.1 Spectral analysis of linear systems

For a linear dynamic system the elastic energy is given in terms of the stiffness matrix $\mathbf{K}$ as $G = \frac{1}{2}\mathbf{u}^T\mathbf{K}\mathbf{u}$. The state-space form of the integrated equation of motion then follows directly from (9.60). In this equation algorithmic damping terms of order $O(h)$ are introduced in the diagonal of the increment matrix. For a linear system without structural damping this leads to the equation of motion in the form

$$\begin{bmatrix} \alpha_1 h\mathbf{K} & \mathbf{M} \\ \mathbf{M} & -\alpha_2 h\mathbf{M} \end{bmatrix} \begin{bmatrix} \Delta\mathbf{u} \\ \Delta\mathbf{v} \end{bmatrix} + \begin{bmatrix} \mathbf{K} & \mathbf{0} \\ \mathbf{0} & -\mathbf{M} \end{bmatrix} \begin{bmatrix} h\bar{\mathbf{u}} \\ h\bar{\mathbf{v}} \end{bmatrix} = \begin{bmatrix} h\,\bar{\mathbf{f}} \\ \mathbf{0} \end{bmatrix}. \tag{9.101}$$

The extra diagonal terms are determined by the fact that only the system matrices $\mathbf{K}$ and $\mathbf{M}$ are available, and the extra terms must be of order $O(h)$. The energy balance equation is obtained by pre-multiplication with $[\Delta\mathbf{u}^T, -\Delta\mathbf{v}^T]$ as

$$\begin{aligned} \big[\tfrac{1}{2}\mathbf{v}^T\mathbf{M}\,\mathbf{v} + \tfrac{1}{2}\mathbf{u}^T\mathbf{K}\,\mathbf{u}\big]_n^{n+1} &= \\ \Delta\mathbf{u}^T\bar{\mathbf{f}} \;-\; \alpha_1\Delta\mathbf{u}^T\mathbf{K}\,\Delta\mathbf{u} \;&-\; \alpha_2\Delta\mathbf{v}^T\mathbf{M}\,\Delta\mathbf{v}. \end{aligned} \tag{9.102}$$

It is seen that both coefficients α_1 and α_2 contribute an energy dissipation term when they are positive. The optimal balance between these two terms is now determined by a spectral analysis.

The spectral analysis is carried out by writing the homogeneous equations in recurrence form,

$$\begin{bmatrix} (\frac{1}{2}+\alpha_1)h\mathbf{K} & \mathbf{M} \\ \mathbf{M} & -(\frac{1}{2}+\alpha_2)h\mathbf{M} \end{bmatrix} \begin{bmatrix} \mathbf{u} \\ \mathbf{v} \end{bmatrix}_{n+1} = \begin{bmatrix} -(\frac{1}{2}-\alpha_1)h\mathbf{K} & \mathbf{M} \\ \mathbf{M} & (\frac{1}{2}-\alpha_2)h\mathbf{M} \end{bmatrix} \begin{bmatrix} \mathbf{u} \\ \mathbf{v} \end{bmatrix}_{n}. \tag{9.103}$$

As explained in detail in the spectral analysis of the Newmark algorithm in Section 9.1, these equations are now applied to a single mode with natural frequency ω and modal coordinates $[u(t), v(t)]$. The resulting equations are expressed in terms of the non-dimensional frequency $\Omega = \omega h$,

$$\begin{bmatrix} (\frac{1}{2}+\alpha_1)\Omega^2 & 1 \\ 1 & -(\frac{1}{2}+\alpha_2) \end{bmatrix} \begin{bmatrix} u \\ hv \end{bmatrix}_{n+1} = \begin{bmatrix} -(\frac{1}{2}-\alpha_1)\Omega^2 & 1 \\ 1 & (\frac{1}{2}-\alpha_2) \end{bmatrix} \begin{bmatrix} u \\ hv \end{bmatrix}_{n}. \tag{9.104}$$

When considering the free response in terms of an amplification factor λ as shown in (9.30), this equation takes the form of the eigenvalue problem

$$\left(\begin{bmatrix} -(\frac{1}{2}-\alpha_1)\Omega^2 & 1 \\ 1 & (\frac{1}{2}-\alpha_2) \end{bmatrix} - \lambda \begin{bmatrix} (\frac{1}{2}+\alpha_1)\Omega^2 & 1 \\ 1 & -(\frac{1}{2}+\alpha_2) \end{bmatrix} \right) \begin{bmatrix} u \\ hv \end{bmatrix} = \begin{bmatrix} 0 \\ 0 \end{bmatrix}. \tag{9.105}$$

The eigenvalue λ is determined by the characteristic equation

$$\begin{aligned} \big[1 + (\tfrac{1}{2}+\alpha_1)(\tfrac{1}{2}+\alpha_2)\Omega^2\big]\lambda^2 \;-\; 2\big[1 - (\tfrac{1}{4} - \alpha_1\alpha_2)\Omega^2\big]\lambda& \\ + \big[1 + (\tfrac{1}{2}-\alpha_1)(\tfrac{1}{2}-\alpha_2)\Omega^2\big] \;=\; 0.& \end{aligned} \tag{9.106}$$

This is a quadratic equation of the form

$$A\,\lambda^2 \;-\; 2B\,\lambda \;+\; C \;=\; 0 \tag{9.107}$$

and the question of complex roots is therefore determined by the discriminant $AC - B^2$. After some algebra it follows that the discriminant is

$$AC - B^2 = \Omega^2 - \tfrac{1}{4}(\alpha_1 - \alpha_2)^2\,\Omega^4. \tag{9.108}$$

If the parameters α_1 and α_2 are different, the sign of the discriminant will change from positive to negative for some finite frequency. This is undesirable, as it would imply that the complex locus of the amplification factor has a bifurcation point on the real axis, and branches along the real axis for high frequencies. Although the algorithm would be stable, this would lead to a non-monotonic damping frequency relation. Thus, the best properties of the algorithm will be obtained by selecting the two parameters equal. It will be convenient to express the parameters in terms of a common parameter α, defined as

$$\alpha_1 = \alpha_2 = \tfrac{1}{2}\alpha. \tag{9.109}$$

The characteristic equation (9.106) then takes the simplified form

$$\begin{aligned}\big[\,1 + \tfrac{1}{4}(1+\alpha)^2\Omega^2\,\big]\lambda^2 \;-\; 2\big[\,1 - \tfrac{1}{4}(1-\alpha^2)\Omega^2\,\big]\lambda \\ +\big[\,1 + \tfrac{1}{4}(1-\alpha)^2\Omega^2\,\big] = 0.\end{aligned} \tag{9.110}$$

It is easily verified that this quadratic equation is the same as (9.32) for the Newmark method when the parameters γ and β are expressed in terms of α according to (9.40), see Exercise 9.7. Thus, the spectral properties of the two algorithms are identical. This implies that α is given in terms of λ_∞ by (9.43), and the low-frequency asymptotic behavior is given by (9.41), (9.42) and (9.44). Also, the spectral behavior illustrated in Fig. 9.3 remains valid for the present algorithm. In spite of identical spectral behavior of the two algorithms the energy balance is different due to differences in the eigenvalue problem defining the free vibration modes $[u, hv]$ of the algorithms.

This algorithm can be modified to high-frequency dissipation by introducing auxiliary state variables, connected to the original state variables by suitable filter equations (Krenk, 2008). The result is an algorithm with frequency properties comparable to the generalized-α method, but without the energy oscillations of the high-frequency components associated with collocation methods.

9.4.2 Linear algorithm with energy dissipation

The algorithm with consistent energy dissipation can now be expressed for linear systems including structural damping $\mathbf{C}$ as Algorithm 9.4. The state-

ALGORITHM 9.4. Linear energy dissipating algorithm.

---

(1) System matrices $\mathbf{K}$, $\mathbf{C}$, $\mathbf{M}$,

$$\mathbf{K}_* = \kappa\Big[\mathbf{K} + \frac{2}{\kappa h}\mathbf{C} + \Big(\frac{2}{\kappa h}\Big)^2\mathbf{M}\Big]$$

(2) Initial conditions $\mathbf{u}_0$, $\mathbf{v}_0$

(3) Displacement increment:

$$\Delta\mathbf{u} = \mathbf{K}_*^{-1}\Big[\mathbf{f}_{n+1} + \mathbf{f}_n - 2\mathbf{K}\mathbf{u}_n + \frac{4}{\kappa h}\mathbf{M}\mathbf{v}_n\Big]$$

(4) State vector update:

$$\mathbf{u}_{n+1} = \mathbf{u}_n + \Delta\mathbf{u}$$

$$\mathbf{v}_{n+1} = \mathbf{v}_n + \Big[\frac{2}{\kappa h}\Delta\mathbf{u} - \frac{2}{\kappa}\mathbf{v}_n\Big]$$

(5) Return to (3) for new time step, or stop

---

space form follows from (9.101) as

$$\begin{bmatrix} \mathbf{C}+\frac{1}{2}\alpha h\mathbf{K} & \mathbf{M} \\ \mathbf{M} & -\frac{1}{2}\alpha h\mathbf{M} \end{bmatrix}\begin{bmatrix} \Delta\mathbf{u} \\ \Delta\mathbf{v} \end{bmatrix} + \begin{bmatrix} \mathbf{K} & \mathbf{0} \\ \mathbf{0} & -\mathbf{M} \end{bmatrix}\begin{bmatrix} h\bar{\mathbf{u}} \\ h\bar{\mathbf{v}} \end{bmatrix} = \begin{bmatrix} h\bar{\mathbf{f}} \\ \mathbf{0} \end{bmatrix}. \quad (9.111)$$

It is most efficient to solve these equations by eliminating the velocity increment $\Delta\mathbf{v}$ from the first equation. The formulation takes on a particularly compact form when introducing the dissipation parameter

$$\kappa = 1 + \alpha. \quad (9.112)$$

The second state-space equation then gives the forward velocity $\mathbf{v}_{n+1}$ via the simple vector relation

$$\kappa\,\Delta\mathbf{v} = \frac{2}{h}\Delta\mathbf{u} - 2\mathbf{v}_n. \quad (9.113)$$

It is observed that the effect of the parameter α is to reduce the velocity increment $\Delta\mathbf{v}$ in the algorithm. In the Newmark algorithm this effect is accomplished by a correction involving the acceleration, while here the effect is obtained by a simple weighting.

The algorithm is most conveniently formulated in terms of the increments $\Delta\mathbf{u}$ and $\Delta\mathbf{v}$. Substitution of $\Delta\mathbf{v}$ from (9.113) into the first of the state-space equations (9.111a) and multiplication by $2/h$ gives the following equation for the displacement increment $\Delta\mathbf{u}$:

$$\kappa\Big[\mathbf{K}+\frac{2}{\kappa h}\mathbf{C}+\frac{4}{(\kappa h)^2}\mathbf{M}\Big]\Delta\mathbf{u} = \big(\mathbf{f}_{n+1}+\mathbf{f}_n\big)-2\mathbf{K}\mathbf{u}_n+\frac{4}{\kappa h}\mathbf{M}\,\mathbf{v}_n. \tag{9.114}$$

In this equation the displacement is determined by the effective stiffness matrix

$$\mathbf{K}_* = \kappa\Big[\mathbf{K}+\frac{2}{\kappa h}\mathbf{C}+\frac{4}{(\kappa h)^2}\mathbf{M}\Big]. \tag{9.115}$$

It is seen that the terms are scaled as if extending the time increment from h to κh, whereby the dynamic terms get less weight.

9.4.3 Non-linear algorithm with energy dissipation

The introduction of algorithmic damping into the energy-conserving algorithm of Section 9.3 consists in the introduction of two damping terms in the state-space equations. The term $-\frac{1}{2}\alpha h\mathbf{M}\Delta\mathbf{v}$ in the second state-space equation is linear and can be introduced directly. Thus, the second state-space equation is the same as (9.111), and $\Delta\mathbf{v}$ is expressed in terms of $\Delta\mathbf{u}$ by (9.113). The linear form of the damping term for the first state-space equation is $\frac{1}{2}\alpha h\mathbf{K}\Delta\mathbf{u}$. This term is proportional to the increment of the internal force vector, and the non-linear equivalent is $\frac{1}{2}\alpha h\dot{\mathbf{g}}$. When this term is inserted into the first state-space equation (9.86a), the following form of the equation of motion is obtained after division by $\frac{1}{2}h$:

$$\begin{aligned}\frac{2}{h}\Big[\mathbf{M}\Delta\mathbf{v} + \big(\mathbf{C}-\tfrac{1}{4}h\Delta\mathbf{K}^g\big)\Delta\mathbf{u}\Big] &+ \\ (1+\alpha)\mathbf{g}_{n+1}+(1-\alpha)\mathbf{g}_n &= \mathbf{f}_{n+1}+\mathbf{f}_n.\end{aligned} \tag{9.116}$$

In this equation the velocity increment $\Delta\mathbf{v}$ is eliminated by use of (9.113), and the damping is represented by the parameter $\kappa = 1+\alpha$ introduced in (9.112), giving an equation in terms of the unknown displacement $\mathbf{u}_{n+1}$:

$$\begin{aligned}\kappa\Big[\frac{4}{(\kappa h)^2}\mathbf{M}+\frac{2}{\kappa h}\mathbf{C}-\frac{1}{2\kappa}\Delta\mathbf{K}^g\Big]\Delta\mathbf{u} &+ \kappa\Delta\mathbf{g} \\ &= \mathbf{f}_{n+1}+\mathbf{f}_n-2\mathbf{g}_n+\frac{4}{\kappa h}\mathbf{M}\,\mathbf{v}_n.\end{aligned} \tag{9.117}$$

For $\alpha = 0$ this equation reproduces the energy-conserving equation of motion (9.86).

The residual of the non-linear equation of motion (9.117) is chosen as the

ALGORITHM 9.5. Non-linear energy-dissipating algorithm.

---

(1) Initial conditions $\mathbf{u}_0$, $\mathbf{v}_0$

(2) Prediction step:

$$\Delta\mathbf{u} = h\,\mathbf{v}_n$$

(3) Residual calculation:

$$\begin{aligned}\mathbf{u}_{n+1} &= \mathbf{u}_n + \Delta\mathbf{u}\\ \mathbf{r} &= \mathbf{f}_{n+1} + \mathbf{f}_n - \big[2\mathbf{g}_n + \kappa\Delta\mathbf{g}\big] + (4/\kappa h)\mathbf{M}\,\mathbf{v}_n\\ &\quad -\kappa\Big[\frac{4}{(\kappa h)^2}\mathbf{M} + \frac{2}{\kappa h}\mathbf{C} - \frac{1}{2\kappa}\Delta\mathbf{K}^g\Big]\Delta\mathbf{u}\end{aligned}$$

(4) Displacement sub-increment:

$$\begin{aligned}\mathbf{K}_* &= \kappa\Big[\mathbf{K}^c_\alpha + \mathbf{K}^g_\alpha + \frac{2}{\kappa h}\mathbf{C} + \frac{4}{(\kappa h)^2}\mathbf{M}\Big]\\ \delta\mathbf{u} &= \mathbf{K}_*^{-1}\mathbf{r}\\ \Delta\mathbf{u} &= \Delta\mathbf{u} + \delta\mathbf{u}\end{aligned}$$

If $\mathbf{r} > \varepsilon_\mathbf{r}$ or $\delta\mathbf{u} > \varepsilon_\mathbf{u}$ repeat from (3)

(5) State vector update:

$$\begin{aligned}\mathbf{u}_{n+1} &= \mathbf{u}_n + \Delta\mathbf{u}\\ \mathbf{v}_{n+1} &= \mathbf{v}_n + \Big[\frac{2}{\kappa h}\Delta\mathbf{u} - \frac{2}{\kappa}\mathbf{v}_n\Big]\end{aligned}$$

(6) Return to (2) for new time step, or stop

---

difference between the right- and left-hand sides,

$$\begin{aligned}\mathbf{r} = \mathbf{f}_{n+1} + \mathbf{f}_n &- \big[2\mathbf{g}_n + \kappa\Delta\mathbf{g}\big]\\ &- \kappa\Big[\frac{4}{(\kappa h)^2}\mathbf{M} + \frac{2}{\kappa h}\mathbf{C} - \frac{1}{2\kappa}\Delta\mathbf{K}^g\Big]\Delta\mathbf{u} + \frac{4}{\kappa h}\mathbf{M}\mathbf{v}_n.\end{aligned} \tag{9.118}$$

The solution is found by iterations using the linearized increment $\delta\mathbf{r}$ of the residual. When the linearized increment is written in the form $\delta\mathbf{r} = -\mathbf{K}_*\delta\mathbf{u}$, the sub-increment $\delta\mathbf{u}$ is determined from the equation

$$\mathbf{K}_*\,\delta\mathbf{u} = \mathbf{r}. \tag{9.119}$$

The effective stiffness matrix $\mathbf{K}_*$ is determined from the energy conservation result in Section 9.3.3,

$$\mathbf{K}_* = \kappa\Big[\mathbf{K}_\alpha^c + \mathbf{K}_\alpha^g + \frac{2}{\kappa h}\mathbf{C} + \frac{4}{(\kappa h)^2}\mathbf{M}\Big]. \tag{9.120}$$

The structure of this formula is identical to (9.115) for the linear case, when introducing the stiffness matrix via the constitutive and geometric stiffness matrices, given by

$$\kappa\,\mathbf{K}_\alpha^c = \mathbf{K}_*^c + \alpha\,\mathbf{K}^c, \qquad \kappa\,\mathbf{K}_\alpha^g = \mathbf{K}_*^g + \alpha\,\mathbf{K}^g. \tag{9.121}$$

In terms of the end-point values these relations are

$$\kappa\,\mathbf{K}_\alpha^c = (\kappa - \tfrac{1}{2})\mathbf{K}_{n+1}^c + \tfrac{1}{2}\mathbf{K}_n^c, \quad \kappa\,\mathbf{K}_\alpha^g = (\kappa - \tfrac{1}{2})\mathbf{K}_{n+1}^g + \tfrac{1}{2}\mathbf{K}^g, \tag{9.122}$$

where the constitutive relation is an approximation as discussed in connection with (9.100).

The implementation of the non-linear energy-dissipating algorithm is illustrated as pseudo-code in Algorithm 9.5. The only difference from the non-linear energy-conserving Algorithm 9.3 is the occurrence of the parameter κ in the residual, the iteration matrices, and the updating formula for the velocity. The effect of algorithmic energy dissipation is illustrated in the following example.

EXAMPLE 9.4. THE ELASTIC PENDULUM – ENERGY DISSIPATION. The motion of the elastic pendulum introduced in Examples 9.1 and 9.2 is integrated by the energy-dissipating algorithm. The mass matrix $\mathbf{M}$, the incremental geometric stiffness matrix $\Delta\mathbf{K}^g$, the internal force vector $\mathbf{g}$ and the external load vector $\mathbf{f}$ are as defined before. The effective constitutive and geometric stiffness matrices defined by (9.122) now take the form

$$\mathbf{K}_\alpha^g + \mathbf{K}_\alpha^c = \frac{\tilde{N}}{l_0}\begin{bmatrix} 1 & 0 \\ 0 & 1 \end{bmatrix} + \frac{EA}{l_0^3}\begin{bmatrix} \tilde{x}\,x & \tilde{x}\,y \\ \tilde{y}\,x & \tilde{y}\,y \end{bmatrix},$$

with $\kappa\tilde{N} = \bar{N} + \alpha N$, and similarly for $\tilde{x}$ and $\tilde{y}$.

The coordinates $x(t)$ and $y(t)$ obtained by the energy-dissipating algorithm with $\alpha = 0.02$ and $h = 0.03$ are shown in Fig. 9.13(a), and the alternative representation of the motion in terms of $l(t)$ and $\varphi(t)$ is shown in Fig. 9.13(b). It is seen that the algorithmic dissipation removes most of the high-frequency oscillations, while retaining the low-frequency oscillation only lightly damped.

Similar results are shown in Fig. 9.14 for the same value of the damping parameter, $\alpha = 0.02$, but the larger time increment $h = 0.1$. It is seen that in this case the algorithmic damping is slightly less efficient in removing

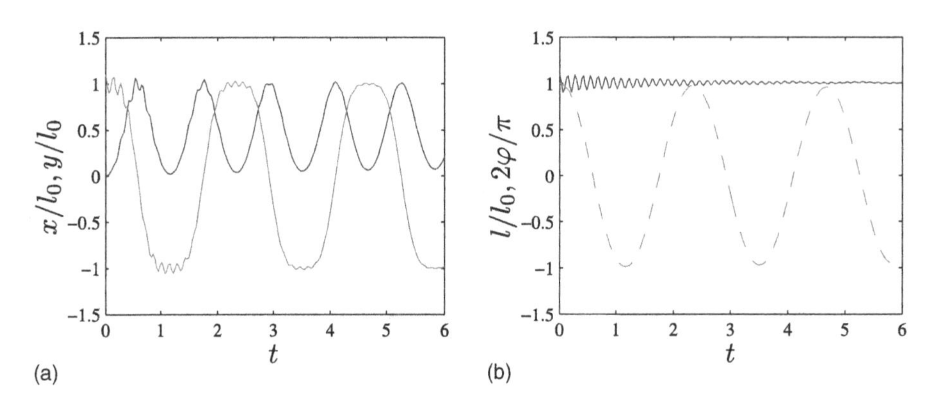

Fig. 9.13. Energy-dissipating integration of pendulum vibrations with $\alpha = 0.02$ and $h = 0.03$. (a) Coordinates $x(t)$ and $y(t)$, (b) length l/l_0 and angle $2\varphi/\pi$.

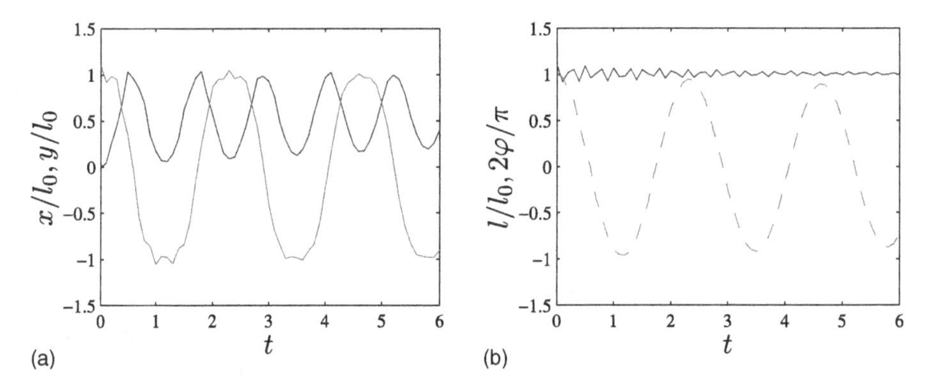

Fig. 9.14. Energy-dissipating integration of pendulum vibrations with $\alpha = 0.02$ and $h = 0.1$. (a) Coordinates $x(t)$ and $y(t)$, (b) length l/l_0 and angle $2\varphi/\pi$.

the high-frequency oscillations, while more damping is introduced into the low-frequency oscillation.

The effect of algorithmic damping on the energy of the system is illustrated in Fig. 9.15, showing the energy development for both time increments $h = 0.03$ and $h = 0.1$. In the initial configuration the potential and kinetic energies are zero, and the total energy is therefore determined by the elastic energy, $E_0/mgl_0 = 1.6357$. As the elastic energy in the pendulum is dissipated, the energy decreases to around zero. However, some energy is also dissipated from the pendulum oscillation, and the energy therefore continues to decrease slightly below zero, but at a lower rate. The energy curves illustrate higher efficiency of the algorithmic damping for $h = 0.03$,

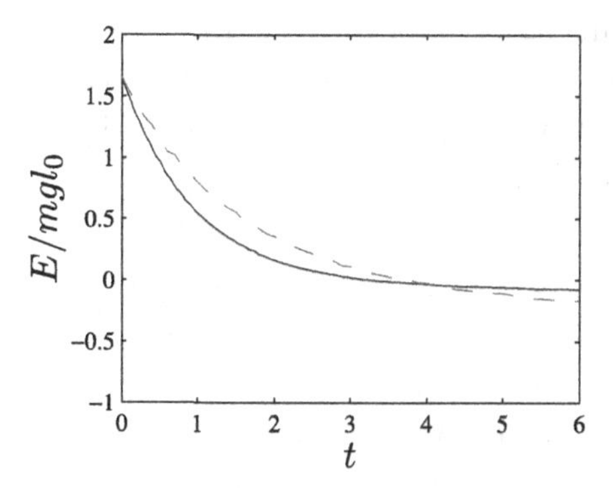

Fig. 9.15. Energy development for $\alpha = 0.02$: $h = 0.03$, —; $h = 0.1$, – –.

as this curve decreases more rapidly in the beginning, and then levels off closer to the zero energy level towards the end of the time interval.

9.5 Summary and outlook

Time integration methods have been presented from two different perspectives: collocation and momentum-based methods. In collocation methods the equation of motion is satisfied at selected points. This provides one equation for the three variables: displacement $\mathbf{u}$, velocity $\dot{\mathbf{u}}$ and acceleration $\ddot{\mathbf{u}}$. Thus, two additional equations are needed. They are typically formed by Taylor expansion-type approximations, and are therefore asymptotically accurate. However, these equations do not represent the conservation properties of the mechanics problem to be solved, except in special cases such as linear problems without damping.

A different approach is used in the momentum-based methods. These methods are developed specifically to represent e.g. conservation properties associated with the underlying problems, see e.g. the general discussion in Hairer *et al.* (2006). These methods found their way into elasto-dynamics via the work of Simo and Tarnow (1992). The equation of motion is considered as a first-order differential equation in time for the momentum of the system. The momentum is defined in terms of the motion by a supplementary first-order differential equation in time. These equations are then integrated to provide two equations for the development over a finite time increment. The central point is to perform this integration in a format that leads to the desired conservation properties. For linear systems the desired properties often follow from use of central differences and mean values, but for non-linear systems more elaborate schemes are needed. The central point is the 'representative average' of the internal forces discussed here in connection with the energy conservation relation (9.64). This equation may be stated in the general form

$$\big[\, G(\mathbf{u}) \,\big]_n^{n+1} \;=\; \int_{V_0} \big[\varphi(\mathbf{E}_{n+1}) - \varphi(\mathbf{E}_n)\big]\, dV_0 \;=\; \int_{V_0} \Delta\mathbf{E}^T \mathbf{S}_*\, dV_0, \qquad (9.123)$$

where the finite time increment of the global $G(\mathbf{u})$ is expressed in terms of the finite strain increment $\Delta\mathbf{E}$ and a representative stress $\mathbf{S}_*$. It was shown in (9.68) that for linear elasticity expressed in terms of the Green strain the representative stress is simply the mean value, i.e. $\mathbf{S}_* = \bar{\mathbf{S}}$, and it was demonstrated that this permits a simple explicit formulation of the effect of the non-linear strain. Two approaches have been used for more general stress–strain relations. Simo and Tarnow (1992) gave a formulation in which the representative stress is constructed from an intermediate state, a scheme later refined by Laursen and Meng (2001). A simpler method is the so-called discrete derivative introduced by Gonzalez (2000) and used, e.g. by Armero and Romero (2001a). The idea is to introduce the gradient of the internal energy in a secant format similar to that used in quasi-Newton methods, discussed in Section 8.4. Thus, the representative stress is defined by the discrete derivative

$$\mathbf{S}_* \;=\; \boldsymbol{\nabla}^*_{\mathbf{E}}\varphi(\mathbf{E}_{n+1/2}) \qquad (9.124)$$

in such a way that multiplication with the finite strain increment $\Delta\mathbf{E}$ produces the correct energy increment,

$$\Delta\mathbf{E}^T\,\mathbf{S}_* \;=\; \Delta\mathbf{E}^T\,\boldsymbol{\nabla}^*_{\mathbf{E}}\,\varphi(\mathbf{E}_{n+1/2}) \;=\; \varphi(\mathbf{E}_{n+1}) \;-\; \varphi(\mathbf{E}_n). \qquad (9.125)$$

This property is obtained as in the BFGS format by a two-term correction. The first term contains the increment of the internal energy, and the second term is a gradient approximation to this difference. The most common format is based on the gradient at the mean state:

$$\boldsymbol{\nabla}^*_{\mathbf{E}}\,\varphi \;=\; \boldsymbol{\nabla}_{\mathbf{E}}\,\varphi_{n+1/2} + \|\Delta\mathbf{E}\|^{-2}\Big[\,\varphi_{n+1} - \varphi_n - \Delta\mathbf{E}^T\boldsymbol{\nabla}_{\mathbf{E}}\,\varphi_{n+1/2}\Big]\Delta\mathbf{E}, \quad (9.126)$$

where the scalar factor in front of the square brackets is a norm of the finite strain increment, $\|\Delta\mathbf{E}\|^2 \;=\; \Delta\mathbf{E}^T\Delta\mathbf{E}$. For small strain increments the contribution from the square brackets becomes of higher order, leaving the gradient $\boldsymbol{\nabla}_{\mathbf{E}}\,\varphi_{n+1/2}$ as the dominating contribution. It is seen that this format satisfies the finite increment condition (9.125). Recently, momentum-conservation algorithms have also been applied to elasto-plasticity, see e.g. Armero (2006).

9.6 Exercises

Exercise 9.1 Determine the period error predicted by the linear theory for the pendulum motion and the oscillations of the elastic bar in Example 9.1.

Exercise 9.2* Demonstrate the effect of algorithmic damping for the pendulum problem of Example 9.1, e.g. by taking $\gamma = 0.52$ and $\beta = \frac{1}{4}(\gamma + \frac{1}{2})^2$. Show by numerical calculations that the non-linear Newmark Algorithm 9.2 becomes unstable for $h \simeq 0.06$, irrespective of application of algorithmic damping.

Exercise 9.3 Formulate the pendulum problem shown in Fig. 9.4 using engineering strain, defined in Chapter 2 as $\varepsilon_E = (l - l_0)/l_0$. (In essence the change in strain corresponds to replacing all occurrences of the initial length l_0 in the denominator with the current length l.)

Exercise 9.4* Use the formulation of the elastic pendulum problem in terms of engineering strain developed in Exercise 9.3 as the basis of a numerical analysis by the energy-conserving Algorithm 9.3. In particular, study the energy history.

Exercise 9.5* In integrated state-space based algorithms the acceleration does not appear directly. However, the acceleration may be of independent interest, either as a result of the analysis or as part of an improved predictor in non-linear problems. In the energy-conserving Algorithm 9.3 the inertial term is represented via the velocity increment $\Delta\mathbf{v}$, which may be related to the acceleration $\mathbf{w}$ by the central difference $\Delta\mathbf{v} = \frac{1}{2}h(\mathbf{w}_{n+1} + \mathbf{w}_n)$. When the initial acceleration $\mathbf{w}_0$ is evaluated from the equation of motion at time $t = 0$, this formula permits subsequent updates of the acceleration of the form $\mathbf{w}_{n+1} = (2/h)\Delta\mathbf{v} - \mathbf{w}_n$. Implement the acceleration vector $\mathbf{w}_n$ in the energy-conserving Algorithm 9.3 and recalculate the elastic pendulum problem of Example 9.2 including the acceleration.

Exercise 9.6* For central differences corresponding to $\gamma = \frac{1}{2}$, $\beta = \frac{1}{4}$ the displacement increment is approximated by (9.8b) as

$$\Delta\mathbf{u} = h\mathbf{v}_n + \tfrac{1}{4}h(\mathbf{w}_n + \mathbf{w}_{n+1}),$$

where $\mathbf{w}$ is the acceleration vector. This formula suggests the following three displacement increment predictors:

$$\Delta\mathbf{u}^A = h\mathbf{v}_n, \quad \Delta\mathbf{u}^B = h\mathbf{v}_n + \tfrac{1}{4}h\mathbf{w}_n, \quad \Delta\mathbf{u}^C = h\mathbf{v}_n + \tfrac{1}{2}h\mathbf{w}_n,$$

The predictor $\Delta\mathbf{u}^A$ disregards the effect of the acceleration, $\Delta\mathbf{u}^B$ corresponds to neglecting the acceleration at the end of the interval, while $\Delta\mathbf{u}^C$ corresponds to the assumption of constant acceleration.

Use the implementation from Exercise 9.5 of the acceleration vector $\mathbf{w}_n$ in the energy-conserving Algorithm 9.3 to investigate the efficiency of the three displacement increment predictors by application to the elastic pendulum treated in Example 9.2.

Exercise 9.7 Show that the characteristic equation (9.110) for the energy-conserving algorithm with dissipation is identical to Ω^2 times the characteristic equation (9.32) for the Newmark method, when the parameters γ and β are expressed in terms of α according to (9.40).

Exercise 9.8 Show that for an elastic Green strain bar element the constitutive and geometric contributions (9.96) and (9.95) to the iteration stiffness matrix are

$$\mathbf{K}_*^c = \int_{V_0} (\overline{\boldsymbol{\nabla}_{\mathbf{u}}\varepsilon})\, E\, (\boldsymbol{\nabla}_{\mathbf{u}}^T\varepsilon)\, dV_0 = \frac{EA_0}{a_0^3}\begin{bmatrix} \bar{\mathbf{a}}\,\mathbf{a}^T & -\bar{\mathbf{a}}\,\mathbf{a}^T \\ -\bar{\mathbf{a}}\,\mathbf{a}^T & \bar{\mathbf{a}}\,\mathbf{a}^T \end{bmatrix}$$

and

$$\mathbf{K}_*^g = \int_{V_0} \bar{\sigma}\,(\boldsymbol{\nabla}_{\mathbf{u}}\boldsymbol{\nabla}_{\mathbf{u}}^T\varepsilon)\, dV_0 = \frac{\bar{\sigma}A_0}{a_0}\begin{bmatrix} \mathbf{I} & -\mathbf{I} \\ -\mathbf{I} & \mathbf{I} \end{bmatrix}.$$

Exercise 9.9* Consider a linear elastic bar element AB described in terms of Green strain ε_G, located in the xy-plane. The initial position is determined by the nodes A at $(-1,0)$ and B at $(1,0)$. There is a concentrated mass $m=1$ at each of the nodes A and B, and the axial element stiffness is $EA_0 = 1$. Formulate the equations of motion in a format similar to that used in Example 9.1. Find the corresponding system matrices and integrate the equations of motion by the energy-conserving Algorithm 9.3 with initial conditions consisting of the initial velocity $\mathbf{v}_0^A = (0,1)$. Evaluate and plot the total energy and the velocity of the center of mass, and compare with the corresponding analytical results. Repeat the numerical analysis using the Newmark Algorithm 9.2 and compare the results.

Exercise 9.10* Let two bar elements AB and BC, connected at B, be located in the xy-plane. The initial position is determined by the nodes A at $(-1,0)$, B at $(0,0)$, and C at $(1,0)$. Let the mass be concentrated at the nodes with magnitude 1, 2 and 1 at A, B and C, respectively, and use the stiffness $EA_0 = 1$. Assemble the system matrices and integrate the equations of motion by the energy-conserving Algorithm 9.3 with initial conditions consisting of the initial velocity $\mathbf{v}_0^A = (0,1)$. In particular, compare the motion of the center of mass of the system with the analytical solution, and evaluate the development of the total energy of the system.

References

Ahadi, A. and Krenk, S. (2000). Characteristic state plasticity for granular materials. Part 2: Model calibration and results, *International Journal of Solids and Structures*, **37**, 6361–6380.

Ahadi, A. and Krenk, S. (2003). Implicit integration of plasticity models for granular materials, *Computer Methods in Applied Mechanics and Engineering*, **192**, 3471–3488.

Al-Rasby, S. N. (1991). Solution techniques in nonlinear structural analysis, *Computers and Structures*, **40**, 985–993.

Argyris, J. (1982). An excursion into large rotations, *Computer Methods in Applied Mechanics and Engineering*, **32**, 85–155.

Argyris, J., Balmer, H., Doltsinis, J. St. *et al.* (1979). Finite element method – The natural approach, *Computer Methods in Applied Mechanics and Engineering*, **17/18**, 1–106.

Argyris, J., Hilpert, H. O., Malejannakis, G. A. and Scharpf, D. W. (1979). On the geometrical stiffness of a beam in space – A consistent V. W. approach, *Computer Methods in Applied Mechanics and Engineering*, **20**, 105–131.

Argyris, J. and Mlejnek, H. P. (1991). *Dynamics of Structures*. North-Holland, Amsterdam, The Netherlands.

Armero, F. (2006). Energy dissipative momentum-conserving time-stepping algorithms for finite strain multiplicative plasticity, *Computer Methods in Applied Mechanics and Engineering*, **195**, 4862–4889.

Armero, F. and Romero, I. (2001a). On the formulation of high-frequency dissipative time-stepping algorithms for non-linear dynamics. Part I: Low-order methods for two model problems and nonlinear elastodynamics, *Computer Methods in Applied Mechanics and Engineering*, **190**, 2603–2649.

Armero, F. and Romero, I. (2001b). On the formulation of high-frequency dissipative time-stepping algorithms for non-linear dynamics. Part II: Second order methods, *Computer Methods in Applied Mechanics and Engineering*, **190**, 6783–6824.

Armstrong, P. J. and Frederick, C. O. (1966). A mathematical representation of the multiaxial Bauschinger effect, Berkeley Nuclear Laboratories, Report RD/B/N731. (Reprinted in *Materials at High Temperatures*, **24**, 1–26, 2007.)

Atluri, S. N. and Cazzani, A. (1995). Rotations in computational solid mechanics, *Archives of Computational Methods in Engineering*, **2**, 49–138.

Bathe, K.-J. (1996). *Finite Element Procedures.* Prentice-Hall, Englewood-Cliffs, NJ.

Bathe, K. J. and Cimento, A. P. (1980). Some practical procedures for the solution of nonlinear finite element equations, *Computer Methods in Applied Mechanics and Engineering*, **22**, 59–85.

Battini, J.-M. (2002). Co-rotational beam elements in instability problems, Department of Mechanics, Royal Institute of Technology, Stockholm, Techn. Dr. thesis.

Belytschko, T., Liu, W. K. and Moran, B. (2000). *Nonlinear Finite Elements for Continua and Structures.* Wiley, Chichester, UK.

Bergan, P. (1980). Solution algorithms for nonlinear structural problems, *Computers and Structures*, **12**, 497–509.

Bergan, P. (1981). Solution by iteration in displacement and load spaces, in *Nonlinear Finite Element Analysis in Structural Mechanics*, ed. W. Wunderlich, E. Stein and K.-J. Bathe. Springer, Berlin, 553–571.

Borup, M. and Hedegaard J. (1995). *DATA REPORT 9403 Baskarp Sand No 15.*, Aalborg University, Denmark.

Cardona, A. and Géradin, M. (1988). A beam finite-element non-linear theory with finite rotations, *International Journal for Numerical Methods in Engineering*, **26**, 2403–2438.

Chaboche, J. L. (1986). Time-independent constitutive theories for cyclic plasticity, *International Journal of Plasticity*, **2**, 149–188.

Chakrabarty, J. (2000). *Applied Plasticity.* Springer, New York.

Chen, W. F. and Han, D. J. (1988). *Plasticity for Structural Engineers.* Springer, New York.

Cheng, H. and Gupta, K. C. (1989). An historical note on finite rotations, *Journal of Applied Mechanics*, **56**, 139–145.

Chung, J. and Hulbert, G. M. (1993). A time integration algorithm for structural dynamics with improved numerical dissipation: The generalized α method, *Journal of Applied Mechanics*, **60**, 371–375.

Christoffersen, J. (1989). When is a moment conservative?, *Journal of Applied Mechanics*, **56**, 299–301.

Clarke, M. J. and Hancock, G. J. (1990). A study of incremental-iterative strategies for non-linear analyses, *International Journal for Numerical Methods in Engineering*, **29**, 1365–1391.

Collins, I. F. (2003). A systematic procedure for constructing critical state models in three dimensions, *International Journal of Solids and Structures*, **40**, 4379–4397.

Collins, I. F. and Houlsby, G. T. (1997). Application of thermomechanical principles to the modelling of geotechnical materials, *Proceedings of the Royal Society of London, A*, **453**, 1975–2001.

Cook, R. D., Malkus, D. S. and Plesha, M. E. (1989). *Concepts and Applications of Finite Element Analysis*, 3rd ed. Wiley, New York.

Corben, H. C. and Stehle, P. (1974). *Classical Mechanics*, 2nd ed. Krieger, New York.

Cowin, S. C. and Mehrabadi, M. M. (1995). Anisotropic symmetries of linear elasticity, *Applied Mechanics Reviews*, **48**, 247–285.

Crisfield, M. A. (1979). A faster modified Newton–Raphson iteration, *Computer Methods in Applied Mechanics and Engineering*, **20**, 267–278.

Crisfield, M. A. (1980). Alternative methods derived from the BFGS formula, *International Journal for Numerical Methods in Engineering*, **15**, 1419–1420.

Crisfield, M. A. (1981). A fast incremental/iterative solution procedure that handles "Snap-through", *Computers and Structures*, **13**, 55–62.

Crisfield, M. A. (1990). A consistent co-rotational formulation for non-linear, three-dimensional, beam elements, *Computer Methods in Applied Mechanics and Engineering*, **81**, 131–150.

Crisfield, M. A. (1991). *Non-Linear Finite Element Analysis of Solids and Structures. Vol. 1, Essentials*. Wiley, Chichester, UK.

Crisfield, M. A. (1997). *Non-Linear Finite Element Analysis of Solids and Structures. Vol. 2, Advanced Topics*. Wiley, Chichester, UK.

Dienes, J. K. (1979). On the analysis of rotation and stress rate in deforming bodies, *Acta Mechanica*, **32**, 217–232.

Drucker, D. C. and Prager, W. (1952). Soil mechanics and plastic analysis or limit design, *Quarterly of Applied Mathematics*, **10**, 157–165.

Dupuis, G. (1969). Stabilité elastique des structures unidimensionelles, *Zeitschrift für Angewandte Mathematik und Physik*, **20**, 94–106.

Erlicher, S., Bonaventura, L. and Bursi, O. S. (2002). The analysis of the generalized-α method for non-linear dynamics problems, *Computational Mechanics*, **28**, 83–104.

Eve, R. A., Reddy, B. D. and Rockafellar, R. T. (1990). An internal variable theory of elastoplasticity based on the maximum plastic work inequality, *Quarterly of Applied Mathematics*, **48**, 59–83.

Forde, B. W. R. and Stiemer, S. F. (1987). Improved arc length orthogonality methods for nonlinear finite element analysis, *Computers and Structures*, **27**, 625–630.

Fujikubo, M., Bai, Y. and Ueda, Y. (1991). Dynamic elastic–plastic analysis of offshore framed structures by plastic node method considering strain hardening effects, in *Proceedings of the 1st ISOPE Conference*, Edinburgh.

Géradin, M. and Cardona, A. (2001). *Flexible Multibody Dynamics. A Finite Element Approach*. Wiley, Chichester, UK.

Géradin, M. and Rixen, D. (1997). *Mechanical Vibrations, Theory and Applications to Structural Dynamics*, 2nd ed. Wiley, Chichester, UK.

Germain, P., Nguyen, Q. S. and Suquet, P. (1983). Continuum thermodynamics, *Journal of Applied Mechanics*, **50**, 1010–1020.

Goldstein, H. (1980). *Classical Mechanics*, 2nd ed. Addison-Wesley, Reading, MA.

Gonzalez, O. (2000). Exact energy and momentum conserving algorithms for general models in nonlinear elasticity, *Computer Methods in Applied Mechanics and Engineering*, **190**, 1763–1783.

Gurtin, M. E. (1981). *An Introduction to Continuum Mechanics*. Academic Press, Orlando, FL.

Hairer, E., Lubich, C. and Wanner, G. (2006). *Geometric Numerical Integration*, 2nd ed. Springer, Berlin.

Halphen, B. and Son, N. Q. (1975). Sur les matériaux standards généralisés, *Journal de Mécanique*, **14**, 39–63.

Hilber, H. M., Hughes, T. J. R. and Taylor, R. L. (1977). Improved numerical dissipation for time integration algorithms in structural dynamics. *Earthquake Engineering and Structural Dynamics*, **5**, 282–292.

Hill, R. (1948). A variational principle of maximum plastic work in classical plasticity, *Quarterly Journal of Mechanics and Applied Mathematics*, **1**, 18–28.

Hill, R. (1950). *The Mathematical Theory of Plasticity.* Oxford University Press, Oxford, UK.

Hill, R. and Rice, J. R. (1973). Elastic potentials and the structure of inelastic constitutive laws, *SIAM Journal of Applied Mathematics*, **25**, 448–461.

Holzapfel, G. A. (2000). *Nonlinear Solid Mechanics.* Wiley, Chichester, UK.

Hughes, T. J. R. (1987). *The Finite Element Method, Linear Static and Dynamic Finite Element Analysis.* Prentice-Hall, Englewood Cliffs, NJ. (Reprinted by Dover, Mineola, NY, 1997.)

Ibrahimbegović, A. (1995). On finite element implementation of geometrically nonlinear Reissner's beam theory: Three-dimensional curved beam elements, *Computer Methods in Applied Mechanics and Engineering*, **122**, 11–26.

Jenkins, J. A., Seitz, T. B. and Przemieniecki, J. S. (1966). Large deflections of diamond-shaped frames, *International Journal of Solids and Structures*, **2**, 591–603.

Jetteur, P. (1986). Implicit integration algorithm for elastoplasticity in plane stress analysis, *Engineering Computations*, **3**, 251–253.

Kouhia, R. (2008). Stabilized forms of orthogonal residual and constant incremental work control path following methods, *Computer Methods in Applied Mechanics and Engineering*, **197**, 1389–1396.

Kouhia, R. and Mikkola, M. (1999a). Some aspects of efficient path-following, *Computers and Structures*, **72**, 509–524.

Kouhia, R. and Mikkola, M. (1999b). Tracing the equilibrium path beyond compound critical points, *International Journal for Numerical Methods in Engineering*, **46**, 1049–1077.

Krenk, S. (1983a). The torsion–extension coupling in pretwisted elastic beams, *International Journal of Solids and Structures*, **19**, 67–72.

Krenk, S. (1983b). A linear theory for pretwisted elastic beams, *Journal of Applied Mechanics*, **50**, 137–142.

Krenk, S. (1994). A general format for curved and non-homogeneous beam elements, *Computers and Structures*, **50**, 449–454.

Krenk, S. (1995a). An orthogonal residual procedure for nonlinear finite element equations, *International Journal for Numerical Methods in Engineering*, **38**, 823–839.

Krenk, S. (1995b). Stiffness of co-rotating elastic beam elements, in *Advances in Finite Element Technology*, ed. N. E. Wiberg. CIMNE, Barcelona, 137–253.

Krenk, S. (1996). A family of invariant stress surfaces, *Journal of Engineering Mechanics*, **122**, 201–208.

Krenk, S. (1998). Friction, dilation and plastic flow potential, in *Physics of Dry Granular Media*, ed. H. J. Herrmann, J.-P. Hovi and S. Ludig. Kluwer, Dordrecht, 255–260.

Krenk, S. (2000). Characteristic state plasticity for granular materials. Part 1: Basic theory, *International Journal of Solids and Structures*, **37**, 6343–6360.

Krenk, S. (2001). *Mechanics and Analysis of Beams, Columns and Cables*, 2nd ed. Springer, Berlin.

Krenk, S. (2006). Energy conservation in Newmark based time integration algorithms, *Computer Methods in Applied Mechanics and Engineering*, **195**, 6110–6124.

Krenk, S. (2007a). The role of geometric stiffness in momentum and energy conserving time integration, *International Journal for Numerical Methods in Engineering*, **71**, 631–651.

Krenk, S. (2007b). A vector format for conservative time integration of rotations, *Multibody Dynamics 2007, ECCOMAS Thematic Conference*, Milan, Italy, June 25–28, 2007.

Krenk, S. (2008). Extended state-space time integration with high-frequency energy dissipation, *International Journal for Numerical Methods in Engineering*, **73**, 1767–1787.

Krenk, S. and Hededal, O. (1995), A dual orthogonality procedure for nonlinear finite element equations, *Computer Methods in Applied Mechanics and Engineering*, **123**, 95–107.

Krenk, S. and Høgsberg, J. R. (2005). Properties of time integration with first order filter damping, *International Journal for Numerical Methods in Engineering*, **64**, 547–566.

Krenk, S., Vissing-Jørgensen, C. and Thesbjerg, L. (1999). Efficient collapse analysis techniques for framed structures, *Computers and Structures*, **72**, 481–496.

Krieg, R. D. and Krieg, D. B. (1977). Accuracies of numerical solution methods for the elastic–perfectly plastic model, *ASME Journal of Pressure Vessel Technology*, **99**, 510–515.

Kuhl, D. and Crisfield, M. A. (1999). Energy-conserving and decaying algorithms in non-linear structural dynamics, *International Journal for Numerical Methods in Engineering*, **45**, 569–599.

Kuhn, H. W. and Tucker, A. W. (1951). Nonlinear programming, in *Proceedings of the Second Berkeley Symposium on Mathematical Statistics and Probability*, ed. J. Neyman. University of California Press, Berkeley, CA, 481–492.

Lade, P. V. and Duncan, J. M. (1975). Elasto-plastic stress–strain theory for cohesionless soil, *ASCE Journal of the Geotechnical Engineering Division*, **101**, 1037–1053.

Lade, P. V. and Kim, M. K. (1995). Single hardening constitutive model for soil, rock and concrete, *International Journal of Solids and Structures*, **32**, 1963–1978.

Laursen, T. A. and Meng, X. N. (2001). A new solution procedure for application of energy-conserving algorithms to general constitutive models in nonlinear elastodynamics, *Computer Methods in Applied Mechanics and Engineering*, **190**, 6309–6322.

Lee, E. H. (1969). Elastic–plastic deformation at finite strains, *Journal of Applied Mechanics*, **36**, 1–6.

Lee, K. L. and Seed, H. B. (1967). Undrained strength characteristics of sand, *ASCE Journal of the Soil Mechanics and Foundations Division*, **93**, 333–360.

Lekhnitskii, S. G. (1963). *Theory of Elasticity of an Anisotropic Body.* Holden-Day, San Francisco.

Lemaitre, J. and Chaboche, J.-L. (1990). *Mechanics of Solid Materials.* Cambridge University Press, Cambridge, UK.

Lo, S. H. (1992). Geometrically nonlinear formulation of 3D finite strain beam elements with large rotations, *Computers and Structures*, **44**, 147–157.

Luenberger, D. G. (1984). *Linear and Nonlinear Programming*, 2nd ed. Addison Wesley, Reading, MA.

Magnusson, A. and Svensson, I. (1998). Numerical treatment of complete load–deflection curves, *International Journal for Numerical Methods in Engineering*, **41**, 955–971.

Malvern, L. E. (1969). *Introduction to the Mechanics of a Continuous Medium.* Prentice-Hall, Englewood Cliffs, NJ.

Markley, F. L. (2008). Unit quaternion from rotation matrix, *Journal of Guidance, Control and Dynamics*, **31**, 440–442.

Marsden, J. E. and Hughes, T. J. R. (1983) *Mathematical Foundations of Elasticity.* Prentice-Hall, Englewood Cliffs, NJ.

Mathiesen, F. (1993). Stability analysis of thin-walled non-symmetric steel beams, Aalborg University, Denmark, Ph.D. thesis.

MATLAB (1992). *Reference Guide.* The MathWorks, Natick, MA.

Matsuoka, H. and Nakai, T. (1985). Relationship among Tresca, Mohr–Coulomb and Matsuoka–Nakai failure criteria, *Soils and Foundations*, **25**, 123–128.

Matthies, H. and Strang, G. (1979). The solution of nonlinear finite element equations, *International Journal for Numerical Methods in Engineering*, **14**, 1613–1626.

Mattiasson, K. (1981). Numerical results from large deflection beam and frame problems analyzed by means of elliptic integrals, *International Journal for Numerical Methods in Engineering*, **17**, 145–153.

Mattiasson, K. (1983). On the co-rotational finite element formulation for large deformation problems, Chalmers University of Technology, Department of Structural Mechanics, Publication 83:1, Göteborg.

Melan, E. (1938), Zur Plastizität des räumlichen Kontinuums, *Ingenieur-Archiv*, **9**, 116–126.

Mises, R. von (1928). Mechanik der plastischen Formänderung von Kristallen, *Zeitschrift für angewandte Mathematik und Mechanik*, **8**, 161–185.

Newmark, N. M. (1959). A method of computation for structural dynamics, *Journal of the Engineering Mechanics Division, ASCE*, **85**, EM3, 67–94.

Ogden, R. W. (1984). *Non-Linear Elastic Deformations.* Ellis Horwood, Chichester, UK. (Reprinted by Dover, Mineola, NY, 1997.)

Oran, C. (1973a). Tangent stiffness in plane frames, *Journal of the Structural Division, ASCE*, **99**, 973–985.

Oran, C. (1973b). Tangent stiffness in space frames, *Journal of the Structural Division, ASCE*, **99**, 987–1001.

Ortiz, M. and Popov, E. P. (1985). Accuracy and stability of integration algorithms for elastoplastic constitutive equations, *International Journal for Numerical Methods in Engineering*, **21**, 1561–1576.

Ortiz, M. and Simo, J. C. (1986). An analysis of a new class of integration algorithms for elastoplastic constitutive relations, *International Journal for Numerical Methods in Engineering*, **23**, 353–366.

Ottosen, N. S. and Ristinmaa, M. (2005). *The Mechanics of Constitutive Modeling.* Elsevier, Amsterdam, The Netherlands.

Pacoste, C. and Eriksson, A. (1997). Beam elements in instability problems, *Computer Methods in Applied Mechanics and Engineering*, **144**, 163–197.

Pecknold, D. A., Ghaboussi, J. and Healey, T. J. (1985). Snap-through and bifurcation in a simple structure, *Journal of Engineering Mechanics*, **111**, 909–922.

Poulsen, P. N. and Damkilde, L. (1996). A flat triangular shell element with loof nodes, *International Journal for Numerical Methods in Engineering*, **39**, 3867–3887.

Powell, G. H. and Chen, P. F. (1986). 3D beam-column element with generalized plastic hinges, *Journal of Engineering Mechanics*, **112**, 627–641.

Prager, W. (1955). The theory of plasticity: A survey of recent achievements, *Institution of Mechanical Engineers*, **169**, 41–57.

Press, W. H., Flannery, B. P., Teukolsky, S. A. and Vetterling, W. T. (1986). *Numerical Recipes, The Art of Scientific Computing.* Cambridge University Press, Cambridge, UK.

Ramm, E. (1981). Strategies for tracing the non-linear response near limit-points, in *Non-linear Finite Element Analysis in Structural Mechanics*, ed. W. Wunderlich, E. Stein, K.-J. Bathe. Springer, Berlin, 63–89.

Reissner, E. (1981). On finite deformations of space curved beams, *Journal of Applied Mathematics and Physics*, **32**, 734–744.

Rice, J. R. (1971). Inelastic constitutive relations for solids: An internal-variable theory and its application to metal plasticity, *Journal of the Mechanics and Physics of Solids*, **19**, 433–455.

Riks, E. (1979). An incremental approach to the solution of snapping and buckling problems, *International Journal of Solids and Structures*, **15**, 529–551.

Schweizerhof, K. H. and Wriggers, P. (1986). Consistent linearization for path following methods in nonlinear FE analysis, *Computer Methods in Applied Mechanics and Engineering*, **59**, 261–279.

Schofield, A. and Wroth, P. (1968). *Critical State Solid Mechanics.* McGraw-Hill, London, UK.

Sewell, M. J. (1987). *Maximum and Minimum Principles*, Cambridge University Press, Cambridge, UK.

Simo, J. C. (1985). A finite strain beam formulation. The three-dimensional dynamic problem. Part I, *Computer Methods in Applied Mechanics and Engineering*, **49**, 55–70.

Simo, J. C. (1988a). A framework for finite strain elastoplasticity based on maximum plastic dissipation and the multiplicative decomposition: Part I. Continuum formulation, *Computer Methods in Applied Mechanics and Engineering*, **66**, 199–219.

Simo, J. C. (1988b). A framework for finite strain elastoplasticity based on maximum plastic dissipation and the multiplicative decomposition: Part II. Computational aspects, *Computer Methods in Applied Mechanics and Engineering*, **68**, 1–31.

Simo, J. C. and Hughes, T. J. R. (1987). General return mapping algorithms for rate-independent plasticity, in *Constitutive Laws for Engineering Materials*, ed. C. S. Desai *et al.* Elsevier, Amsterdam, The Netherlands.

Simo, J. C. and Hughes, T. J. R. (1998). *Computational Inelasticity.* Springer, New York.

Simo, J. C. and Miehe, C. (1992). Associative coupled thermoplasticity at finite strains: Formulation, numerical analysis and implementation, *Computer Methods in Applied Mechanics and Engineering*, **98**, 41–104.

Simo, J. C. and Ortiz, M. (1985). A unified approach to finite deformation elastoplastic analysis based on the use of hyperelastic constitutive equations, *Computer Methods in Applied Mechanics and Engineering*, **49**, 221–245.

Simo, J. C. and Tarnow, N. (1992). The discrete energy-momentum method. Conserving algorithms for nonlinear elastodynamics, *Zeitschrift für angewandte Mathematik und Physik*, **43**, 757–792.

Simo, J. C. and Taylor, R. L. (1985). Consistent tangent operators for rate-independent elastoplasticity, *Computational Methods in Applied Mechanics and Engineering*, **48**, 101–118.

Simo, J. C. and Taylor, R. L. (1986). A return mapping algorithm for plane stress elastoplasticity, *International Journal for Numerical Methods in Engineering*, **22**, 649–670.

Simo, J. C., Taylor, R. L. and Pister, K. S. (1985). Variational and projection methods for the volume constraint in finite deformation elasto-plasticity, *Computational Methods in Applied Mechanics and Engineering*, **51**, 177–208.

Simo, J. C. and Vu-Quoc, L. (1986). A three dimensional finite strain rod model. Part II: Computational aspects, *Computer Methods in Applied Mechanics and Engineering*, **58**, 79–116.

Simo, J. C. and Vu-Quoc, L. (1988). On the dynamics in space of rods undergoing large motions – a geometrically exact approach, *Computer Methods in Applied Mechanics and Engineering*, **66**, 125–161.

Simo, J. C. and Wong, K. K. (1991). Unconditionally stable algorithms for rigid body dynamics that exactly preserve energy and momentum, *International Journal for Numerical Methods in Engineering*, **31**, 19–52.

Skallerud, B. and Amdahl, J. (2002). *Nonlinear Analysis of Offshore Structures*, Research Studies Press, Baldock, Hertfordshire, UK.

Spurrier, R. A. (1978). Comment on "Singularity-free extraction of a quaternion from a direction-cosine matrix", *Journal of Spacecraft*, **15**, 255.

Truesdell, C. (1991), *A First Course in Rational Continuum Mechanics*, 2nd ed. Academic Press, San Diego, CA.

Timoshenko, S. P. and Gere, J. M. (1961). *Theory of Elastic Stability.* McGraw-Hill, New York.

Ueda, Y. and Yao, T. (1982). The plastic node method: A new method of plastic analysis, *Computer Methods in Applied Mechanics and Engineering*, **34**, 1089–1104.

Wallin, M., Ristinmaa, M. and Ottosen, N. S. (2003). Kinematic hardening in large strain plasticity, *European Journal of Mechanics, A/Solids*, **22**, 341–356.

Washizu, K. (1974). *Variational Methods in Elasticity and Plasticity*, 2nd ed. Pergamon Press, Oxford, UK.

Williams, F. W. (1964). An approach to the nonlinear behavior of the members of a rigid jointed plane framework with finite deflections, *Quarterly Journal of Mechanics and Applied Mathematics*, **17**, 451–469.

Wood, D. M. (1990). *Soil Behaviour and Critical State Soil Mechanics*. Cambridge University Press, Cambridge, UK.

Wood, W. L., Bossak, M. and Zienkiewicz, O. C. (1981). An alpha modification of Newmark's method, *International Journal for Numerical Methods in Engineering*, **15**, 1562–1566.

Yang, Y.-B. and Kuo, S.-R. (1994). *Theory and Analysis of Nonlinear Framed Structures*. Prentice-Hall, New York.

Yang, Y.-B. and Leu, L.-J. (1991). Constitutive laws and force recovery procedures in nonlinear analysis of trusses, *Computer Methods in Applied Mechanics and Engineering*, **92**, 121–131.

Ziegler, H. (1959). A modification of Prager's hardening rule, *Quarterly of Applied Mathematics*, **17**, 55–65.

Ziegler, H. (1963). Some extremum principles in irreversible thermodynamics with application to continuum mechanics, in *Progress in Solid Mechanics*, Vol. 4, ed. I. N. Sneddon, R. Hill. North-Holland, Amsterdam.

Ziegler, H. (1977). *Principles of Structural Stability*, 2nd ed. Birkhäuser, Basel.

Zienkiewicz, O. C. and Taylor, R. L. (2000). *The Finite Element Method*, 5th ed., Vol. 1–3. Butterworth-Heinemann, London, UK.

Index

additive elastic energy, 230
additive internal energy, 201, 211
algorithmic complementary energy, 239
algorithmic damping ratio, 304
algorithmic energy dissipation, 325
amplification factor, 301, 305
angle of friction, 244
angular velocity, 68, 163
anisotropy from kinematic hardening, 231
arc-length methods, 260, 270–281
  displacement sub-increments, 271, 274
  flexibility parameter, 275, 279
  general constraint formulation, 272–274
  hyperplane constraint, 274–278
  hyperplane constraint condition, 275
  hypersphere constraint, 278–281
  hypersphere constraint condition, 279
  load sub-increments, 271
  orthogonal arc-length algorithm, 277
  quadratic constraint condition, 278
  sub-increment selection strategy, 271
Armstrong–Frederick flow potential, 232
assembly of elements, 31
associated plasticity, 212
  evolution equations, 214
axial stiffness, 3, 127, 140

back stress, 218
bar element, 18–28, 39
Bauschinger effect, 217, 230
beam element 2D implementation, 112
beam element 3D implementation, 134, 137
beam tangent stiffness matrix, 110, 130
beam-column element, 140
  shortening, 140
  tangent stiffness, 141
bending mode
  anti-symmetric, 103, 110, 127
  symmetric, 102, 109, 127
BFGS in product format, 286
BFGS matrix inverse, 285
BFGS matrix update, 285
biaxial extension, 152
bifurcation, 117, 269
bifurcation point, 30, 267
bulk modulus, 199, 247

Cam-Clay theory, 241
  friction assumption, 245
  hardening rule, 247
Cauchy elasticity, 191
Cauchy stress, 158–160
  Truesdell rate, 162
Cauchy stress tensor, 158
Cayley rotation representation, 71
Cayley–Hamilton equation, 194
characteristic equation, 193, 194, 327
co-factor, 195
co-rotating formulation, 100
co-rotating frame, 101, 118
co-rotation stiffness matrix, 106, 122
  non-symmetric, 124
  symmetric, 125, 126
co-rotation tangent stiffness, 104–107, 120–126
collocation methods, 292
combined isotropic–kinematic hardening, 230–234
complementary energy, 94
complementary energy density, 191, 196, 238
complementary internal variable, 236
complementary internal variables, 252
complementary virtual work, 108
compound rotation matrix, 104, 119
conjugate element forces, 102, 118
conjugate variables, 204
  external, 205
  in elasto-plastic iteration, 238
  internal, 205, 230, 235
conservative moments, 59
conservative time integration, 310
  displacement increment, 317
  velocity increment, 317
consistency condition, 212
consistent tangent matrix, 240
constitutive beam stiffness matrix, 111, 130
constitutive stiffness, 168

constitutive stiffness matrix, 107
constrained extremum problem, 208
continuum mechanics, 146
contraction phase, 249
critical state soil mechanics, 241
current coordinates, 148
cut-off frequency, 303

deformation gradient
  factored format, 250
  vector format, 171
deformation gradient tensor, 148, 151, 201
deformation mode stiffness matrix, 106, 122
deformation stiffness matrix, 107
determinant of stress components, 194
deviatoric contour, 242
deviatoric plane, 197
deviatoric sections, 242
deviatoric strain, 229
  equivalent, 229
  invariants, 198
deviatoric stress, 195, 199, 229
  components, 197
  contour, 243
  equivalent, 229
  invariants, 243
dilation, 249
direction control, 261
discrete derivative, 334
discretized hardening evolution equation, 237
discretized strain evolution equation, 237
displacement control, 4, 7
displacement gradient
  tensor, 149
  vector format, 171
displacement increment, 258
displacement increment control, 263
displacement sub-increment, 258, 265
dissipation parameter, 328
dissipation rate, 207, 211, 236, 245
  friction mechanism, 245
divergence theorem, 155
Drucker–Prager flow potential, 236
Drucker–Prager yield surface, 234
dual flow potential, 216

effective stiffness, 262
effective stiffness matrix, 329
eigenvalue problem, 193
elastic flexibility matrix, 200
elastic incremental stiffness, 205
elastic pendulum, 307, 320, 331
elastic potential, 312
elastic solids, 190–203
elastic stiffness matrix, 200
elastic stiffness tensor, 199
elastic strain, 204
elastica, 77
  constitutive stiffness, 82
  curvature components, 79
  curvature parametrization, 80
  element stiffness, 87–89
  equilibrium equations, 77
  geometric stiffness, 83–85
  strain components, 79
elasto-plastic flow
  evolution equations, 212–216
element stiffness matrix, 106
element tangent stiffness matrix, 33
energy balance equation
  finite increment form, 312, 326
  linear form, 297
  Newmark algorithm, 298
  non-linear form, 310
energy dissipating time integration
  displacement increment, 329
energy dissipating time integration algorithm, 330
energy fluctuations, 309
energy-conserving integration, 316
energy-conserving time integration algorithm, 319
energy-dissipating time integration
  velocity increment, 328
energy-dissipating time integration algorithm, 328
engineering strain, 3, 18, 23, 40, 44, 335
equation of motion, 291, 293, 305
  recurrence format, 326
  state space increment format, 316, 326, 328
  state-space format, 311
  state-space increment format, 311
equilibrium, 3, 18
equilibrium equation, 155
equilibrium force system, 103
equilibrium iterations, 9–14
equilibrium path, 258, 267
equipotential curve, 245
equivalent bending stiffness, 96
equivalent elastic stress, 226
equivalent plastic strain, 220
equivalent shear stiffness, 95
equivalent stiffness matrix, 298
equivalent stress, 219
error estimate, 224
Euler explicit method, 8
Eulerian formulation, 147
evolution equations, 210
  combined hardening, 233
  elasto-plastic flow, 212–216
  reduced format, 214
explicit elasto-plastic integration, 222–225
explicit integration, 222
explicit method, 224
extension mode, 102
external nodal forces, 170, 180
external virtual work, 22, 58, 104, 120, 167, 174, 180

factored format
  deformation gradient, 250
  elastic potential increment, 313, 314, 334

Green strain increment, 313
finite displacement theory, 145
finite strain, 252
finite strain elasticity, 200–203
finite strain plasticity, 249–252
first variation, 33
flow potential, 210, 244
Armstrong–Frederick, 232
forward Euler procedure, 223
free motion of solid frame, 322

generalized shear, 244
generalized strain vector, 79
generalized stress vector, 79
geometric beam stiffness matrix, 111, 131
geometric stiffness, 25, 168
geometric stiffness increment, 316
global frame, 101, 118
global tangent stiffness matrix, 34
granular materials, 241–249
Green deformation tensor, 201, 250
Green elasticity, 191
Green strain, 19, 40, 145, 148–151, 312, 334
Green strain tensor, 150, 201, 250, 312, 313
components, 150
variation, 151

hardening modulus, 214, 226
granular material, 247
high-frequency algorithmic damping, 325
homogeneous function, 199
hyper-elasticity, 191
hyperplane, 209

implicit integration, 225
implicit method, 227, 237
incompressible strain increments, 220
increment of internal virtual work, 168
increment of rotation variation, 55–58, 121
incremental rotation vector, 56
incremental stiffness equation, 43
inertial forces, 291
initial conditions, 292
initial stress, 25
integrated state space equations, 311
internal energy density, 191, 204, 252
internal force, 3
internal nodal force, 172
internal porosity, 235
internal porosity variable, 235
internal variables, 204
conjugate, 206
original, 206
internal virtual work, 22, 167, 172, 174
invariants, *see* stress invariants
irreversible deformation, 207
isochoric deformation, 201
isostatic compression, 241, 249
isotropic hardening, 216
isotropic material, 192
isotropic scaling, 201
iteration sub-increments, 239

Jacobian determinant, 159, 201
Jaumann stress increment, 180
Jaumann stress rate, 163
Cauchy stress, 164
Kirchhoff stress, 164

kinematic hardening, 217
exponential decay, 233
Kirchhoff stress, 159
Truesdell rate, 161
Kuhn–Tucker conditions, 210

Lagrangian formulation, 147
large displacement theory, 145
lateral buckling of beam, 132
left-hand decomposition, 152
Legendre transformation, 192
Lie derivative, 162
limit point, 30, 260
linearized equilibrium equation, 258
linearized evolution equations, 237
linearized stability theory, 169
load step, 258
load step control, 262
load sub-increment, 258
local beam geometric stiffness, 130
local director basis, 78
local force transformation matrix, 104, 120
local frame, 101
local geometric stiffness, 109, 110, 128
local geometry, 151
local orthonormal basis, 77
logarithmic strain, 187
logarithmic strain, 19, 40

material formulation, 147
material particle, 147
material point, 204
maximum dissipation rate, 208–212
mean rotation by quaternions, 69
mean strain, 229
mean stress, 195, 199, 229, 314
mid-point integration, 223
mode shape representation, 299
modified mass matrix, 295
modified Newton–Raphson method, 13–14, 37, 256
modified tangent stiffness matrix, 307
momentum-based methods, 292, 333
momentum-based time integration, 310–333

natural deformation modes, 102, 119
Newmark algorithm, 293
alpha-modifications, 305
characteristic equation, 302
energy balance equation, 298
equivalent modal stiffness, 300
integral approximations, 295
modal energy equation, 300

non-linear form, 306
prediction step, 295
recurrence format, 301
stability diagram, 299
state space format, 297
unconditional stability, 299
Newton iteration, 227
Newton iterations, 305
Newton–Raphson method, 9–12, 237, 256, 259
nodal residual force, 173, 182
non-associated flow, 215, 234–236
non-associated plasticity, 212
evolution equations, 214
non-linear solution strategies, 259

objective stress rate, 160
orthogonal residual algorithm, 265
orthogonal residual method, 263–266

parameter representation of incremental rotation, 60–63
period error, 303
asymptotic low-frequency, 304
permutation symbol, 51, 64
Piola–Kirchhoff stress, 155–157
beam theory, 157
in finite strain plasticity, 252
Piola–Kirchhoff stress tensor, 312
components, 156
first kind, 156
second kind, 155
plastic deformation gradient, 250
plastic hinges, 141
plastic multiplier, 212, 220, 232
finite increment, 240
plastic rotation, 251
plastic strain, 252
plasticity theory, 203
Poisson's ratio, 200
polar decomposition theorem, 151
potential energy, 7, 29, 43, 145, 169, 308
principal deviatoric components, 242
principal direction, 193
principal strain invariants, 198
principal stress, 193
principal stress invariants, 193
principal stress space, 197, 219, 242
principle of virtual work, *see* virtual work
projection tensor, 53
proper orthogonal tensor, 50, 151
pseudo stress, 218
pseudo time, 204
pull-back, 80
push-forward, 80

quasi-Newton condition, 284
quasi-Newton methods, 13, 284–287
factored form with scaling, 287
inverse with stiffness scaling, 286
stiffness scaling factor, 286
quasi-stationary equilibrium, 210
quaternion addition of rotations, 66
quaternion parameters, 64
beam elements, 133
extraction of, 64
increment, 134
update, 134

radial return algorithm, 225–229
rate-independent plasticity, 203
reduced integration, 94
reduced plastic dilation, 236
reference configuration, 146
representative internal force, 314, 316
residual deformation gradient, 250
residual force, 167, 257, 305
residual force methods, 259
residual orthogonality condition, 264
return algorithm, 226
reversible deformation, 205
right-hand decomposition, 152
rigid body modes, 102
rigid body motion, 160
Rodrigues' formula, 50
Rodrigues' vector, 70
roll-up of cantilever, 114
rotation increment by quaternions, 68
rotation into specified direction, 53–55
rotation of external moment, 58, 86
rotation of internal moment, 85
rotation of local basis, 135
rotation parameter increments, 57
rotation pseudo vector, 70
rotation pseudo-vector, 50
rotation tensor, 49–51, 151
rotation transformation tensor, 63

scalar invariant, 6
scaled load increment, 263
secant condition, 284
secant-Newton method, 287
second variation, 33
second variation of rotation, 58
semi-tangential moment, 59
sequence of equilibrium states, 257
shallow truss dome, 268
shape functions
bar element, 26
beam element, 87, 109, 129
solid element, 170, 180
shear at constant volume, 153
Jaumann stress rate, 164
Truesdell stress rate, 162
shear flexibility, 108, 127
shear locking, 91–98
shear modulus, 199, 247
Sherman–Morrison formula, 285
six-component tensor format, 171, 181, 192
skew-symmetric vector component matrix, 50
smooth triangular contour, 243
snap-through, 4, 117
spatial components, 146

spatial formulation, 147
spectral radius, 301, 305
spin tensor, 163
state vector, 292
state-space variables, 297, 301, 311
stiffness matrix, 273, 284
  constitutive, 318
  elastica, 87–89
  geometric, 318
  non-symmetric, 318
  total Lagrangian, 173
  updated Lagrangian, 183
strain, *see* Green strain
strain energy density, 42, 169, 191, 192
strain invariants, 192
strain rate, 163
  elastic, 208
  plastic, 208
  total, 208
stress, *see* Cauchy stress, Kirchhoff stress, Piola–Kirchhoff stress
stress free volumetric strain, 235
stress invariants, 192, 194
  deviatoric, 196
stress rates, 160–165
stretch tensor, 187
sub-increment, 223, 237
summation convention, 146
symmetric non-associated iteration equations, 239
symmetry correction of stiffness matrix, 125

tangent stiffness, 4
  algorithmic, 240
  constitutive, 90, 316, 319
  elasto-plastic, 221, 222
  geometric, 316, 318
  modified, 307
  non-symmetric, 180
  symmetric, 169
tangent stiffness matrix, 9, 24
tensor analysis, 146
three-component tensor basis, 195
time increment limit, 300
total displacement components, 166
total Lagrangian formulation, 18, 37, 165–170
trace, 194
trace of rotation tensor, 51, 64
triaxial compression, 243
triaxial tension, 243
Truesdell stress increment, 175, 178, 180, 182
Truesdell stress rate, 160
  Cauchy stress, 162
  Kirchhoff stress, 161
truss model, 2, 28, 266, 281

uncoupled materials, 211
undrained test, 249
uniaxial compression, 241
uniaxial yield stress, 219
updated Lagrangian formulation, 18, 37, 174–180

velocity gradient tensor, 161
virtual displacement, 6, 32, 78, 105, 180
virtual Green strain, 158, 166
virtual nodal displacement, 39–41
virtual rotation, 56, 58, 78
virtual strain, 32, 39, 40
virtual strain components, 177
virtual work
  equation, 155, 166, 177
  increment, 105, 121, 176, 177, 179, 182
  principle, 6, 14, 22, 40–42, 55, 72, 78, 80
virtual work of elastica, 78–81
visco-plasticity, 203
volume-preserving deformation, 201
volumetric strain, 198
von Mises plasticity, 218–222
  elasto-plastic tangent stiffness, 222
  hardening modulus, 221
  plane stress, 229
  uniaxial stress–strain relation, 221
  yield function, 219
  yield surface, 219, 232

yield function, 205, 209, 219
  convex, 207
yield potential, 209
yield surface, 206, 219
Young's modulus, 200